D0821237

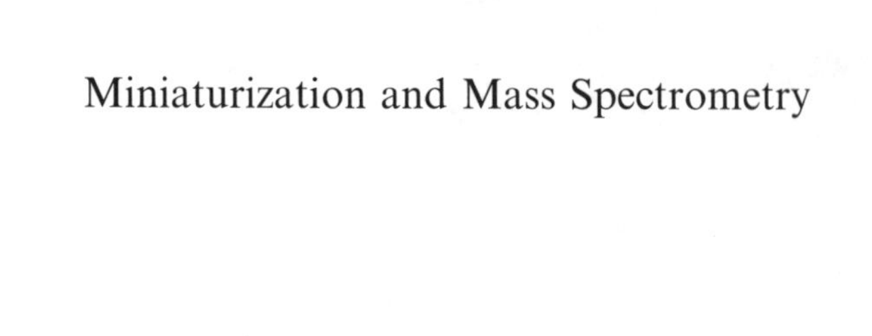

Miniaturization and Mass Spectrometry

Miniaturization and Mass Spectrometry

Edited by

Séverine Le Gac and Albert van den Berg
BIOS, University of Twente, Enschede, The Netherlands

RSC Publishing

ISBN: 978-0-85404-129-9

A catalogue record for this book is available from the British Library

Published by The Royal Society of Chemistry,
Thomas Graham House, Science Park, Milton Road,
Cambridge CB4 0WF, UK

Registered Charity Number 207890

For further information see our web site at www.rsc.org

Preface

Dear Reader,

With this book we want to illustrate how two quickly growing fields of instrumentation and technology, both applied to life sciences, mass spectrometry and microfluidics (or microfabrication) naturally came to meet at the end of the last century and how this marriage impacts on several types of applications.

Since the mid-20th century, the techniques of mass spectrometry (MS) have seen tremendous growth for both biochemical detection and analysis purposes. In particular, new developments regarding MS techniques have been strongly driven by the explosive expansion in the field of biological sciences since the elucidation of the structure of DNA by Watson and Crick in 1953. Quickly thereafter, the quest for understanding biological paths and processes and the need for characterizing an increasing amount of new and unidentified biological species resulted in a demand for new and powerful analytical techniques and tools. MS quickly turned out to be one key technique to perform such studies and research on biomolecules. MS capability became broadened with more recent technological developments such as (i) the discovery of two soft ionization techniques facilitating the analysis of large (bio)molecules and (ii) MS coupling to a liquid-based separation for analysis of more complex samples. In parallel, the capability of MS techniques has been enhanced and has evolved so as to be suitable for handling small volumes of samples down to the low microlitre range. These different evolutions turned MS into a mature technique for the analysis of a wide range of compounds and that of small and complex biological samples. The growth of MS techniques is still best illustrated by the current large interest in proteomics, for which MS is by far the most preferred analytical technique.

Later in the 20th century, biology also impacted on the field of analytical chemistry, especially for biological, medical and diagnostic purposes. The rapid expansion of this field resulted in a need for fast, integrated, portable, more reliable and more sensitive tools that could handle reduced-size samples. This

Miniaturization and Mass Spectrometry
Edited by Séverine Le Gac and Albert van den Berg

Published by the Royal Society of Chemistry, www.rsc.org

gave rise to a new trend in miniaturization for a number of applications and the increasing use of microfabrication techniques to produce micrometer-sized devices for chemical processing, analytical chemistry, biological analysis and diagnostic purposes. However, a new challenge quickly arose from this extreme miniaturization: the detection of the analytes and compounds of interest. Analysts decided to have recourse to MS as a consequence of the strong potential of this detection and analysis technique, together with its suitability to work on low-concentration and reduced-size samples. Additionally, MS was already prevailing for proteomic analysis, and the first microfluidics-to-MS connection was reported for on-line analysis of protein samples processed on a microfluidic device.

The use of MS in combination with microfabricated devices extended to other fields of application, not only biological and medical analysis, but also organic chemistry and the study of organic reactions with their on-line analysis at the outlet of a microfluidic system. For the latter application, this microfluidic-MS hyphenated technique exhibits novel performances and provides a new insight in to reactions compared to conventional tools in organic chemistry.

Other sectors have been reached since then by the combined use of microfabrication and MS techniques, such as forensics and homeland safety. For instance, microfabrication techniques are now also used in view of the miniaturization of the mass spectrometer itself and its implementation on a microchip to "kill" the current paradox of combining tiny devices for sample preparation to bulky and almost room-sized instrumentation. From such ongoing development, one can expect soon the appearance of fully integrated and portable devices for on-site analysis, with both the implementation of the microfluidic-based sample preparation step and the MS analysis on a single device of a few inches in size.

In spite of this "matching" marriage between miniaturization and MS and the fruitful ongoing research into various applications using this unique combination, no book has yet aimed at highlighting the potentials and benefits of this hyphenated tool. Until now, only a number of reviews have focused on one particular subfield of application without giving a really comprehensive coverage of this field of "miniaturization and mass spectrometry" or describing the real status of the field. With this book we hope to give to the reader a better and complete overview of this combined field and to convince people who are new in the field of the potential and the capability of allying mass spectrometry to the trend in miniaturizing chemical processing and analytical tools.

The book covers several subtopics of miniaturization and mass spectrometry. It combines (i) technological developments in the quest for miniaturization of sample preparation and how to connect micrometer-sized devices to a mass spectrometer, (ii) various illustrations of fields that benefit from such a hyphenated technique and (iii) technological developments for the miniaturization of the mass spectrometer. Additionally, the book is not restricted to one ionization technique as is often the case for many reviews, but it reports on efforts for both the ESI and MALDI ionization techniques. After an introduction to miniaturization and mass spectrometry, the book is divided in three sections that respectively concern (i) ESI-MS applications, (ii) MALDI-MS applications and in every section both the connection of a micrometer-size system to MS and applications of this coupling are illustrated and (iii) the miniaturization of the mass spectrometer.

The first chapter gives a comprehensive introduction to the combination of miniaturization and mass spectrometry by approaching several aspects of this coupling. After a short recall of MS analysis and of the two soft ionization techniques of ESI and MALDI, we mention a number of motivations that encouraged people to use microfabrication techniques in the field of MS analysis and we describe also how they have done it and for which fields this unique combination is applied.

The first section of the book focuses on ESI-MS techniques and reports on various strategies to connect a microfluidic or microfabricated device to ESI-MS and on some applications of such a microfluidics-to-ESI-MS coupling. Early coupling strategy between a microfluidic system and ESI-MS using a conventional ionization source and a transfer capillary is described by Daniel Figeys. The following chapters review microfabricated approaches for ESI-MS coupling with ionization sources integrated or not (yet) on a microfluidic system and using various materials, silicon as well as different types of polymers. Gary Schultz presents a conventional silicon and dry-etching-based approach to produce ESI Chip™ that includes 100 individual capillary-shape ionization sources that are operated by a robotic interface making the link from a 96-well plate to an ESI Chip™. Polydimethylsiloxane (PDMS), which is one of the most popular materials nowadays for microfluidics applications, has also been used to produce nanoESI sources, and Kim and Knapp describe here the production of PDMS-based nanoESI sources using three different machining routes as well as their testing. An original design for nanoESI tips that resembles the shape and the functioning of a fountain pen is reported by Le Gac *et al.* through several generations of prototypes either based on SU-8, a photopatternable resist, or polysilicon so as to achieve smaller dimensions and enhanced ionization performance. The last nanoESI microsystem described by Yang *et al.* is made from cyclo-olefin polymer and includes a parylene-based nanoESI interface. The performance of multiplexed devices comprising of a two- or a four-channel tip array is demonstrated, notably for bioanalysis applications and the detection of a small drug in crude urine samples.

The ESI section continues with three contributions on the application of microfluidics and ESI-MS coupling for proteomic analysis or on-line chemical investigation. Iulia Lazar gives a first example of a fully integrated glass-based microfluidic system applied for biomarker discovery and proteomic analysis; the system presents both horizontal and vertical integration and is comprised of six independent devices each including an electroosmotic flow-based pump, a liquid chromatography separation system with a pre-injection/concentration step and an inserted nanoESI capillary source. A second illustration in the field of proteomics is provided by Ghitun *et al.* with a polymer-based microfluidic platform for multidimensional chromatography and on-line nanoESI-MS analysis using an integrated nanoESI source: they especially illustrate the capability of their integrated platform combining ion exchange and reverse phase chromatography for the identification of low-abundance species starting from crude cellular extracts. The last application of a microfluidics-and-ESI-MS combination focuses on chemical investigations. For that purpose, Brivio *et al.* describe and discuss two alternative approaches for coupling a microfluidic system to nanoESI analysis, with or without the use of an ionization capillary source and they notably illustrate the performance of their systems for on-line monitoring of both supramolecular interactions and organic reactions.

The MALDI-MS section is comprised of three chapters, reporting on different aspects of the combination of microfluidics and MALDI-MS analysis. In the first chapter, Musyimi *et al.* discuss the pros and cons of off-line *vs.* on-line analysis using microfluidics and MALDI-MS and particularly describe a very original interface based on a rotating ball to couple a microfluidic sample preparation step to an on-line analysis using MALDI-MS. Thereafter, Brivio *et al.* demonstrate another approach for MALDI-MS on-line analysis after microfluidic processing of chemical or biochemical samples; liquid actuation is triggered and driven by the vacuum environment present in the ionization source of the mass spectrometer and the microchip includes an open area (outlet reservoir or detection window) where samples can be irradiated by a laser. Finally, Nichols and Gardeniers describe a microfluidic dedicated to MALDI-MS analysis and kinetic studies of an enzymatic reaction; fluids are actuated and mixed extremely fast using the electrowetting-on-dielectric (EWOD) principle and analysis is performed off-line after quenching of the enzymatic reaction.

The last chapter of this book gives an example of the miniaturization of the analysis instrumentation, *i.e.* the mass spectrometer. Cotter *et al.* illustrate here the conception of a miniaturized MALDI-TOF mass spectrometer and its applications for homeland safety and clinical diagnostics purposes.

We hope you will enjoy this book, and we wish you lots of pleasure and inspiration reading this volume!

Séverine Le Gac and Albert van den Berg

Contents

Miniaturization and Mass Spectrometry
Edited by Séverine Le Gac and Albert van den Berg

Published by the Royal Society of Chemistry, www.rsc.org

Section 1 ESI-MS

Early Couplings

Micromachined Source

Chapter 4 Microfabricated Multichannel Electrospray Ionization Emitters on Polydimethylsiloxane (PDMS) Microfluidic Devices
Jin-Sung Kim and Daniel R. Knapp

Chapter 5 Microfabricated Nanoelectrospray Emitter Tips based on a Microfluidic Capillary Slot
Séverine Le Gac, Steve Arscott and Christian Rolando

CHAPTER 1
Introduction

SÉVERINE LE GAC AND ALBERT VAN DEN BERG

BIOS the Lab-on-a-Chip Group, University of Twente, PO Box 217, 7500 AE Enschede, The Netherlands

The development of miniaturized analytical or chemical processing systems for both biological and chemical applications is a fast growing field because such systems enable the performance of a series of successive operations at scales which are not easily handled by human experimenters. A key challenge arising from this continuous system miniaturization towards the micrometer scale, or even smaller, lies in the ability to sensitively detect low molecular concentrations in reduced sample volumes. Additionally, such analytical systems must be coupled to microfluidic devices with minimal loss of analytes and information. The last issue is the scalability of the detection technique, as the detection is performed on small sample sizes. The ideal technique for microfluidic detection would therefore present an enhanced sensitivity upon downscaling. The dream of the users in the (bio)chemical field would be a fully integrated and portable device that includes (micro)systems for sample handling, preparation and detection. Conventional detection systems are still bulky instruments, resulting in the paradox of coupling a smaller and smaller analytical or processing device to room-sized instrumentation for the detection.

On-chip detection firstly relied on optical techniques, such as ultraviolet (UV) absorbance, fluorescence or laser-induced fluorescence (LIF).[1,2] The latter technique in particular has a sensitivity in the (sub)micromolar range which is suitable for microfluidic applications. Besides optical techniques, electrical-based techniques are also widely used for on-chip detection due to their sensitivity, *e.g.* detection based on conductivity,[3] electrochemistry,[4] electrochemiluminescence,[5] *etc*. The main advantage of these techniques is that they

Miniaturization and Mass Spectrometry
Edited by Séverine Le Gac and Albert van den Berg

Published by the Royal Society of Chemistry, www.rsc.org

are fully integrated on the microdevice via the introduction of electrodes; they do not rely on the use of complex and bulky instrumentation as is the case for optical techniques. More exotic techniques are also used in combination with microfluidics, such as nuclear magnetic resonance (NMR)[6] and Raman spectroscopy.[7] These techniques are less popular but are currently developing at a rapid rate. Since the late 1990s mass spectrometry (MS) has also been used for the detection stage for microfluidic processing systems;[8] this combination is particularly striking if one considers the size of a mass spectrometer compared to that of a microchip! MS has rapidly replaced other techniques due to its very high sensitivity and other advantages such as a high selectivity compared to optical-based techniques, for instance. Consequently, it turned out that MS analysis could also benefit from the use of microfluidic systems for sample preparation prior to analysis. As a consequence, the field of microfluidics and MS has been rapidly growing with the appearance of dedicated products within the last decade.

In the first part of this introductory chapter, we briefly introduce the technique of mass spectrometry as well as two ionization methods, namely ESI (electrospray ionization) and MALDI (matrix-assisted laser desorption ionization), commonly used for the analysis of biological/biochemical samples or for organic chemistry purposes. The second part highlights the advantages brought by the miniaturization and coupling of microfabricated devices to MS, and how this marriage benefits both on-chip detection and the MS analysis. The third part of this chapter focuses on the different approaches adopted for coupling microfabricated systems to ESI-MS or MALDI-MS and on the miniaturization of the mass spectrometer itself. In the final part different fields of applications of miniaturization for MS analysis are presented. Moreover, the different technological developments and applications that are treated in greater detail in separate chapters in this book about miniaturization and mass spectrometry are reviewed.

1.1 Brief Introduction to MS Techniques and the ESI and MALDI Ionization Techniques

Mass spectrometry is an analysis technique that detects substances as a function of their molecular weight, or, more precisely, that detects substances as ions as a function of their mass-to-charge ratio (m/z). The analysis starts with the ionization of the molecules, which are subsequently separated in an analyzer according to their size (m/z ratio) before they reach the detector. A mass spectrum is composed of a series of peaks at given m/z values, indicating the presence of ionic species characterized by these mass-to-charge ratio values.

The key part of the connection between microfabricated/microfluidic devices and a mass spectrometer is the ionization of the analyte, as molecules are introduced as ions for the analysis. Subsequently, they must be ionized on the chip or at the outlet of the chip to be detected. Ionization is achieved using many different techniques, depending on the molecule properties. However, we will focus here on two ionization techniques, ESI and MALDI. These two techniques prevail nowadays in the field of MS analysis and they consist of the two main ionization methods used in combination with microfluidic analysis, although two other techniques, atmospheric pressure chemical ionization (APCI) and atmospheric pressure photoionization (APPI), have also recently been reported for on-chip detection.[9,10]

MALDI and ESI ionization techniques are known as soft ionization techniques: they are suitable for the ionization and the analysis of large molecules (MW > 1 kDa) such as polymers, proteins, peptides, nucleic acids and poly- or oligosaccharides without fragmenting them. During the ionization process, the molecules acquire enough energy to be transferred to the gas phase, but the amount of energy remains low enough not to induce any fragmentation of rearrangement of the molecules. MALDI and ESI are the two mostly used ionization techniques because of their compatibility with the analysis of large molecules, and especially biomolecules, and their routine sensitivity in the low picomole down to the femtomole range. This popularity of ESI and MALDI has mainly been caused by the explosion of analytical needs in the fields of biology and biochemistry. The discovery of these two soft ionization techniques in the mid-1980s is strongly linked to the prosperity of MS, and the recent growth of proteomics analysis marks in particular the golden age of the techniques of ESI-MS and MALDI-MS. As a consequence, the discovery of both ESI and MALDI was rewarded in 2002 with the Nobel prize to their inventors, Fenn and Tanaka, respectively.

More recent technological developments in the field of MS have strengthened this prosperity and the potential of MS as an analysis tool for complex samples, "real-world" biological samples as well as complex synthetic mixtures. For instance, coupling MS to a separation technique such as capillary electrophoresis (CE) or liquid chromatography (LC) enables one to reduce the sample complexity and to successively analyze the species present therein. Also, tandem (MS^2), or even MS^n, helps to elucidate the structure of substances and the sequence of biopolymers such as peptides.

These points, the potential of MS and the recent developments in the field of MS account for the increasing popularity of (ESI- and MALDI)-MS and explains why MS has naturally been associated with the recent miniaturization trend and the recent appearance of microfabricated and microfluidic systems.

1.1.1 ESI Technique

Electrospray ionization was first reported by Fenn in 1984, and further developed in 1988.[11] This technique relies on the generation of a spray from a liquid upon application of a high voltage. Typically, the sample to be analyzed is introduced in a liquid phase in a capillary. A strong electric field is created (high voltage of several kilovolts) between the liquid and a counter electrode (*i.e.* inlet of the mass spectrometer) placed some centimeters in front of the capillary. Upon application of the electric field, the liquid breaks into a gas of highly charged droplets whose size depends on different parameters such as the capillary inner diameter, the flow rate of the liquid and the applied voltage. During transport, the droplets evolve to the ultimate stage of ions in the gas phase. The solvent (typically, an organic solvent such as methanol, ethanol or acetonitrile) evaporates, and the droplets thereby shrink until they reach the so-called Rayleigh limit where surface tension exactly compensates electrical forces, as expressed by

$$q = ze = 8\pi(\varepsilon_0 \gamma r)^{1/2} \tag{1.1}$$

where q is the droplet charge, r its radius, γ the surface tension and ε_0 the permittivity of vacuum. Beyond this limit, the equilibrium is lost and droplets break into smaller droplets that undergo the same process. This cycle proceeds until the stage of ions in the gas phase is reached, and these ions in the gas phase enter the mass spectrometer to be analyzed.

The technique of ESI gives rise to multi-charged species as long as there are several protonation sites on the analytes. A typical ESI mass spectrum presents a group of peaks for one analyte species, corresponding to the different charge states with a Gaussian distribution. One advantage of this is that the analyzer and detector of the mass spectrometer work with a reduced mass range as ions are analyzed and detected as a function of m/z and not of their molecular weight. Mass spectra are of course more difficult to interpret compared to MALDI mass spectra, although many deconvolution techniques are now routinely available to transform raw data into reconstructed spectra presenting peaks for mono-charged species. A limitation of the technique of ESI, especially compared to MALDI, is the low analysis throughput and its tedious and cumbersome preparation with the manual introduction of the sample in the capillary source.

1.1.1.1 ESI vs. nanoESI

There are now two commonly defined regimes for electrospray analysis. These regimes are distinguished by the inner diameter of the capillary source, the liquid flow rate and the applied ionization voltage, and as already mentioned above these three parameters dictate the size of the generated droplets. These two regimes are

- classical ESI with a capillary inner diameter of *ca* 100 μm, a flow rate of 1–20 μL min^{-1} and ionization voltage of 3–4 kV; and
- nanoESI with a capillary inner diameter of *ca* 1–10 μm, a flow rate below 1 μL min^{-1} and ionization voltage of 1 kV.

The miniaturization of the technique started in 1994 with the description of microESI by the group of Caprioli.[12,13] A further step towards miniaturization was reported by Wilm and Mann in the 1990s with the development of nanoESI.[14] The miniaturization of the technique is driven by the improvement it offers. ESI is sensitive only to the concentration of species and not to the total amount of analytes; as a consequence, its miniaturization does not affect the sensitivity of the analysis but rather enhances it, as explained below.

Working with smaller sources appears to be an advantage as it leads to the formation of smaller droplets, and this in turn gives a number of improvements. The increased surface-to-volume ratio of the droplets brings two main advantages: it favors evaporation phenomena allowing for using a higher relative amount of water in solution; and it yields an increased surface charge density, which promotes coulombic fission of the droplets. The evaporation–fission cycle becomes more efficient and gives more ions in the gas phase, and this leads all together to an increase of the ionization yield, from 10^{-9} for classical ESI up to 10^{-4} for nanoESI. This is illustrated by Equation (1.2) that gives the charge concentration as a function of the droplet size (radius): the smaller the droplet, the higher the charge concentration, and the higher the ionization probability for a molecule in the droplet:

$$\frac{q}{V} = 3\left(\frac{\varepsilon_0\gamma}{2r^3}\right)^{1/2} \tag{1.2}$$

where q/V represents the charge volume concentration in a droplet, r the radius of a droplet and γ the surface tension. Moreover, this increased charge concentration limits ion suppression phenomena; the competition between molecules to acquire a charge is lower, and as droplets are smaller there are also obviously fewer analytes per droplet. One particular consequence is a higher tolerance of the analysis to the presence of salts or other solution contamination; this is of great interest for the technique of ESI where the presence of salts usually fully hinders the process of ionization.

Additionally, miniaturization of the sources also leads to a decrease of the sample flow rate and the use of a lower ionization voltage. Ionization conditions are smoother, and the ionization source can be placed closer to the mass spectrometer inlet. Consequently, not only more ions are formed, but also more ions enter the mass spectrometer for their analysis.

The enhancement in performance brought about by nanoESI compared to classical ESI is crucial for some fields of applications, such as bioanalysis, and especially proteomics. Miniaturizing ESI provides increased sensitivity of the analysis and enables its application for the analysis of complex samples with a wide range of analyte concentrations. When working with nanoESI the probability of ionizing molecules is higher, and this is of great importance for the analysis of complex real-world samples where some compounds are present as traces.

As ESI works on a continuous flow of liquid, it has quickly been coupled to LC or other liquid-phase separation techniques as an alternative to optical detection.[15] Mass spectrometry gives more information on the eluted compound, and the resulting hyphenated technique enables one to decrease the complexity of samples before their analysis by MS. High performance liquid chromatography (HPLC) is coupled to conventional ESI-MS while nanoLC is connected to nanoESI-MS for a better match in the flow-rate values.

1.1.2 MALDI Technique

Matrix-assisted laser desorption ionization has simultaneously been developed by Karas and Hillenkamp[16] in Germany and by Tanaka *et al.*[17] in Japan in 1985. With this technique molecules are ionized via laser irradiation of the sample and with the help of other small organic molecules, called the matrix. The matrix strongly absorbs the light of the laser and transfers it together with a charge, mostly a proton, to the analytes. Thereby, analytes reach the gas phase as ions that are ultimately analyzed by the mass spectrometer.

A MALDI-MS analysis proceeds as follows. A solution of matrix is mixed with the sample to be analyzed. A droplet of the resulting mixture is deposited on a plate and allowed to dry; the evaporation of the solvents leads to the co-crystallization of the analytes with the matrix. Once placed under vacuum, the spots of crystal are irradiated with the laser; the laser energy is absorbed by the molecules of the matrix, and subsequently transferred to the analytes that are desorbed from the surface. Simultaneously, the analytes capture a proton from the matrix molecules and become ions. Analytes finally reach the state of ions in the gas phase after desolvation of the matrix molecules.

As for ESI, the MALDI technique depends neither on the properties of the analytes nor on the absorption properties of the molecule as energy transfer proceeds via molecules of matrix. Besides, the mass or the size of the molecules does not influence the ionization and desorption process so that it can be applied to any molecule. Contrary to ESI, analyzed species are mono-charged with MALDI; this simplifies the interpretation of mass spectra but this imposes working with a detector covering a wider mass (m/z) range.

The laser wavelength can be either in the UV or infrared range, with the former being most commonly used. Therefore, matrix molecules are aromatic compounds that can absorb UV light and present a carboxylic acid moiety for the protonation of the analytes. Matrix molecules are simultaneously detected together with the analytes, and give peaks in the low mass range, *i.e.* below m/z 600. Consequently, MALDI-MS analysis is often limited to compounds with a molecular weight above m/z 600. Compared to ESI, the MALDI technique has a higher tolerance to the contamination of samples, and especially to the presence of salts that strongly hinders analyte ionization in ESI-MS. On-target sample cleaning is widely used in case of a high level of contamination. Another advantage of MALDI compared to ESI is the higher throughput of the analysis and its possible automation.

While ESI works on continuous flows of samples, MALDI analysis is performed on droplets or discrete amounts of liquid. Yet, recent developments have aimed at coupling a separation step relying on liquid chromatography to MALDI-MS analysis. For that purpose, the liquid eluted from the chromatography column is deposited in a continuous and automated way on a MALDI target.[18,19] Other improvements concern the MALDI target to alleviate the use of a matrix. The matrix can be covalently attached on the target plate surface to avoid its desorption together with the analytes,[20] or targets based on porous silicon[21] are used that does not require the addition of a matrix and that gives enhanced analysis sensitivity. Lastly, an obvious improvement for the process of ionization comes from on-target concentration of the analytes, as will be discussed later. An easy way to obtain sample concentration is to confine it on a smaller surface area. Subsequently, a major breakthrough recently was obtained with the appearance of smart targets consisting of a uniform hydrophobic area patterned with small hydrophilic spots where the sample is deposited.[22]

1.2 Why Couple Microfabricated Chips to MS? Chip-MS: Matching and Improvement

Coupling a microfluidic or microfabricated system to MS appears to be fruitful for both the detection on the chip (microchip point of view) and for the MS analysis (MS point of view). On the one hand, MS is a powerful technique for on-chip detection due to its sensitivity and the amount of information it provides on the sample; on the other hand, by using microfluidics prior to the MS analysis, new opportunities for the field of MS are created as it provides better MS capabilities compared to conventional sample preparation techniques.

1.2.1 Microchip Point of View

A first interest of MS when used in combination with microfabricated structures, or at the outlet of microfluidic devices, is the match in the volume of liquid handled. A typical MS analysis requires less than 1 μL of liquid, for ESI-MS as well as for MALDI-MS techniques. When working with a continuous flow of liquid and ESI-MS, the MS performance is even more enhanced for flow rates down to 50–100 nL min^{-1}: the lower the flow rate, the better the MS analysis. This flow-rate range corresponds to flow-rate values observed in microfluidic devices. Consequently, the technique of MS is easily scalable and exhibits an enhanced response when the sample size is decreased. This is not the case for instance for other detection techniques, such as UV absorbance or amperometry: these two techniques require large detection area or volume, which is the opposite of the quest of microfluidics. This first advantage of MS compared to other technique goes together with its high sensitivity.

Mass spectrometry is a fairly universal technique: it is a label-free detection technique and can be applied to any molecule as long as it can be ionized (*i.e.* that it presents a protonation site for instance). Optical techniques (*e.g.* UV absorbance and fluorescence-based techniques) require that the analytes present given properties to be detected. They must absorb or emit in the UV or in the fluorescence range. If they do not, they must be coupled to an external moiety (aromatic group, fluorophore) that presents such absorption/emission properties, and this imposes an additional derivatization of the analytes before their detection, but after their separation for instance on a microfluidic device. On the contrary, MS is widely used without any further derivatization step of the analytes for a large variety of species with different sizes: small organic compounds, inorganic compounds such as macromolecular complexes, any kind of biological samples, from metabolites up to large native proteins via oligosaccharides, nucleic acids and peptides. Lastly, MS enables the simultaneous analysis of different compounds, and consequently the analysis of complex samples. The technique presents a large dynamic range for the analysis and covers a wide range of concentrations, also in a single complex sample without any compound separation.

An obvious advantage of MS compared to other techniques is the amount of information a single MS analysis can provide about a sample. Optical techniques give a binary response regardless of the analyzed compound; absorption/emission or not. It is therefore difficult to identify the analytes as these techniques do not yield any further information on the nature or the structure of the analytes. Mass spectrometry does. Firstly, analytes are detected as a function of their molecular weight and the precision of the mass determination lies in the region of 10^{-4} amu. From this basic information, the size of a molecule can be deduced, and the molecule can be identified. The precision of the molecular weight of the species is particularly crucial for biological analysis, and the characterization and identification of large compounds such as peptides and proteins. Secondly, MS analysis is nowadays routinely performed in the tandem (MS^2) or even MS^n mode, where molecules are analyzed 2 or n times, respectively. For instance, for a MS^2 analysis, molecules are analyzed a first time to provide their molecular weight. Thereafter, they are individually selected, fragmented and the resulting fragments analyzed anew. The characterization of the resulting fragments provides structural information on the analyzed species. Tandem MS is widely used for instance in the fields of proteomics, glycomics, metabolomics or organic chemistry. For the two former fields of applications, fragmenting the analytes enables one to derive their sequences. In particular, for proteomics analysis, the peptide sequences are used for database mining and the identification (name, nature and origin) of proteins present in the analyzed samples. The fragmentation of smaller compounds such as small synthetic organic molecules or metabolites gives the precise structure of the compounds, and for instance the localization of substituted groups. NMR analysis provides the same type of structural information, but the latter requires a much larger amount of material (several milligrams) for a single analysis.

Lastly, MS is compatible with high-throughput analysis. Each analysis, for both ESI-MS and MALDI-MS, is fast and lasts for less than a minute. As already mentioned, MALDI-MS analysis can easily be automated with dedicated software so that hundreds of samples can be analyzed without human intervention. ESI-MS lends itself less to high-throughput analysis as sample preparation and introduction in the ionization capillary are more tedious and time consuming. Still, automation is fully conceivable when coupled to LC for instance.

1.2.2 MS Point of View

1.2.2.1 Gain in Analysis Quality (Reproducibility and Sensitivity)

The coupling to microfabricated structures or a microfluidic-based preparation of samples also benefits the MS analysis. Replacing conventional ionization tools by microfabricated structures brings a tremendous enhancement of the analysis quality. As discussed above, both ESI-MS and MALDI-MS can exhibit enhanced performance when using smaller ionization sources for introducing the samples or smaller spots for sample deposition, respectively. In both cases, the smaller the structures, the higher the ionization yield and the less the amount of sample required and consumed. Moreover, using microfabricated structures guarantees their reproducibility, and thereby the reproducibility of the analysis. NanoESI in particular suffers from the poor quality and a lack of reproducibility of the ionization sources whose fabrication relies on heating and pulling technique to realize apertures in the range of a few micrometers. For ESI-MS, the use of microfabrication techniques reliably produces small sources with decreased characteristic dimensions, leading to enhanced ionization phenomena. Moreover, the electrical contact for applying the ionization high voltage can be fabricated using robust microfabrication processes and integrated into the sources providing better and smoother conditions for high-voltage application. Similarly, microfabricated targets for MALDI-MS analysis would present smaller patterns for sample deposition, leading thereby to on-target sample concentration.

1.2.2.2 Analysis Automation and Multiplexing

The chip format is fully compatible with the use of robots and the automation of the analysis, leading to high-throughput analysis or screening. The connection to a robotic interface is already routinely done for MALDI-MS analysis, and is especially interesting for ESI-MS for which the preparation of the samples is tedious and time consuming, and requires a skilled operator for the introduction of the sample in the ionization source. In that case, the sample preparation step represents a limitation for high-throughput analysis. This is no longer required if a microfluidic system is coupled to a robotic and automated interface. MALDI-MS analysis benefits also from the use of an automated and robotic interface for sample deposition on a target. Lastly, microfluidic systems lend themselves also to analysis multiplexing; as they are fabricated using microtechnology techniques, each single microchip can include several independent analysis microsystems ("horizontal integration"). This possible integration on a microchip is another step towards high-throughput analysis especially if the different systems "work" simultaneously.

1.2.2.3 Integration of Sample Preparation

One strong point of microfluidic systems is their high level of integration. This integration can be "horizontal" when a microfluidic chip presents several identical analytical systems in parallel, but also "vertical" when several subsequent steps of a single process can be performed on-line using a single system. Sample preparation prior to MS analysis usually proceeds in several steps of sample treatment (*e.g.* reaction and digestion), separation and especially purification (*e.g.* desalting) and concentration. In a conventional protocol these different steps are performed manually by an experimenter and off-line, and this implies extensive and time-consuming manipulation of the samples, and as a result increased risk of sample loss and contamination, and human error. For instance, preparation of proteomic samples proceeds as follows. Proteins are extracted from a crude biological matrix, the proteins are processed (denaturation, disulfide bond reduction, *etc.*), digested, and the resulting peptides separated and purified using LC techniques. Particularly, the last step of sample separation and purification limits the analysis throughput, as a liquid-based separation roughly lasts one hour, which is much longer than the acquisition time for a single mass spectrum. On the contrary, if carried out on a microfluidic device, the preparation of the samples is fully integrated and much faster, because of the reduced size of the microsystems and the absence of human intervention, than when using a conventional methodology and gives rise to few or none of the issues listed above. Besides the factor of time, microfluidic systems lend themselves better to the control of low flow rates that leads to better ESI-MS analysis conditions. Lastly, using a fully integrated microsystem appears to be a cheaper alternative to bulky and conventional devices (*e.g.* nanoLC columns).

1.2.3 Towards a Single Chip for a Full Analysis

The next step to get fully integrated analytical tools that perform a total sample analysis from sample preparation to sample detection is to miniaturize the detection instrumentation and to integrate it into the same miniaturized device. Such efforts have already been described and successfully achieved for optical detection techniques. In the field of MS, the miniaturization of the instrumentation (mass spectrometer) started much before the miniaturization trend for organic and biological applications. Such development is driven by the need to have portable detection systems for on-site analysis for various fields of applications, among which, for instance, are spatial or environmental analysis and homeland safety. The mass spectrometer itself is often not sufficient for performing such analysis, especially when dealing with complex mixtures. Consequently, coupling the MS analysis with a front-end sample preparation step appears to be necessary so as to purify, concentrate or even separate species present in the crude sample. Thereby low-abundance species can be detected from complex samples that present a high level of contamination.

Additionally, as already mentioned earlier in this chapter, coupling a miniaturized device to MS is a sort of paradox if one compares the difference in sizes of the two parts of the analysis setup: while the microsystem has a size in the millimeter range and is portable, the mass spectrometer can be up to several meters in size and some hundreds of kilograms in weight! As a consequence, the analysis is confined to dedicated laboratories. Coupling microfabricated systems to a miniaturized and portable mass spectrometer having a size comparable to that of a microsystem to give a fully integrated and portable analysis setup would also benefit, for instance, the fields of biological and medical analysis.

1.3 Coupling Microfabricated Structures and Microfluidic Devices to MS

The combination of miniaturized devices with MS can be divided in several categories, first according to the ionization technique which is used (ESI *vs.* MALDI), and second depending whether the device is fluidic (microfluidic device) or static (microfabricated structure or device), as a large number of miniaturized devices for MALDI-MS analysis mainly consist of smart MALDI targets without any integrated microfluidics. Fluid handling is in that case carried out by either an external experimenter or an automated and robotic interface. We will distinguish here these different "categories".

1.3.1 Microsystems and ESI-MS

Electrospray ionization works on a continuous flow of liquid, wherever the sample is introduced in the mass spectrometer by infusion or by using a pumping system that controls the flow rate (*e.g.* connection to liquid-based chromatography). Therefore, coupling a microsystem appears to be easier and more natural for ESI-MS analysis than for MALDI-MS, as is the case for a LC-MS connection. The microfluidic chip can be seen in that case as a miniaturized and more integrated LC system that is connected to a mass spectrometer. However, in the case of a microsystem the optimization of the coupling represents a bigger challenge, especially to avoid any kind of dead volume, as for the smaller systems the dead volumes are larger!

Coupling microfluidic systems to ESI-MS has been achieved via three different technological approaches. In the first approach, the electrospray is generated at the outlet of the microchannel, directly from the edge of the chip. This approach was firstly reported by Ramsey and Ramsey in 1997.[23] The ionization voltage was applied at the inlet reservoir of the chip. Nonetheless, this simple coupling approach was not seen to be very successful as the liquid had a tendency to spread around the channel outlet instead of forming a spray. Two improvements were subsequently introduced, a pneumatic assistance to force the formation of an electrospray and a hydrophobic coating on the side surface of the chip to alleviate any spreading phenomena.

The second alternative relies on the use of a conventional capillary ionization source which is connected at the outlet of a microchannel, the capillary being directly inserted into the outlet channel or connected to the chip via a transfer capillary.[24] The use of a capillary enables one to focus the liquid and direct it into the mass spectrometer. Using this configuration the spray quality is roughly the same as for conventional ESI-MS analysis and several other issues arise from this type of coupling. Capillaries are inserted manually which prevents any mass production of integrated systems. Besides, the connection between the microfluidic channel and the capillary inserted therein must be optimized to minimize any dead volume.[25] Lastly, the inserted capillary is glued on the chip to maintain it in place, and care must be taken that the glue does not dissolve in the sample, which would induce background noise on the mass spectra.[26]

The third approach is to micromachine the ionization source as a sharp structure and to integrate it onto the microfluidic chip. This last alternative apparently does not suffer from any of the aforementioned issues. The ionization source is fabricated at the same time as the rest of the microsystem, and should exhibit better characteristics than commercial capillary-based sources (smaller dimensions and reproducibility). With this approach, the production of multiplexed microsystems with separate sources for any individual device to avoid any cross-contamination issue is conceivable and easier. This microfabrication-based route to produce integrated microsystems or only ionization sources has already given rise to several devices based on different designs and materials. Structures can be classified according to two main types of designs, whether the ionization source reproduces a capillary-type cylindrical structure[27,28] or whether the ionization source is planar and placed horizontally at the end of the outlet microchannel.[29] Microfabricated sources have been made from different materials such as silicon,[28] and also from polymer-based materials such as polydimethylsiloxane (PDMS),[29] poly(methyl methacrylate),[30] SU-8,[31] polyimides,[32] poly(ethylene terephthalate),[33] parylene,[34] polycarbonate,[27] *etc*. The choice of the material has two main impacts on the structure and the analysis quality: it should be compatible with the production of reduced-sized structures in the micrometer range and should not give rise to any contamination issue, via degradation or dissolution in the solvent used for the analysis for instance. Another improvement to microfabricated sources consists of integrating the electrode for application of the ionization high voltage;[33] this has been reported to give smoother and enhanced ionization conditions. A number of these microfabricated structures, demonstrated in stand-alone conditions or integrated in a more complex microfluidic device, are described in this book, in the chapters presenting ESI-related microfluidic development.

1.3.2 Microsystems and MALDI-MS

MALDI ionization mostly takes place in vacuum in spite of the recent development of atmospheric pressure ionization sources. Therefore, coupling a microfluidic system to MALDI-MS implies working under vacuum on the chip or performing the MS analysis off-line, *i.e.* introducing the microchip in the MALDI-MS once the fluidic operations are finished. Consequently, the first miniaturized developments for MALDI-MS analysis only concerned the microfabrication of enhanced or smart MALDI targets that do not include any fluidic component.

Microsystems for MALDI-MS analysis can be divided into three groups: microfabricated MALDI target plates, microsystems for off-line preparation of samples and microsystems that integrate both the sample preparation steps and the MALDI targets, with an on-line or off-line analysis of the samples.

The first category comprises micromachined MALDI targets that aim at concentrating and eventually cleaning samples on site using passive structures. One limitation in MALDI analysis lies in the fact that samples spread on the surface of the target upon deposition, and this results in a decrease of analysis sensitivity. To circumvent this liquid spreading, MALDI targets can be patterned so as to decrease the MALDI spot size and thereby to yield local concentration of the analytes. Two main approaches have been reported. The first one consists of machining nanovials where samples are deposited. These vials have an inverted pyramid structure, at the bottom of which the sample is concentrated.[35] Such systems were first realized in silicon,[35] and more recently their realization has been demonstrated in a polymer material.[36] The analysis time is also decreased as the experimenter does not need to search for the sample on a large spot of crystal. The second approach relies on a chemical patterning of the surface. Typically, the whole surface of the MALDI target is made hydrophobic (via appropriate coating, a monolayer or the use of a hydrophobic material) and small hydrophilic spots are created therein.[22] Thereby, the sample is confined to the hydrophilic spots, and the smaller the spots, the higher the sensitivity enhancement. Both approaches give a sensitivity improvement down to the attomole range. Another major technological development in this direction is the implementation of the purification step on such a chemically patterned MALDI target.[20] For that purpose, the hydrophilic areas of the MALDI target are functionalized with a given stationary phase aiming at retaining a specific class of compounds. These devices have in particular been developed by the company Ciphergen and are now commercially available under the name ProteinChip® devices.[20]

The second category of microsystems for MALDI-MS analysis focuses on the sample preparation steps before deposition on a (conventional) MALDI target. MALDI-MS analysis is subsequently done off-line and not on the same device. Such microsystems comprise two parts, one for the sample treatment (digestion, purification, separation) and the other for sample dispensing onto the MALDI target. The first part consists of a microfluidic device working on a continuous flow of liquid. The second part aims, as before, at limiting sample spreading on the MALDI target; however, instead of structuring a MALDI target, a robotic interface is used to improve the deposition of the sample on a confined area on a (conventional) MALDI target. Sample dispensing has been described using various types of technologies: a spotting technology, piezoelectric actuation or by spraying the sample as a thin film. The spotting technology[18] borrows much from the interface which is used for coupling LC to MALDI-MS, but works here for smaller amounts of liquids. Another technique uses a piezoelectric dispenser[35] that enables continuous deposition of sample on a MALDI plate via sub-nanolitre droplets of liquid. As soon as the droplet reaches the surface, solvent evaporates so that the sample does not spread on the surface. This dispensing technique has been coupled to a microreactor for the tryptic digestion of proteins and to a nanovial-based MALDI plate.[35] An original dispensing method developed by the group of Murray is based on a rotating ball placed at the outlet of a microfluidic system on which samples are separated;[37,38] the rotating ball makes the interface between the zone at atmospheric pressure where samples are prepared and the zone under vacuum where MALDI ionization takes place. This technique is described in more detail in Chapter 10. The last reported technology to couple sample preparation steps with a continuous flow to MALDI-MS consists of spraying the sample into a uniform film onto the MALDI target surface.[39] This technology provides more homogeneous sample crystals, whose size is dictated by the distance between the spraying head and the MALDI surface. Spots down to 170 μm diameter with 150 μm spacing have been realized using the spraying technique. Sample deposition has been demonstrated after digestion of protein samples inside the spraying device using immobilized trypsin. One major advantage of these devices is their ability to be multiplexed and to work in parallel without any risk of contamination from one sample to another.

The third and last group of microfluidic systems for MALDI analysis comprise integrated systems that include both steps of the analysis, *i.e.* sample preparation and the MALDI target. The advantage of these systems is that they are fully integrated and hardly need any external intervention for sample analysis. While for most of these devices sample preparation is carried out off-line (*i.e.* not in the mass spectrometer), one unique system works "under vacuum" and enables on-line sample treatment and analysis. In this 3rd category, the first microfluidic device was developed by Gyros AB and gave rise to a commercial product, the Gyrolab®. The Gyrolab® is dedicated to biological sample analysis using affinity chromatography, and particularly targets the field of proteomics applications.[40] This device has the shape of a compact disk that includes a series of radial microfluidic networks with MALDI targets placed at the external edge. Sample preparation is performed off-line on a dedicated workstation, with a robot making the link between conventional 96-well plates and Gyrolab® devices that contain 96 independent analysis networks. Thereafter the device is cut and the MALDI target parts are placed in the mass spectrometer for the analysis. Liu *et al.*[41] have developed an alternative system whose channels are not covered with a lid for accessibility to the laser. In those open microchannels, peptide and oligosaccharide samples to which a matrix solution has first been added are separated using CE and dried in place under vacuum. MALDI analysis is done on the crystals formed in the channels by scanning along the whole channel with the laser. EWOD (electrowetting on a dielectric) is particularly suitable for MALDI-MS analysis, where liquids are displaced as droplets, *i.e.* the format which is conventionally used for MALDI analysis. Consequently, several integrated microsystems for MALDI-MS analysis rely on this pumping principle that enables sample treatment and mixing of the sample with the matrix solution. Wheeler *et al.* have reported

EWOD-based microfluidic systems for MALDI-MS analysis. In a first system,[42] the principle of EWOD is exploited to displace a sample droplet, to mix it with a droplet of matrix solution and to place it on a Teflon-based MALDI target. They further improved their system by including a purification step of the sample;[43] a droplet of sample is displaced onto a Teflon-coated site where analytes and contamination adsorb. This spot is washed with deionized water to remove contamination, and later matrix is added to the purified analytes. In both cases, the system is opened at the end of sample preparation and spots are dried under vacuum before the system is introduced in the mass spectrometer. Chapter 12 presents an alternative system based on EWOD which is dedicated to kinetics studies;[44] its performance is illustrated here for the investigation of an enzymatic reaction. The functioning principle of this device relies on the successive mixing within a few milliseconds of liquid droplets: (i) between the enzyme and the substrate solutions, (ii) the reaction mixture and a quenching agent, and finally (iii) the addition of the matrix solution to the quenched reaction mixture. A last and unique integrated microsystem for MALDI-MS analysis enables the on-line preparation of the samples within the mass spectrometer. This system which is described in Chapter 11 benefits from the vacuum environment inside the ionization source of the MALDI mass spectrometer for fluid actuation. Detection is done via the irradiation of the sample, either in the outlet reservoir of the device[45] or through a detection window added for this purpose in the outlet channel of the chip.[46] This system has successfully been demonstrated for the investigation of both biochemical and chemical reactions.

1.4 Main Fields of Applications of Miniaturization for MS Analysis

The combination of miniaturization to MS analysis finds two main fields of applications: biological/biochemical analysis and organic chemistry. These are two important fields of applications of MS analysis. In both cases, by using microfabricated devices for MS analysis or by combining microfluidics to MS, novel capabilities are reached and new applications are found.

1.4.1 Biological Analysis: Medical Analysis, Proteomics

The main field of application of microfluidics-to-MS coupling is the field of biological and medical diagnostics, and especially proteomics or protein analysis. Microfluidic developments in this field of analysis are driven by the large number of advantages provided by the miniaturization of the analysis and the coupling of microfabricated devices to MS analysis. Most of these advantages are already listed in Section 1.2.2. These advantages can be divided in three categories: (i) the gain in analysis quality and sensitivity, (ii) performance improvement by the high level of integration of sample treatment and preparation and (iii) the suitability of such microfabricated structures for high-throughput analysis. Using microfabricated tools and structures bring an enhancement in the analysis sensitivity. For ESI-MS analysis, this is accounted for by the improved ionization efficiency which is reached by decreasing the size of the ionization source, while MALDI benefits from both the miniaturization of the spots of analytes and the use of a microfabricated interface for sample deposition. Moreover, the miniaturization of the analysis may also result in a greater tolerance to sample contamination. This is of major interest for specific subfields of biological analysis where compounds cannot be amplified as is the case for nucleic acid analysis. For metabolomics applications, analytes of interest may be present as traces in the samples and crude samples consist of complex biological matrix. Proteomics analysis suffers from the same issues: proteins present a large dynamic range and proteins of interests are mostly of low abundance. Besides that, for proteomics applications analysis resolution is also an issue to guarantee the accurate identification of proteins using database mining. This is now made possible with the development and improvement of FT-MS instruments. Sample treatment and preparation prior to MS analysis can be fully integrated on a microfluidic device limiting sample manipulation. This high level of integration results in faster analysis and higher sample quality due to decreased loss and contamination. Moreover, more steps can be integrated on a single chip, and for proteomics applications a single chip can be used for performing sample preparation from the protein digestion step to the ionization of a purified and separated peptide mixture. Lastly, due to the reduced sample preparation and analysis times by using miniaturized systems and the possibility to integrate a (large) number of independent devices on a single system, microchips are a potential tool for high-throughput and multiplexed analysis. This is all the more reinforced by the easy connection of microfluidic systems to an automated and robotic interface.

1.4.2 Organic Chemistry, Biochemical Analysis and on-Line Analysis

The connection of microfluidics to mass spectrometry is also widely used for the investigation of biochemical and organic chemistry reactions. Chapters 9, 11 and 12 illustrate such couplings for both ESI-MS and MALDI-MS. The advantages of direct coupling between a microfluidic system on which a reaction is carried out and a mass spectrometer are manifold. Firstly, it should be emphasized that with a microfluidic system the reaction mixture is always more homogeneous than in conventional glassware. Therefore, on-line analysis using a microsystem is always done on a homogeneous medium. Consequently, MS analysis from a microsystem gives precise snapshots and information on the progression of a given reaction in real time, as long as the reaction is "stopped" before the reaction mixture is introduced in the mass spectrometer. For ESI-MS, this is easily achieved by simply coupling a microfluidic system or simply short capillary tubing to a mass spectrometer and by taking care to avoid any dead volume. For MALDI-MS, unless the reaction is performed within the mass spectrometer and analyzed on-line, care should be taken to quench the reaction at given time points. The real-time characterization of reactions makes possible novel investigation of reaction kinetics, as reported for instance in Chapter 12 for the study of enzymatic reactions. Moreover, these continuous and real-time snapshots also enable one to study and identify reaction intermediates as long as their lifetime is long enough for an MS analysis. A last and novel application of such microfluidics-to-MS coupling concerns high-throughput analysis and the screening for instance of libraries of compounds using multiplexed microfluidic systems. This novel analytical tool in this context appears to be complementary to combinatorial chemistry studies that give rise to huge amounts of compounds to be characterized and identified.

1.4.3 Mass Spectrometer Miniaturization

A brief history of mass spectrometer miniaturization is given in Chapter 13. Developments for the miniaturization of MS instrumentation have been especially driven until now by a few fields of application that require on-site analysis and portable analysis instruments. However, novel fields of analysis would much benefit from fully integrated and portable analysis tools as we already discussed in Section 1.2.3.

One main field of applications of miniaturized mass spectrometers is related to space exploration.[47] In such a context, the instrumentation must fulfill a number of requirements, starting from its size, weight and power consumption and going to its resistance to harsh environmental conditions (microgravity, exposure to shock and vibrations, for instance) and including of course its analytical performance and ability to detect trace-level species in complex samples. In that field, miniaturized mass spectrometers are mainly developed for two types of analysis: determining the composition of planetary atmospheres and monitoring the air quality in aircraft during long-duration manned space missions, with especially the detection of volatile organic compounds (VOCs) in the air.[48]

A second important application of miniaturized mass spectrometers which is currently growing rapidly in the USA is homeland safety and the prevention of civil (bio)terrorism. For such purposes, the miniaturized mass spectrometer can also include a system for wireless data transmission for remote site sensing.[49]

More recently, handheld mass spectrometers have also been successfully used for the study of ion/molecule reactions[50] as well as for biological analysis, and the detection of peptides, oligonucleotides and biological spores.[51]

Recent achievements in the miniaturization of MS instrumentation gave rise to the first instruments with characteristic dimensions in the micrometer range[52–54] as in the case of microsystems dedicated to analysis and chemical processing. Such achievements should lead in the near future to the development of a fully integrated setup including the steps of both sample preparation and detection.

1.5 Content of the Book

In this book we endeavor to cover most of the topics highlighted in the introduction, and to present (i) technological developments achieved for the coupling of microfluidic systems to MS, both for the ESI and MALDI ionization techniques, (ii) relevant applications of microfluidics-to-MS couplings and (iii) research aiming at miniaturizing the mass spectrometer itself. We hope thereby to give a complete and comprehensive view to the reader of the "miniaturization and mass spectrometry" field with this collection of chapters.

The book is divided into three sections, dealing respectively with ESI-MS coupling, MALDI-MS analysis and the miniaturization of mass spectrometers. The ESI-MS section starts by illustrating early couplings of microfluidic systems to nanoESI-MS using a transfer capillary. Subsequent chapters describe the development of novel microfabricated sources for nanoESI-MS analysis, these sources being integrated on a microfluidic system or not (yet), using various and eventually new geometries, and using various materials (silicon, parylene, SU-8, PDMS, polyimide). Finally, the application of on-line nanoESI-MS analysis after a microfluidic step of sample handling is demonstrated in the fields of (i) proteomics with the digestion/separation/purification of samples performed on a microchip and analyzed on-line by ESI-MS and (ii) organic chemistry and biochemical analysis. The second section of the book reports on MALDI-MS analysis after microfluidics through three chapters reporting a novel microfluidic-to-MALDI target interface and two fully integrated microfluidic systems. The first chapter describes an original interface for coupling a microfluidic device for proteomic sample preparation to a MALDI target, for both off-line and on-line analysis. In the second chapter of the section, on-line monitoring of reactions carried out within a mass spectrometer and using a microfluidic chip with vacuum-driven actuation of liquid is discussed. Lastly, the functioning of an EWOD-based microfluidic system is illustrated for real-time kinetic studies of an enzymatic reaction. The last section of the book gives an example of a micro-mass spectrometer (micro-MALDI-TOF-MS) and its applications for homeland security and clinical diagnostics.

References

1. K. Uchiyama, H. Nakajima and T. Hobo, *Anal. Bioanal. Chem.*, 2004, **379**, 375.
2. M. A. Schwarz and P. C. Hauser, *Lab Chip*, 2001, **1**, 1.
3. E. X. Vrouwe, R. Luttge and A. van den Berg, *Electrophoresis*, 2004, **25**, 1660.
4. P. F. Gavin and A. G. Ewing, *J. Am. Chem. Soc.*, 1996, **118**, 8932.
5. A. Arora, A. J. de Mello and A. Manz, *Anal. Commun.*, 1997, **34**, 393.
6. H. Wensink, F. Benito-Lopez, D. C. Hermes, W. Verboom, H. Gardeniers, D. N. Reinhoudt and A. van den Berg, *Lab Chip*, 2005, **5**, 280.

7. R. Keir, E. Igata, M. Arundell, W. E. Smith, D. Graham, C. McHugh and J. M. Cooper, *Anal. Chem.*, 2002, **74**, 1503.
8. I. M. Lazar, J. Grym and F. Foret, *Mass Spectrom. Rev.*, 2006, **25**, 573.
9. P. Ostman, S. J. Marttila, T. Kotiaho, S. Franssila and R. Kostiainen, *Anal. Chem.*, 2004, **76**, 6659.
10. T. J. Kauppila, P. Ostman, S. Marttila, R. A. Ketola, T. Kotiaho, S. Franssila and R. Kostiainen, *Anal. Chem.*, 2004, **76**, 6797.
11. J. B. Fenn, M. Mann, C. K. Meng, S. F. Wong and C. M. Whitehouse, *Science*, 1989, **246**, 64.
12. P. E. Andren, M. R. Emmett and R. M. Caprioli, *J. Am. Soc. Mass Spectrom.*, 1994, **5**, 867.
13. M. R. Emmett and R. M. Caprioli, *J. Am. Soc. Mass Spectrom.*, 1994, **5**, 605.
14. M. Wilm and M. Mann, *Anal. Chem.*, 1996, **68**, 1.
15. B. Mehlis and U. Kertscher, *Anal. Chim. Acta*, 1997, **352**, 71.
16. F. Hillenkamp and M. Karas, *Int. J. Mass Spectrom.*, 2000, **200**, 71.
17. K. Tanaka, H. Waki, Y. Ido, S. Akita, Y. Yoshida, T. Yoshida and T. Matsuo, *Rapid Commun. Mass Spectrom.*, 1988, **2**, 151.
18. E. Nagele and M. Vollmer, *Rapid Commun. Mass Spectrom.*, 2004, **18**, 3008.
19. K. K. Murray, *Mass Spectrom. Rev.*, 1997, **16**, 283.
20. N. Tang, P. Tornatore and S. R. Weinberger, *Mass Spectrom. Rev.*, 2004, **23**, 34.
21. R. A. Kruse, X. L. Li, P. W. Bohn and J. V. Sweedler, *Anal. Chem.*, 2001, **73**, 3639.
22. T. Redeby, J. Roeraade and A. Emmer, *Rapid Commun. Mass Spectrom.*, 2004, **18**, 1161.
23. R. S. Ramsey and J. M. Ramsey, *Anal. Chem.*, 1997, **69**, 1174.
24. D. Figeys, C. Lock, L. Taylor and R. Aebersold, *Rapid Commun. Mass Spectrom.*, 1998, **12**, 1435.
25. J. J. Li, P. Thibault, N. H. Bings, C. D. Skinner, C. Wang, C. Colyer and J. Harrison, *Anal. Chem.*, 1999, **71**, 3036.
26. D. M. Pinto, Y. B. Ning and D. Figeys, *Electrophoresis*, 2000, **21**, 181.
27. K. Q. Tang, Y. H. Lin, D. W. Matson, T. Kim and R. D. Smith, *Anal. Chem.*, 2001, **73**, 1658.
28. G. A. Schultz, T. N. Corso, S. J. Prosser and S. Zhang, *Anal. Chem.*, 2000, **72**, 4058.
29. J. S. Kim and D. R. Knapp, *Electrophoresis*, 2001, **22**, 3993.
30. M. Schilling, W. Nigge, A. Rudzinski, A. Neyer and R. Hergenroder, *Lab Chip*, 2004, **4**, 220.
31. S. Le Gac, S. Arscott and C. Rolando, *Electrophoresis*, 2003, **24**, 3640.
32. N. F. Yin, K. Killeen, R. Brennen, D. Sobek, M. Werlich and T. V. van de Goor, *Anal. Chem.*, 2005, **77**, 527.
33. J. S. Rossier, N. Youhnovski, N. Lion, E. Damoc, S. Becker, F. Reymond, H. H. Girault and M. Przybylski, *Angew. Chem. Int. Ed.*, 2003, **42**, 54.

34. Y. N. Yang, C. Li, J. Kameoka, K. H. Lee and H. G. Craighead, *Lab Chip*, 2005, **5**, 869.
35. S. Ekstrom, D. Ericsson, P. Onnerfjord, M. Bengtsson, J. Nilsson, G. Marko-Varga and T. Laurell, *Anal. Chem.*, 2001, **73**, 214.
36. G. Marko-Varga, S. Ekstrom, G. Helldin, J. Nilsson and T. Laurell, *Electrophoresis*, 2001, **22**, 3978.
37. H. K. Musyimi, J. Guy, D. A. Narcisse, S. A. Soper and K. K. Murray, *Electrophoresis*, 2005, **26**, 4703.
38. H. Orsnes, T. Graf, H. Degn and K. K. Murray, *Anal. Chem.*, 2000, **72**, 251.
39. Y. X. Wang, Y. Zhou, B. M. Balgley, J. W. Cooper, C. S. Lee and D. L. DeVoe, *Electrophoresis*, 2005, **26**, 3631.
40. M. Gustafsson, D. Hirschberg, C. Palmberg, H. Jornvall and T. Bergman, *Anal. Chem.*, 2004, **76**, 345.
41. J. Liu, K. Tseng, B. Garcia, C. B. Lebrilla, E. Mukerjee, S. Collins and R. Smith, *Anal. Chem.*, 2001, **73**, 2147.
42. A. R. Wheeler, H. Moon, C. J. Kim, J. A. Loo and R. L. Garrell, *Anal. Chem.*, 2004, **76**, 4833.
43. A. R. Wheeler, H. Moon, C. A. Bird, R. R. O. Loo, C. J. Kim, J. A. Loo and R. L. Garrell, *Anal. Chem.*, 2005, **77**, 534.
44. K. P. Nichols and H. Gardeniers, *Anal. Chem.*, 2007, **79**, 8699.
45. M. Brivio, R. H. Fokkens, W. Verboom, D. N. Reinhoudt, N. R. Tas, M. Goedbloed and A. van den Berg, *Anal. Chem.*, 2002, **74**, 3972.
46. M. Brivio, N. R. Tas, M. H. Goedbloed, H. Gardeniers, W. Verboom, A. van den Berg and D. N. Reinhoudt, *Lab Chip*, 2005, **5**, 378.
47. P. T. Palmer and T. F. Limero, *J. Am. Soc. Mass Spectrom.*, 2001, **12**, 656.
48. B. J. Shortt, M. R. Darrach, P. M. Holland and A. Chutjian, *J. Mass Spectrom.*, 2005, **40**, 36.
49. L. Gao, Q. Y. Song, G. E. Patterson, R. G. Cooks and Z. Ouyang, *Anal. Chem.*, 2006, **78**, 5994.
50. H. W. Chen, R. F. Xu, H. Chen, R. G. Cooks and Z. Ouyang, *J. Mass Spectrom.*, 2005, **40**, 1403.
51. M. C. Prieto, V. V. Kovtoun and R. J. Cotter, *J. Mass Spectrom.*, 2002, **37**, 1158.
52. D. Cruz, J. P. Chang, M. Fico, A. J. Guymon, D. E. Austin and M. G. Blain, *Rev. Sci. Instrum.*, 2007, **78**, 015107.
53. D. E. Austin, D. Cruz and M. G. Blain, *J. Am. Soc. Mass Spectrom.*, 2006, **17**, 430.
54. M. G. Blain, L. S. Riter, D. Cruz, D. E. Austin, G. X. Wu, W. R. Plass and R. G. Cooks, *Int. J. Mass Spectrom.*, 2004, **236**, 91.

SECTION 1
ESI-MS

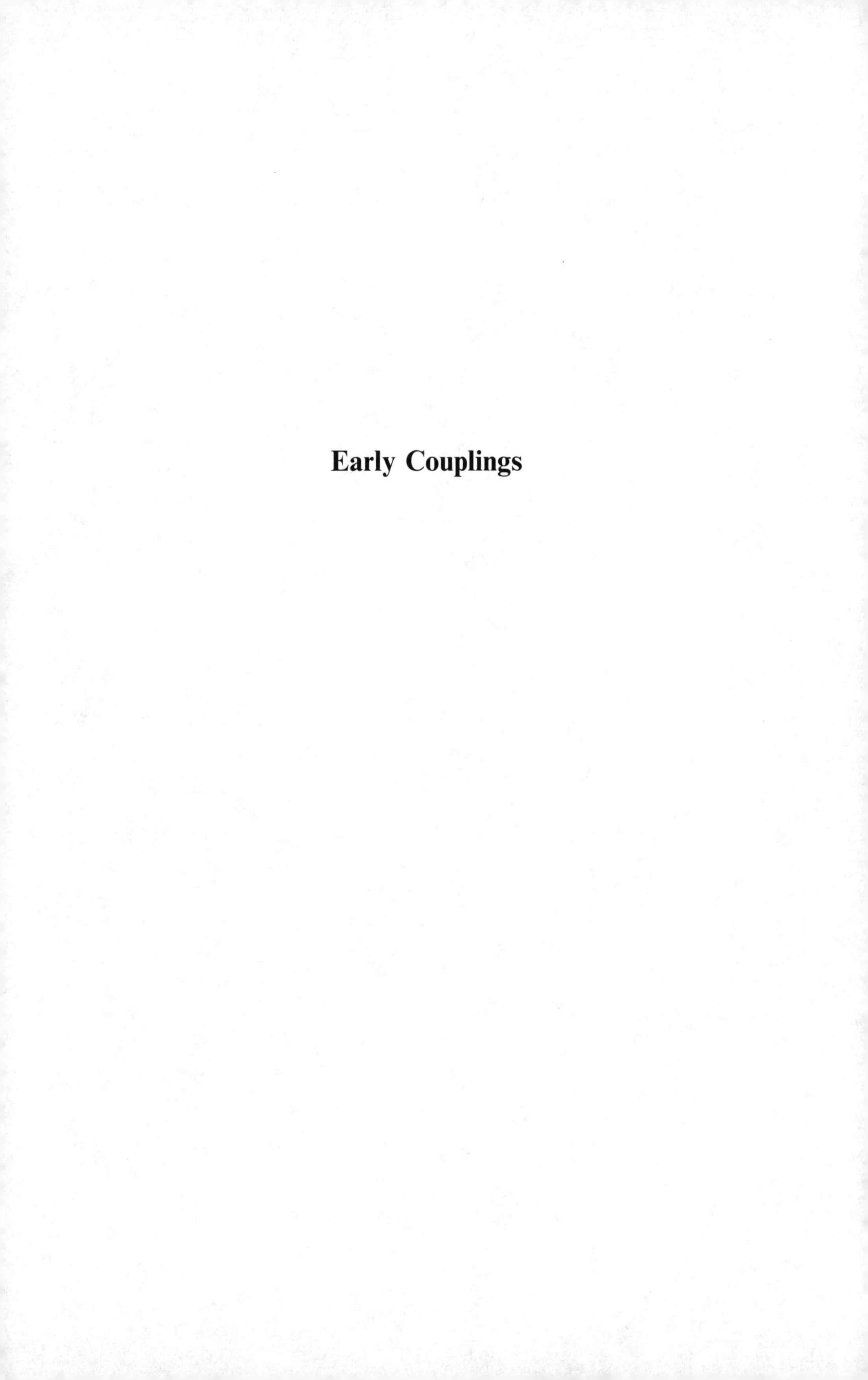

Early Couplings

CHAPTER 2

The Coupling of Microfabricated Fluidic Devices with Electrospray Ionization Mass Spectrometers

DANIEL FIGEYS[a] AND RUEDI AEBERSOLD[b]

[a] Ottawa Institute of Systems Biology, University of Ottawa, 451 Smyth Road, Ottawa, Ontario, Canada K1H 8M5; [b] Institute of Molecular Systems Biology, ETH Zürich, and Faculty of Science, University of Zürich, Switzerland, and Institute for Systems Biology, Seattle, WA 98103, USA

2.1 Introduction

The early days of proteomics were driven by the need to develop novel analytical techniques for the rapid and sensitive identification of gel-separated proteins. At that time, two-dimensional (2D) gel electrophoresis was the predominant technique for separating complex protein mixtures. Furthermore, mass spectrometry was increasingly used for protein analysis. However, serious challenges arose from the use of these technologies. Mass spectrometers and HPLC were developed primarily to analyze small molecules; however, they were not adapted to consider the full nature of proteomic samples, including sample volume, sample lost and sample contamination.

The adaptation or invention of other technologies was therefore incorporated to overcome the restrictions of 2D gel electrophoresis and mass spectrometers. For example, interfaces for performing micro- and nanoelectrospray and separation techniques for proteomics were used, although they were

Miniaturization and Mass Spectrometry
Edited by Séverine Le Gac and Albert van den Berg

Published by the Royal Society of Chemistry, www.rsc.org

rudimentary and still experimental at best. HPLC and capillary electrophoresis coupled to electrospray mass spectrometry were also actively pursued for the sensitive analysis of proteins and peptides. HPLC systems often consisted of high flow rate HPLC being flow split to lower flow rate using home-made devices and proper auto-samplers to handle proteomic samples were inexistent. Therefore, there were still clear issues with robustness, throughput and handling of small volumes of sample.

Microfabricated fluidic devices were an attractive alternative for these restrictions because they are compatible with low sample volumes and provide multiplexing possibilities, reductions in sample handling, reductions in sample contamination and, finally, the potential for the complete processing of protein samples on a single device.[1,2] At that time, microfluidic systems were primarily used with *in situ* detection devices in which the analytes remained in the system during the analysis.[3] However, this was not an option for the field of proteomics because the analytes had to be transferred to a mass spectrometer for analysis. Therefore, researchers in the field of proteomics were faced with the challenge of how to couple a microfabricated planar fluidic device to the outside world. Their options were limited. They could either electrospray straight off the chip or transfer the analytes through some sort of a connection to an electrospray device.

Over the years, many different approaches based on these two basic principles have been developed.[4–9] We decided to focus on developing an approach to transfer analytes by coupling capillary tubing with electrospray ionization devices. From this basic design principle, we were able to develop a simple three-position device for the analysis of proteomic samples by mass spectrometry.[7,10] We developed this principle further into an automated nine-position device,[6] and to perform frontal analysis separations of peptides.[11] This chapter reviews these early developments in coupling microfabricated devices to mass spectrometers.

2.2 Interfacing Microfabricated Fluidic Devices to the Outside World

We successfully developed two interfaces which couple microfluidic devices to electrospray ionization mass spectrometers.[7,10] Both interfaces were based on separating the microfluidic system from the microelectrospray interface and the mass spectrometer. A transfer capillary was used to couple the microfluidic device to a microelectrospray interface. This coupling produced a stable and permanent positioning of the microelectrospray interface and generated an optimal electrospray process achieved through pulled capillary tubing.

The first interface design[10] was based upon a butt-end connection between the transfer capillary and the end of the device's main channel (Figure 2.1A). The primary challenges were to align a capillary tubing in three dimensions with the exit of the microfludic device and then to seal them together. We opted for an approach that consisted of using multiple Teflon sleeves to align the capillary tubing with the edge of the device. Briefly, the microfabricated device was connected to a microsprayer through a 12 cm piece of 75 μm i.d. × 150 μm

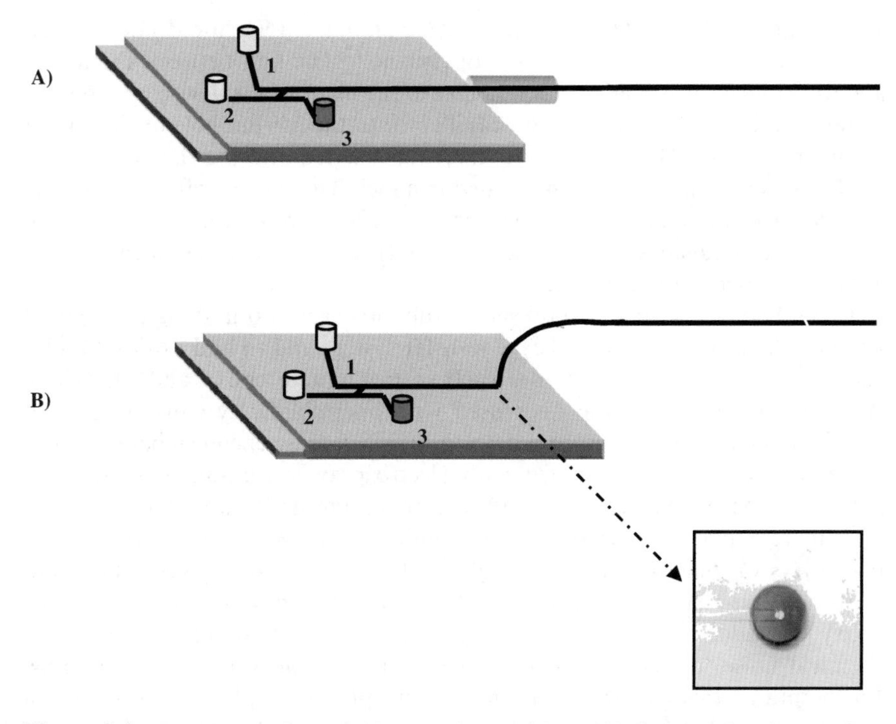

Figure 2.1 Two types of connections to the outside world that were implemented on our devices. (A) A butt-end connection was established at the end of a channel at the edge of the device. A transfer capillary was aligned (x, y, z) using a sleeve in a sleeve approach. (B) A connection through the top of the device to the main channel was established using a sleeve in a sleeve approach. In this design, the alignment was preset on the plan of the device (x, y) while the alignment on the z axis (perpendicular to the device) was easily performed by hand. The inset shows the cross-section of the capillary with the main channel viewed from the bottom.

o.d. fused silica capillary tubing which was derivatized on the inner surface with 3-aminopropylsilane.[12] The link between the device and the capillary was made by gluing a 1.6 mm i.d. × 3.2 mm o.d. Teflon sleeve to the edge of the device. Another Teflon sleeve of 250 μm i.d. × 1.6 mm o.d. was inserted into the first one until it made contact with the device. Finally, the capillary was inserted into the inner Teflon sleeve until it cleanly aligned with the etched channel of the microfabricated device. The whole assembly was held in place and was stabilized by a Teflon tube dual-shrink placed on top of the two Teflon sleeves. The alignment of this interface was critical and very time consuming.

Our second interface design[6] made the connection through the top of the device (Figure 2.1B). This design includes a guiding internal channel which aligns the transfer capillary with the end of the main channel of the microfluidic system. Here again, we used a sleeve in a sleeve system with a hand-tight fitting to seal the system. Briefly, the device was connected to a microESI source via a

15 cm long fused silica transfer capillary (75 μm i.d. × 150 μm o.d.) derivatized on the inner surface with 3-aminopropylsilane.[12] The link between the device and the capillary was made perpendicular to the plane of the device by inserting a 200 μm i.d. × 350 μm o.d. sheath capillary into the 350 μm hole at the end of the main channel. The transfer capillary was inserted into the sheath capillary so that its end reached into the etched channel. The whole transfer assembly was sealed and stabilized by a "dual shrink" Teflon tube placed over the sheath capillary. The advantage of this second design was that the alignment on two axes was preset in the cover device.

These two approaches were successful; however, both designs required glue to attach the sleeve in a sleeve transfer system and to add external buffer reservoirs, *i.e.* pipette tips. Glues such as five-minute epoxy and ultraviolet (UV)-curable glue (like the ones used with microfluidic systems coupled to UV or laser-induced fluorescence detectors) gave intense chemical backgrounds when the devices were coupled with electrospray ionization mass spectrometry (ESI-MS). Therefore, we had to explore alternative approaches to join the different parts of the device. We settled on an approach that employs two layers of glue: a heat-curable glue, which when cured gives a hard but brittle solvent-resistant seal, on top of which is placed a layer of five-minute epoxy for added durability. Although this approach significantly reduces chemical noise, the construction of the final assembly was time consuming. Furthermore, this particular assembly was prone to plugging because of glue seeping into the channel while curing and from loose particles of the cured glue.

We therefore redesigned the microfluidic system and the coupling interface to completely remove the glue and to incorporate a filtration system in the reservoir (Figure 2.2).[13] First, the glass microfabrication processes were modified to increase the total glass thickness to 3.2 mm instead of 1.2 mm, and the reservoir diameters were expanded. This made the device more robust, and it expanded the volume of the reservoir from 0.4 to 5 μL. For most samples, this procedure removes the need to add external reservoirs. For larger volumes, an expansion reservoir can also be press-fitted by hand to the glass reservoir. Second, a filter made of inert Teflon membrane was incorporated into every reservoir and exit. This reduces the possibility of channel plugging. Channel plugging is a particular problem when dealing with real biological samples which are not easily filtered off-line without sample and volume losses. Finally, a Plexiglas mount (with no glue) was constructed to hold the device in place, to align the transfer capillary into the guiding channel and to seal the interface. The end product was a microfluidic system that did not use glue, it incorporated solvent-resistant Teflon filters for filtration and it was easy to use.

These techniques represented our full attempts at connecting microfabricated fluidic devices to the outside world. We had pushed as far as we could with our limited engineering resources in order to develop the interface. At that point, it was time for industries to also get into the game of coupling microfabricated devices to mass spectrometers.

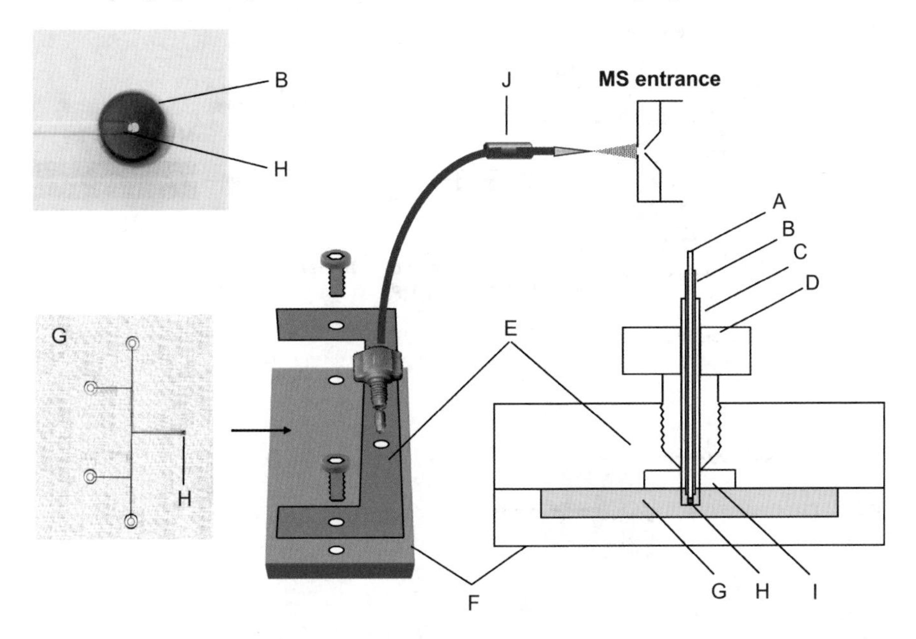

Figure 2.2 Microfabricated device and the coupling system to an ESI-MS-MS mass spectrometer. A, transfer capillary (15 cm); B, guide capillary (200 μm i.d. × 360 μm o.d.); C, Teflon tubing; D, finger-tight fitting and ferrule; E, mount—top plate; F, mount—bottom plate; G, μFAB device; H, main channel exit; I, silicone gasket; J, liquid-junction ESI interface. The inset shows the union of the device and a 350 μm o.d. and 50 μm i.d. capillary tubing at the end of the main channel. (Adapted with permission from Ref. 13).

2.3 Sample Handling and Continuous Flow Infusion to a Mass Spectrometer

2.3.1 Three-Position Device

Our first device was a simple three-position microfabricated device that was used to illustrate the principle that these devices could be coupled to a mass spectrometer (Figure 2.3A).[7] It consisted of the actual micromachined device which contained channels, sample reservoirs and electrodes, a fused silica connection and a microelectrospray ion source. The device also included conducting pads and wires connected to the individual reservoir. These pads and wires were incorporated during the microfabrication process. High voltages could then be directly applied to the individual pads. The analytes were brought from the reservoirs to the mass spectrometer using electroosmotic pumping. The inner surface of the 12 cm fused silica capillary was derivatized using 3-aminopropylsilane as previously described. At pH = 3.0, the permanent positive charge generated a strong electroosmotic flow toward the anode. The channel walls on the microfabricated device were not derivatized, and at

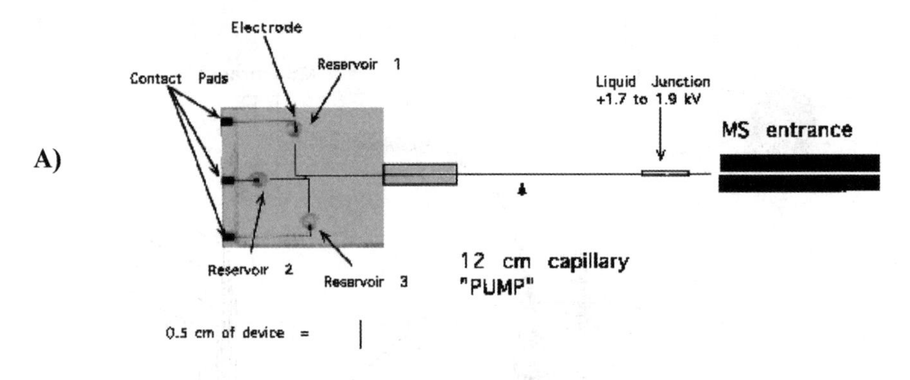

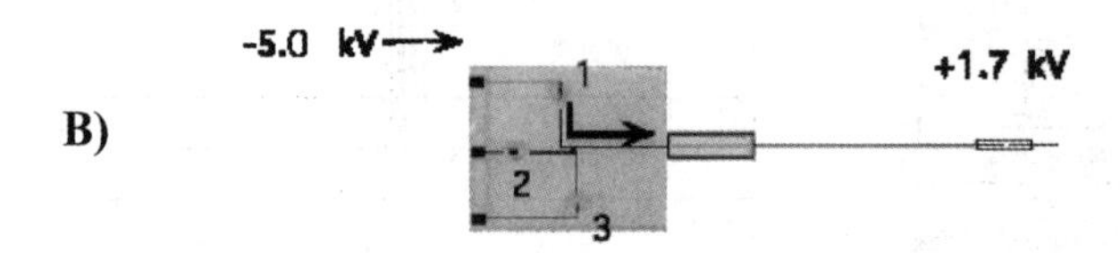

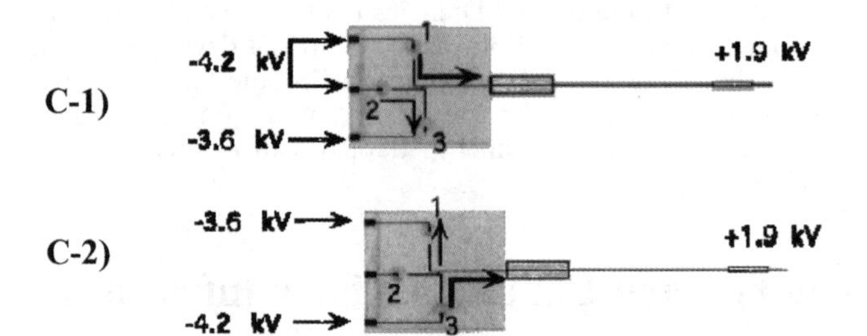

Figure 2.3 (A) Schematic illustration of a microfabricated three-position analytical system. Shown are the microfabricated device, the coated capillary electroosmotic pump and the microelectrospray interface. The channels on the device are graphically enhanced to make them more visible. (B, C) Flow diagram of experiments performed. The potentials applied and the induced electroosmotic flows are indicated for the operation of the system with a single sample (B) and in the sequential sample mobilization mode (C). The electroosmotic flows are indicated by the wide line and an arrow. (B) A potential of −5 kV was applied to reservoir 1, while reservoirs 2 and 3 were floated, and the microsprayer was at + 1.7 kV. This resulted in a constant sample flow from reservoir 1 to the MS. (C-1) A potential of −4.2 kV was applied to reservoirs 1 and 2, −3.6 kV was applied to reservoir 3 and + 1.9 kV was applied to the microsprayer. This resulted in a sample flow from reservoir 1 to the MS and a smaller flow of electrophoresis buffer from reservoir 2 to reservoir 3. (C-2) A potential of −3.6 kV was applied to reservoir 1, −4.2 kV was applied to reservoir 3, reservoir 2 was floating and + 1.9 kV was applied to the microsprayer. This resulted in a main sample flow from reservoir 3 to the MS and in a smaller flow from reservoir 3 to reservoir 1. (Adapted with permission from Ref. 7).

pH = 3.0, they generated a small electroosmostic flow towards the anode. The main portion of the total electroosmotic flow was therefore generated by the 12 cm long capillary, which is designated "pump" in Figure 2.3A. The differential in electroosmotic pumping between the microfabricated device and the capillary pump required a tight seal at the interface. The equilibrium liquid flow through the micromachined device was established by viscous dragging of the liquid. Because the dimensions of the etched channels were different from the dimensions of the capillary pump, different linear flow velocities were generated on the device and in the capillary. A volume flow rate of 200 nL min^{-1} was obtained with −4 kV applied to reservoir 1 and + 1.7 kV applied to the microelectrospray.

It is important to realize that up to this point the field of microfabricated fluidic devices had been predominantly focused on the use of *in situ* detection of analytes, typically by laser-induced fluorescence.[14–16] The selectivity of fluorescence detection relaxed the requirements for sample purity compared to a more general mode of detection. ESI-MS, a general and sensitive detection method, limited the choice of materials which could be used for the construction of the system. In particular, because the channel surfaces and the connections were exposed to organic solvents during normal operation, leakage of small amounts of contaminants from the system was a concern. We found that the main source of contamination was the glue used to form connections during the assembly of the system. Initially, the expansions over the reservoirs and the Teflon sleeves were glued in place using rapidly hardening (5 min) epoxy. From these connections, the solvents extracted high levels of small-molecule and polymeric contaminants, which limited the sensitivity of detection to picomole per microlitre levels. Replacing the 5 min epoxy with a one-phase high molecular weight epoxy significantly reduced this problem.

We first set out to determine the limit of detection (LOD) achievable with this system. A standard peptide, fibrinopeptide A, was used to establish the LOD of the system. Briefly, different solutions of increasing concentrations of fibrinopeptide A were analyzed on the device using the configuration described in Figure 2.3B. A sample aliquot containing fibrinopeptide A was applied to reservoir 1, and an electrophoresis buffer was introduced in reservoirs 2 and 3. The peptide was directed from reservoir 1 to the spectrometer by applying −5.0 kV to reservoir 1 and + 1.7 kV at the microsprayer end of the pump while reservoirs 2 and 3 were left floating. The spectrometer was scanned in MS mode from 400 to 1000 Da. When the $(M + 2H)^{2+}$ ion of fibrinopeptide A was detected at m/z 850.6, the instrument was switched manually to MS-MS mode, and collision-induced dissociation (CID) spectra were generated at different energies until sufficient information was generated. The average of the spectra was used to search a small composite database containing 2413 protein sequences with the Sequest program. The device was thoroughly washed between each experiment with acetonitrile, methanol and buffer consecutively, until the fibrinopeptide A was no longer observed by the mass spectrometer. The experiments were repeated three times with independently prepared solutions. Figure 2.4 shows the average of tandem mass spectra of fibrinopeptide A

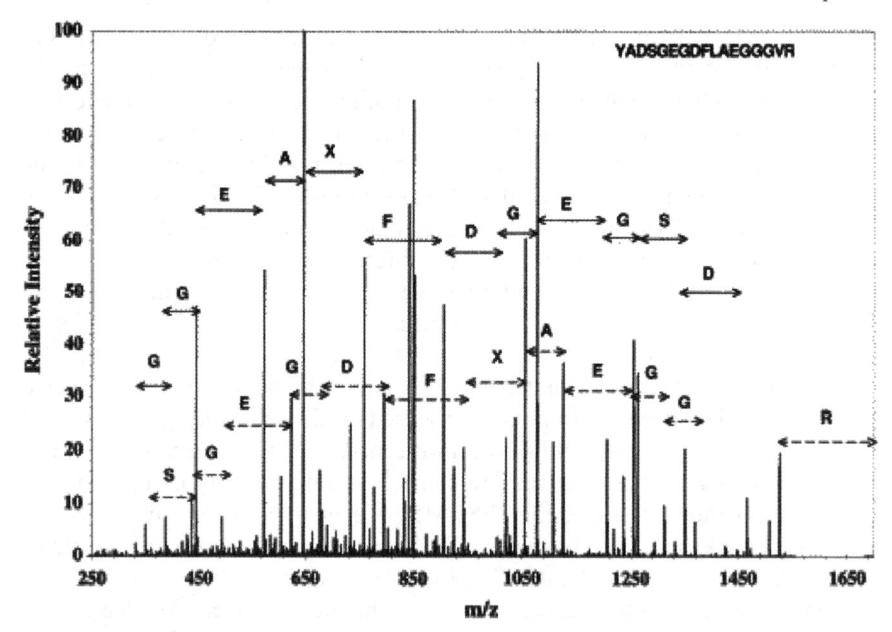

Figure 2.4 Tandem mass spectrum obtained for fibrinopeptide A. A solution of fibrinopeptide A at 33 fmol μL^{-1} was infused at a flow rate of 200–300 nL min^{-1} from reservoir 1 using the conditions indicated in Figure 2.3A. The buffer was 10 mM acetic acid, pH = 3.0, containing 10% (v/v) methanol. The Y ion series is indicated by full lines. The B ion series is indicated by broken lines. The letters in the right upper corner indicate the fibrinopeptide A amino acid sequence in the one-letter code. (Adapted with permission from Ref. 7).

obtained at 33 fmol μL^{-1} which were infused at an estimated flow rate of 200–300 nL min^{-1}. An amount of 2 fmol of fibrinopeptide A was used to generate this spectrum. We estimated that the LOD of this device running in continuous infusion mode and in MS mode was in the region of 2 fmol μL^{-1}.

We next demonstrated that different proteomic samples could be concurrently applied to the microfabricated device, sequentially mobilized and analyzed (Figure 2.3C). Since there are no mechanical valves or gates on the device, we were particularly interested in whether individual samples could be electrophoretically contained in their reservoirs while another sample was analyzed, thus avoiding cross-contamination. The configuration of the system for this experiment is shown in Figure 2.3C. Briefly, tryptic digests of carbonic anhydrase (CA) at a concentration of 290 fmol μL^{-1} and bovine serum albumin (BSA) at a concentration of 130 fmol μL^{-1} were applied to reservoirs 1 and 3, respectively, while an electrophoresis buffer was applied to reservoir 2. Initially, the CA tryptic digest was mobilized towards the spectrometer by applying a potential of −4.2 kV at reservoir 1 and a potential of + 1.9 kV at the microsprayer (Figure 2.3C–1). During the analysis of the CA sample, the BSA tryptic digest was contained by applying a potential of −4.2 kV to reservoir 2 and a potential of −3.6 kV to reservoir 3. The potential difference between reservoirs 2 and 3 generated a slow flow of buffer towards reservoir 3 which effectively isolated the analyte in reservoir 3. This was followed by the analysis of the BSA tryptic digest from reservoir 3. The sample was electroosmotically pumped toward the spectrometer by applying a potential of −4.2 kV at reservoir 3 and a potential of + 1.9 kV at the microsprayer (Figure 2.3C–2), while the remainder of the CA sample was contained by applying a potential of −3.6 kV to reservoir 1. Reservoir 2 was left floating. These conditions resulted in a small flow of BSA sample from reservoir 3 to reservoir 1, while most of the BSA flowed towards the spectrometer without cross-contamination from the CA tryptic digest. Although the design of the device was not optimum for this type of application, it was sufficient enough to provide a proof of principle that such an experiment is possible.

Overall, the development of the three-position microfabricated device coupled to ESI-MS-MS was a success. However, the throughput of the device was limited to one sample at a time. Therefore, our next step was to explore different designs for multiplexed devices.

2.3.2 Nine-Position Device

We decided to develop a more complex sample handling device. Our goals were to demonstrate that the process of sample analysis could be fully automated and that multiple samples could be simultaneously handled by the device. The system is shown schematically in Figure 2.5. It consists of a microfabricated device for sequential sample delivery, an ESI-MS-MS instrument for the structural analysis of analytes, an array of high-voltage relays controlling direction, origin and magnitude of the electroosmotic flow, and a computer workstation which controls the relays, the MS instrument and analyzes the generated data.

The microfabricated device consists of nine reservoirs connected via channels to a common transfer capillary. Reservoirs and channels were etched in glass using a photolithographic mask and an isotropic etching process with hydrofluoric acid. The samples were applied to the reservoirs, and one sample at a time was mobilized by inducing an electroosmotic flow from the specific reservoir through the transfer capillary to the spectrometer. The flow was controlled by a high-voltage electrode which was connected via a computer-controlled set of high-voltage relays to a high-voltage power supply. The microESI ion source at the end of the transfer capillary interfaced the sample delivery module with the spectrometer in which selected peptide ions were subjected to CID. The resulting CID spectra were recorded and searched against a protein or DNA database using Sequest.[17,18] This cycle was automatically repeated by mobilizing the sample in the next reservoir until all the samples initially applied to the device were analyzed.

We first evaluated the performance of the device described in Figure 2.5 for the analysis of protein digests. Different standard protein tryptic digests

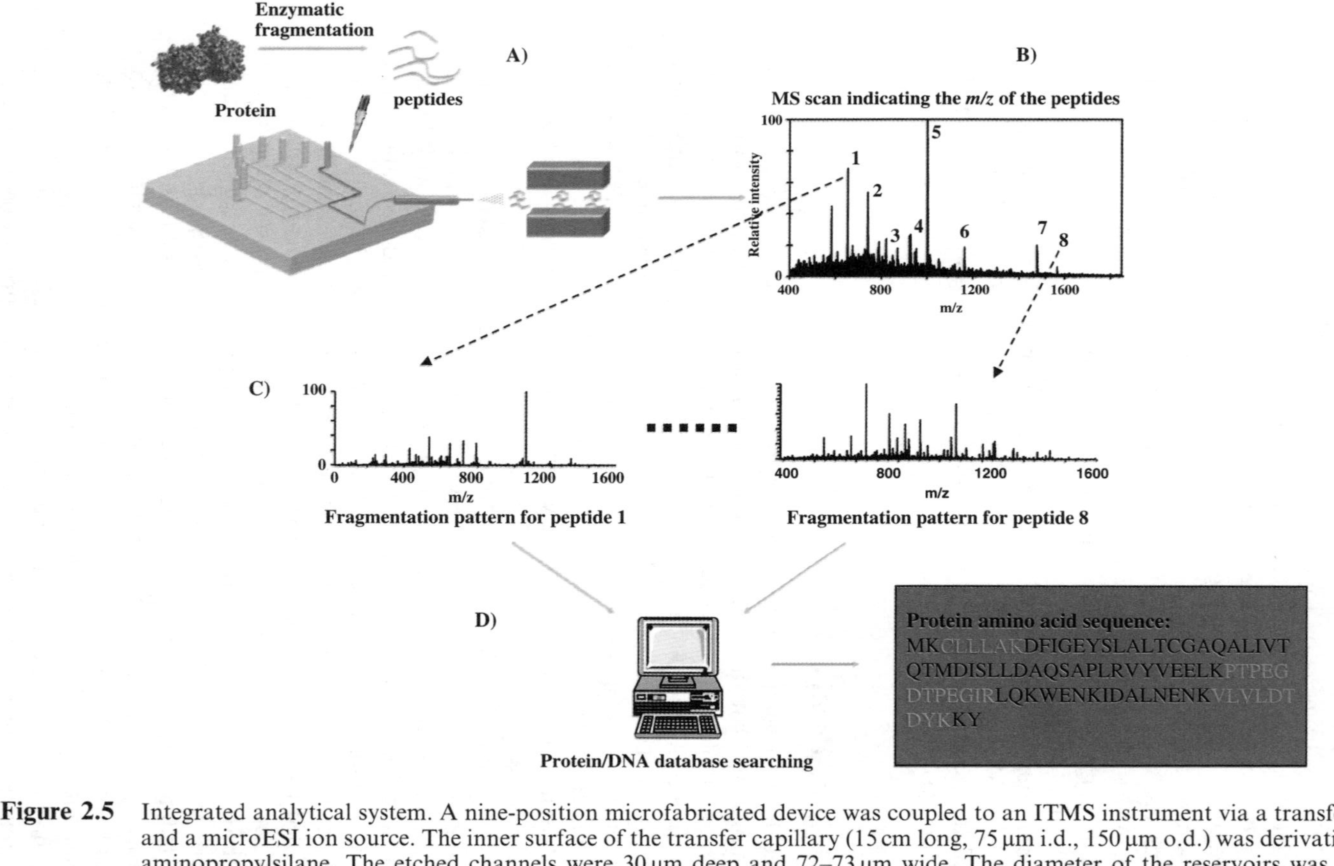

Figure 2.5 Integrated analytical system. A nine-position microfabricated device was coupled to an ITMS instrument via a transfer capillary and a microESI ion source. The inner surface of the transfer capillary (15 cm long, 75 μm i.d., 150 μm o.d.) was derivatized with 3-aminopropylsilane. The etched channels were 30 μm deep and 72–73 μm wide. The diameter of the reservoirs was 1 mm. The sample flow was controlled by an array of computer-controlled high-voltage relays which are also schematically represented. The software controlled the sample flow from the different reservoirs, the generation of MS spectra, the selection of potential peptides, the generation of MS-MS spectra and the matching of the MS-MS spectra against a protein sequence database. (Adapted with permission from Ref. 6).

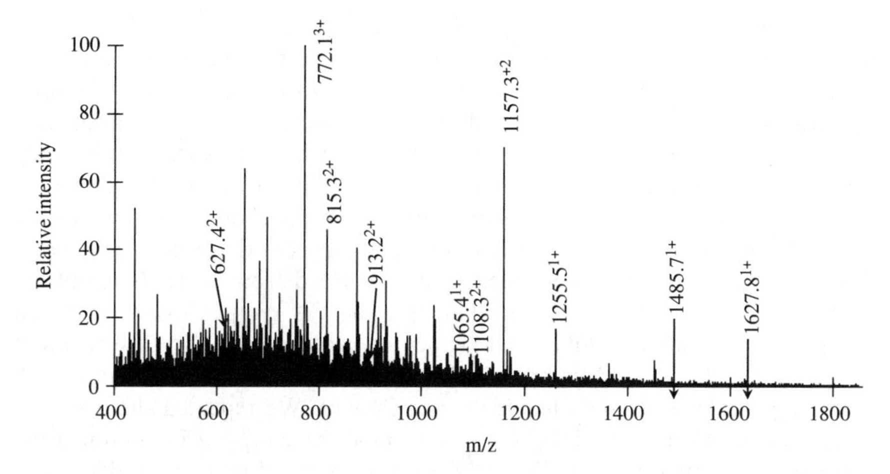

Figure 2.6 Automated analysis of BSA tryptic digest at 26 fmol μL^{-1} from position 3 on the device. MS-MS spectra were generated for the observed peaks. The MS-MS spectra for the labeled peaks were successfully matched to BSA by searching a protein sequence database using the Sequest software. The numbers indicate the measured *m/z* ratio and the charge state of the peptide ions.

(BSA, β-lactoglobulin (β-lac) and CA) were tested on the system. Figure 2.6 shows the mass spectra and the identified peptides from BSA tryptic digest at 26 fmol μL^{-1} infused from reservoir 3. All the labeled peaks were identified by generating MS-MS spectra. The concentration LOD was established using a tryptic digest of β-lac. Briefly, different concentrations of β-lac tryptic digests calibrated to concentrations between 160 amol μL^{-1} and 2.6 pmol μL^{-1} were successively applied in reservoir 2, mobilized by electroosmotic pumping, and analyzed by MS. The LOD was established by monitoring selected peptides by ESI-MS and ESI-MS-MS. The β-lac peptides exceeding a concentration of 5 fmol μL^{-1} were detectable in MS mode. To determine the LOD in MS-MS mode we used the cross-correlation factor (X_{corr}), which was calculated by Sequest software when the generated CID spectra were searched against a bovine protein database. The factor indicates the quality of the correlation with a value of 2 empirically being considered significant. X_{corr} values exceeding 2 were achieved at sample concentrations between 160 amol μL^{-1} (for peptide at *m/z* 771.7) and 1 fmol μL^{-1} (for peptides at *m/z* 458.8 and 533.3).

In summary, the results show that β-lac-derived peptides were conclusively identified at subfemtomole per microlitre concentration. This sensitivity is comparable to the sensitivity achieved by nanoESI-MS,[19] by sample delivery through fused silica capillaries, and from a simple micromachined device.[7]

The next step was to demonstrate that multiple samples could be present on the device and successively delivered to the mass spectrometer with limited cross-contamination. To demonstrate the feasibility of this approach, standard tryptic digests of β-lac, CA and BSA were applied to reservoirs 1, 2 and 3, respectively, and then sequentially mobilized by manually switching the high-voltage relays. To even make minor sample-to-sample cross-contamination detectable, these analyses were done at concentrations of $\sim$200 fmol μL^{-1} which exceeded the LOD by a factor of at least 50. MS spectra were only acquired after 14 minutes of continuous infusion of a new sample to allow time for the new sample to replace the previous one in the shared segments of the flow path. Analysis of the resulting CID spectra indicated that in the case of a β-lac sample, 13 peptides were identified as derived from β-lac. No peptides from either BSA or CA were detected. For the CA sample, five CA-derived peptides were identified while one β-lac peptide was detected. No BSA peptides were detected. In the BSA sample, 12 BSA peptides were identified and no CA or β-lac peptides were detected (data not shown). We repeated these experiments at least 17 times and we found that on average 0.7 ± 0.7 contaminating peptides per sample were analyzed. Typically, contaminating peptides had a low signal in MS mode. These results indicate that samples concurrently present on the device could be sequentially analyzed with minimal sample-to-sample cross-contamination.

The manual procedure for sample mobilization and analysis was automated by implementing a computer program that coordinately controlled the sample flow from the appropriate reservoirs to an ion trap mass spectrometer (ITMS) and for generating and analyzing CID spectra of selected ions. Human intervention was thus limited to inserting the sample into the reservoirs and starting the experiment. It is important to note that the manipulation of high voltages requires special high-voltage relays, and special care must be taken to avoid potential electrical hazards with these devices.

The performance of the fully automated device for the analysis of 2D gel-separated proteins was then evaluated (Figure 2.7). Yeast proteins were separated by 2D gel electrophoresis, excised and in-gel digested. The tryptic digests were introduced in individual reservoirs, and the software controlling the automation was activated. Protein identities were established by searching the yeast sequence database with the generated CID spectra using the Sequest search algorithm. One cycle of the system took 30 min, and the search of the yeast database 15–30 min. This means that the database search was finished before the data were generated for the next sample. The amount of protein present in the spots was estimated to be in the range from 12 pmol (spot 1) to 180 fmol (spot 3), and sample concentrations ranged from an estimated 800 (spot 1) to 12 fmol μL^{-1} (spot 3). The majority of the yeast proteins were successfully analyzed by the fully automated device.

At that point, we had achieved our goal to demonstrate that a microfabricated device can be coupled to a mass spectrometer and efficiently used for the rapid and sensitive identification of proteins separated by 2D gel electrophoresis. Although we toyed with the idea of more complex devices (we even designed a 42-position device which is still sitting on the author's desk), we quickly decided instead to move towards expanding the types of sample manipulation that could be done on and off the device.

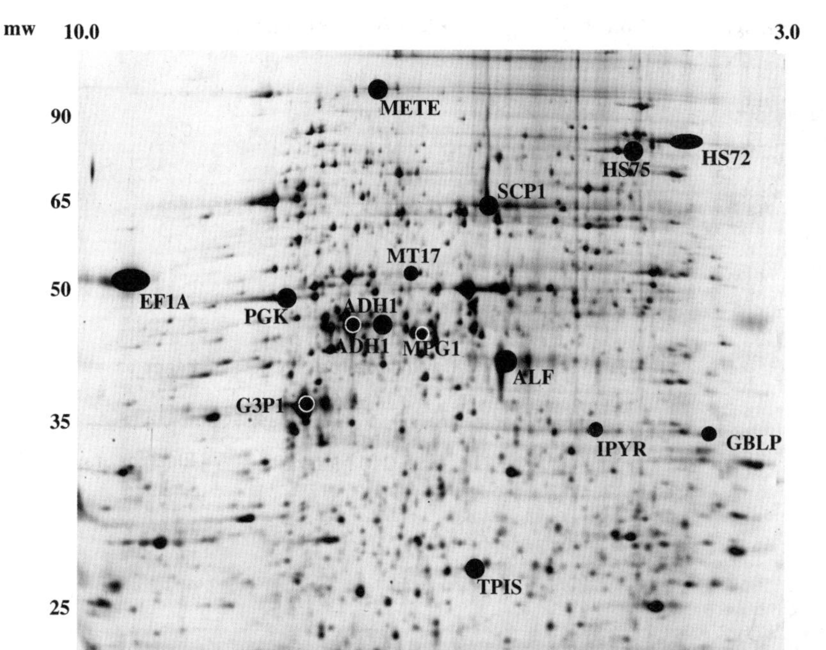

Figure 2.7 2D gel electropherogram of yeast proteins. Aliquots of total yeast (*S. cerevisea*) lysate containing 40 μg of protein were separated by 2D electrophoresis (pI range 4–7). Separated proteins were detected by silver staining and selected spots were subjected to tryptic digestion and automated analysis on the nine-position device. The abbreviations of the names of the proteins identified by the system are next to the spots on the 2D gel electropherogram. (Adapted with permission from Ref. 6).

2.4 Gradient Generation from Devices

We decided to test the idea of using microfluidic devices for solvent handling and gradient generation instead of for sample handling. Microfabricated fluidic devices are well suited for generating solvent gradients since multiple reservoirs and flow paths can be present on a single device and individually addressed. Electroosmotically generated flows from specific reservoirs can be joined in a controlled fashion to generate solvent gradients. The magnitude and the direction of the electroosmotic flow are controlled by the potentials applied to the reservoirs.

We developed an approach based on the system described in Figure 2.8 in which ramping potentials are applied to reservoirs 1 and 3 for establishing electroosmotic pumping in order to produce a solvent gradient. Briefly, for these experiments, a bare fused silica capillary was used as transfer capillary. Reservoir 1 was filled with 10 mM acetic acid/10% (v/v) methanol and reservoir 3 was filled with 65% (v/v) acetonitrile/3 mM acetic acid. The difference in potential between the reservoirs and the ESI interface generated an electroosmotic flow toward the

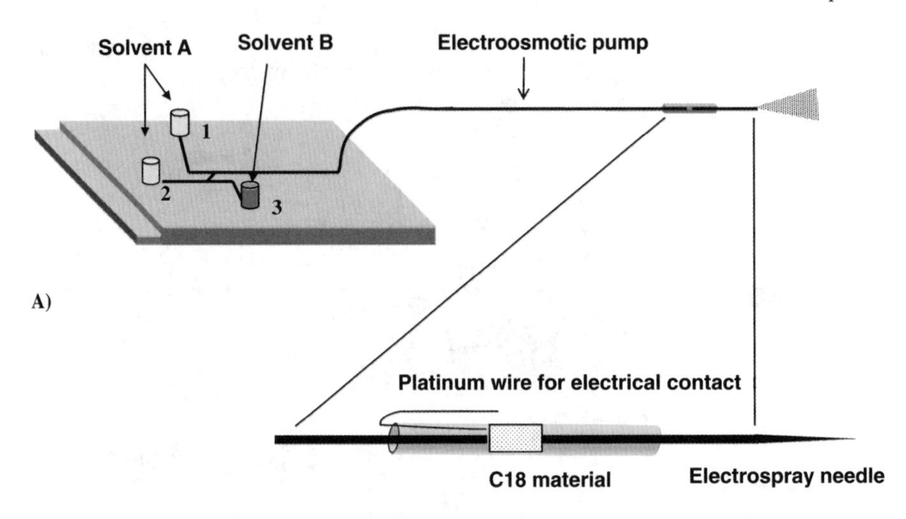

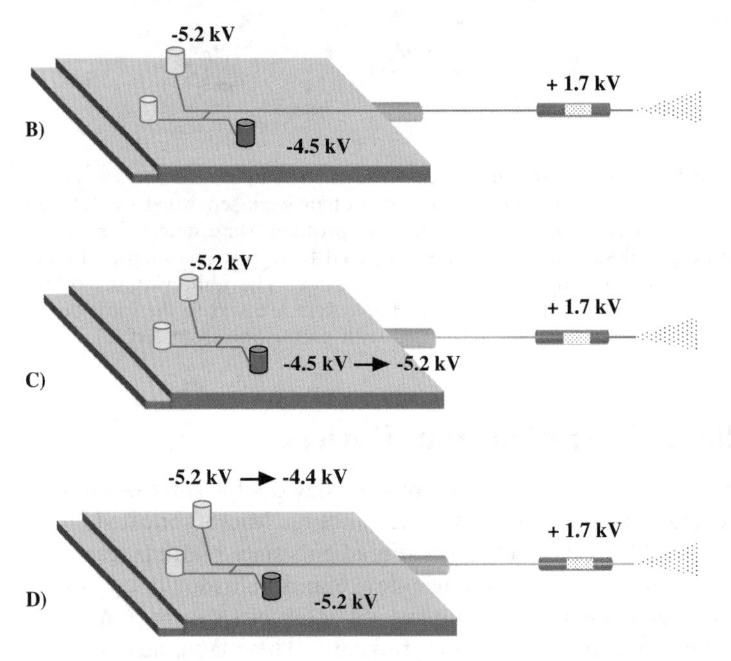

Figure 2.8 (A) Device combining three reservoirs for the generation of solvent gradients, a transfer capillary and an ESI interface that incorporates a small C18 cartridge to concentrate peptides. (B–D) Three steps for the generation of an aqueous/organic solvent gradient. (B) The flow from reservoir 1 towards the MS and back in reservoir 3. (C) The flow from reservoirs 1 and 3, respectively, towards the MS. (D) The flow from reservoir 3 towards the MS and back to reservoir 1. The aqueous phase was 10 mM acetic acid/10% (v/v) methanol. The organic phase was 65% (v/v) acetonitrile/3 mM acetic acid. (Adapted with permission from Ref. 11).

ESI interface. Therefore, solvent gradients were generated without the need for a pump. We had to use two high-voltage power supplies to provide the appropriate potentials to reservoirs 1 and 3. Each power supply was controlled through a digital to analog converter board (DAC) and by a Labview procedure written in-house. This procedure controlled the potential (0–10 V) applied to two analog outputs of the DAC which in turn controlled the high-voltage power supply. The Labview program ramped the potential which was applied to the DAC output according to the preset initial and final voltages and ramping times.

We then monitored the gradient by using the change in the MS signal which was generated by the increase of acetonitrile. A signal representing the clustering of acetonitrile in the spectrometer was monitored. The signal increased as the acetonitrile concentration increased. The course of a typical experiment is illustrated in Figure 2.8B–D. At the start of an experiment, reservoir 1 was held at −5.2 kV, and reservoir 3 was held at −4.5 kV (Figure 2.8B). These potentials generated a flow from reservoir one towards the ESI interface. It also generated a small secondary flow from reservoir 1 to reservoir 3, ensuring that only the buffer from reservoir 1 reached the ESI interface.

The gradient was then started by ramping the potential applied to reservoir 3 from −4.5 to −5.2 kV over 14 min (Figure 2.8C). Simultaneously the potential on reservoir 1 was held constant. While the potential was ramped on reservoir 3, the flow from reservoir 1 to reservoir 3 slowly decreased. At a certain potential value, at which the potential on reservoir 3 was higher than the potential at the junction of the flow path from reservoirs 1 and 3, no flow was apparent from reservoir 3. At higher potential values an increasing flow from reservoir 3 toward the ESI interface was generated. The net effect of this step was an increase in the acetonitrile concentration flowing towards the spectrometer. Finally (Figure 2.8D), the potential on reservoir 1 was slowly increased from −5.2 to −4.2 or −4.4 kV over 14 min and the potential on reservoir 3 was kept constant at −5.2 kV. During this ramping, the flow from reservoir 1 slowly decreased. At a certain point, the flow from reservoir 1 stopped and then reversed into reservoir 1 from reservoir 3.

A typical solvent gradient observed by this approach is illustrated in Figure 2.9A. The gradient is well described with an exponential fit. The gradient started at 28 min and finished at 50 min, thus providing 22 min for a gradient to develop from 0 to 65% acetonitrile.

Interestingly, the change in the solvent composition over the course of an experiment was also apparent from the current delivered by the power supply. During the formation of a gradient, the ionic composition of the solution in the channels of the device and in the transfer capillary changed. The current supplied by the ESI power supply was the summation of the current from the microfabricated device and the current for the electrospray ionization process. Changes in ionic composition in the channels of the microfabricated device and in the transfer capillary were therefore reflected by changes in the current provided by the MS high-voltage power supply. In the first phase of the experiment, the current provided by the MS power supply was stable, reflecting a stable ionic composition of aqueous buffer in the channels and in the transfer

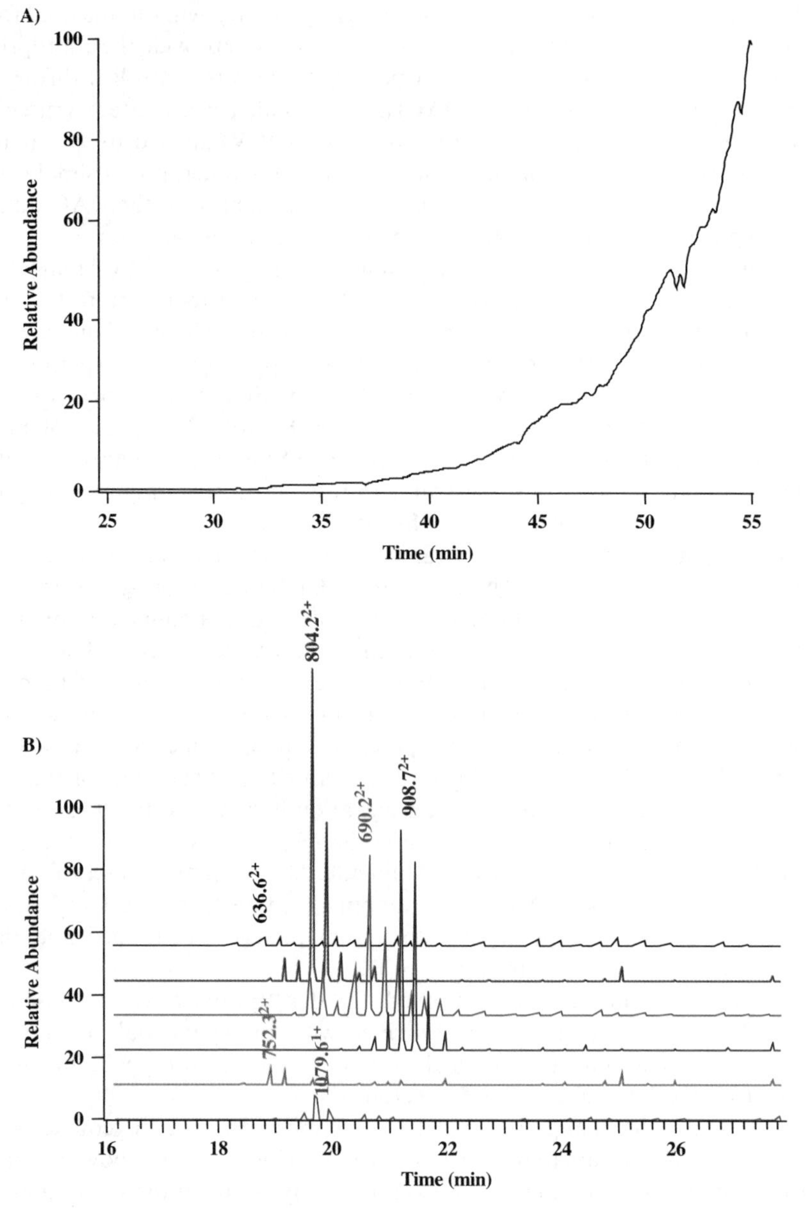

Figure 2.9 (A) Gradient profile generated using a 12 cm long uncoated 50 µm i.d. capillary. The gradient was generated using the device described in Figure 2.8A and the procedure described in the text. The gradient was monitored in the MS following a signal generated by acetonitrile clustering. (B) Analysis of 74 fmol of a myoglobin tryptic digest by gradient frontal analysis-MS-MS. The procedure described in Figure 2.8B–D was used to generate the gradient and a (3-aminopropyl) silane-coated capillary was used as the transfer line (gradient profile different from Figure 2.9A). Myoglobin (10 µL) at a concentration of 7.4 fmol μL^{-1} was injected off-line. The trace for each peptide identified is displayed. (Adapted with permission from Ref. 11).

capillary. In the second phase, the current provided by the ESI power supply started to drop due to the difference in conductivity between the solutions in reservoirs 1 and 3, indicating an increasing concentration of acetonitrile present in the channels and transfer capillary. In the third phase, the current continued to drop, and it eventually reached a plateau reflecting a stable acetonitrile concentration in the channels and transfer capillary.

2.5 Frontal Analysis of Protein Digests

2.5.1 Standard Protein Digests

The introduction of a small C18 cartridge in the ESI interface allows the accumulation of protein digests. Furthermore, the coupling of this C18 cartridge with the microfabricated device created a system well suited for the concentration and frontal analysis of protein digests by MS. Briefly, in this system, the protein digests of interest is first pressure-injected on the C18 cartridge through a transfer capillary. The transfer capillary is connected to the microfluidic device which is then activated for gradient generation.

Our first experiment on this combined system aimed to establish the LOD for the analysis of a calibrated protein digest. Briefly, a standardized solution of myoglobin digested with trypsin was used to establish the LOD. Figure 2.9B illustrates the analysis of a 10 μL aliquot of 7.4 fmol μL^{-1} (total 74 fmol) digested myoglobin on the gradient system equipped with a derivatized transfer capillary. The sample was pressure loaded (6–9 psi) off-line. The transfer capillary was then reconnected to the microfluidic device, and the gradient was developed. As the peptides eluted from the C18 cartridge, they were detected by the spectrometer. If a specific peptide ion exceeded a predetermined intensity, the instrument automatically switched to MS-MS mode and the resulting CID spectra were recorded. These spectra were used in conjunction with Sequest to search a horse protein sequence database to identify the origin of the peptides. Six peptides were identified as being derived from myoglobin. Other peptides from myoglobin were also present. These peptides did not generate CID spectra of good enough quality for unambiguous identification or they were small peptides with a 1^+ charge. Such spectra were not assigned by Sequest. Using the signal from the peptide ion at m/z 804.2^{2+} and a noise value calculated as three times the average background signal in a mass window 20 Da below the detected peptide ion, we determined a LOD of 1 fmol and a concentration LOD of 100 amol μL, assuming that 10 μL of sample was applied to the C18 cartridge.

2.5.2 One- and Two-Dimensional Gel Electrophoresis

Our next step was to establish the general applicability of the microfluidic device for real biological samples. Proteins obtained from 1D and 2D gel separation of *Saccharomyces cerevisiae* were used to evaluate the device.

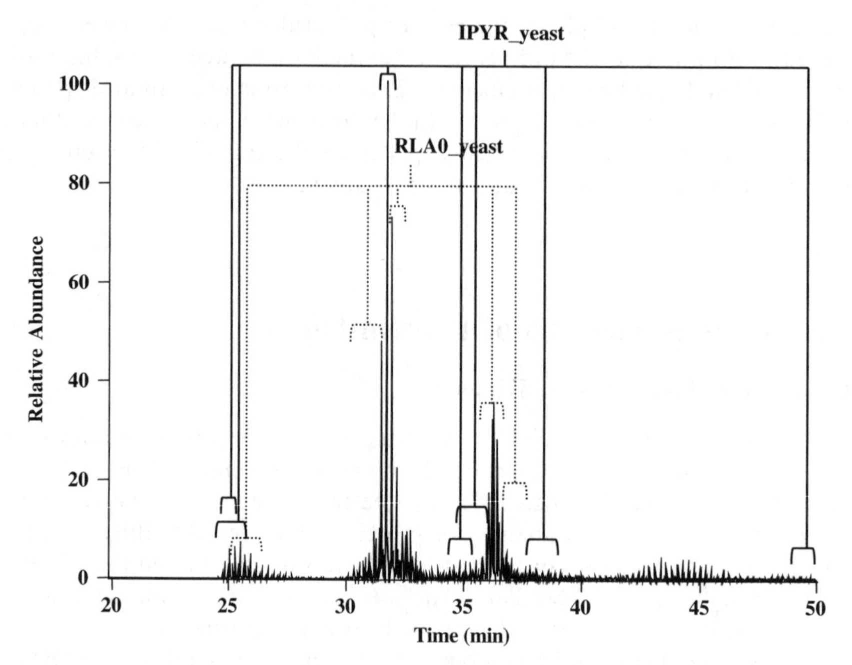

Figure 2.10 Analysis of a band of yeast proteins separated by 1D gel electrophoresis of yeast total cell lysate and digested with trypsin. The proteins migrating to a band with an apparent molecular mass of 34 kDa were digested with trypsin and the extracted peptide mixture was analyzed by frontal analysis-ESI-MS-MS. Six different proteins were identified from the mixture: G3P1 yeast, G3P2 yeast, G3P3 yeast, IPYR yeast, RLAO yeast and BMH1 yeast. The position of the peptides for only two of the proteins is illustrated for clarity. The procedure described in Figure 2.8B–D was used to generate the gradient, and an uncoated capillary was used as the transfer line. The sample was loaded off-line. (Adapted with permission from Ref. 11).

We first analyzed tryptic digests of protein bands obtained from a 1D gel electrophoresis separation of total yeast lysate. Figure 2.10 shows the frontal analysis of a tryptic digest of a sample representing a single band migrating at 34 kDa in the gel. The sample was pressure loaded off-line on the C18 cartridge. The transfer capillary was then reconnected to the microfabricated device, and the gradient was generated. CID spectra of the eluted peptides were generated on the fly. The signal intensity was a few orders of magnitude above the LOD. Six different proteins were identified in this band with up to seven peptides identifying a specific protein.

We next analyzed proteins obtained from a 2D gel separation of a total yeast lysate. Four peptides were identified from HXKA yeast along with one peptide from keratin and four peptides from trypsin. All the peptides identified had an X_{corr} higher than 2.0 except for the ion MH^+ at m/z 630.4 which was a small peptide observed as a 1^+ ion. The amount of protein present in the gel was estimated to be 200–300 fmol.

2.6 Concluding Remarks

The coupling of microfabricated devices to mass spectrometers has been a great success through the efforts of many laboratories around the world. The depth of applications has been limited by our ability to perform protein chemistry and biochemistry on a scale compatible with microfluidic devices. To date, microfluidic devices have been generally used to handle small amounts of sample. The next frontier for this field of research will be the development and integration of *in situ* chemical and biochemical processing.

References

1. D. Figeys and D. Pinto, *Electrophoresis*, 2001, **22**, 208.
2. D. Figeys and D. Pinto, *Anal. Chem.*, 2000, **72**, 330A.
3. C. L. Colyer, T. Tang, N. Chiem and D. J. Harrison, *Electrophoresis*, 1997, **18**, 1733.
4. Q. Xue, F. Foret, Y. M. Dunayevskiy, P. M. Zavracky, N. E. McGruer and B. L. Karger, *Anal. Chem.*, 1997, **69**, 426.
5. Q. Xue, Y. M. Dunayevskiy, F. Foret and B. L. Karger, *Rapid Commun. Mass Spectrom.*, 1997, **11**, 1253.
6. D. Figeys, S.P. Gygi, G. McKinnon and R. Aebersold, *Anal. Chem.*, 1998, **70**, 3728.
7. D. Figeys, Y. Ning and R. Aebersold, *Anal. Chem.*, 1997, **69**, 3153.
8. R. S. Ramsey and J. M. Ramsey, *Anal. Chem.*, 1997, **69**, 1174.
9. N. Xu, Y. Lin, S. A. Hofstadler, D. Matson, C. J. Call and R. D. Smith, *Anal. Chem.*, 1998, **70**, 3553.
10. D. Figeys, C. Lock, L. Taylor and R. Aebersold, *Rapid Commun. Mass Spectrom.*, 1998, **12**, 1435.
11. D. Figeys and R. Aebersold, *Anal. Chem.*, 1998, **70**, 3721.
12. D. Figeys, A. Ducret, J. R. Yates III and R. Aebersold, *Nat. Biotechnol.*, 1996, **14**, 1579.
13. D. M. Pinto, Y. Ning and D. Figeys, *Electrophoresis*, 2000, **21**, 181.
14. C. S. Effenhauser, G. J. Bruin and A. Paulus, *Electrophoresis*, 1997, **18**, 2203.
15. D. J. Harrison, K. Fluri, K. Seiler, Z. Fan, C. S. Effenhauser and A. Manz, *Science*, 1993, **261**, 895.
16. S. C. Jacobson, L. B. Koutny, R. Hergenroeder, A. W. J. Moore and J. M. Ramsey, *Anal. Chem.*, 1994, **66**, 3472.
17. J. Eng, A. L. McCormack and J. R. I. Yates, *J. Am. Soc. Mass Spectrom.*, 1994, **5**, 976.
18. J. R. Yates III, J. K. Eng, A. L. McCormack and D. Schieltz, *Anal. Chem.*, 1995, **67**, 1426.
19. M. Wilm, A. Shevchenko, T. Houthaeve, S. Breit, L. Schweigerer, T. Fotsis and M. Mann, *Nature*, 1996, **379**, 466.

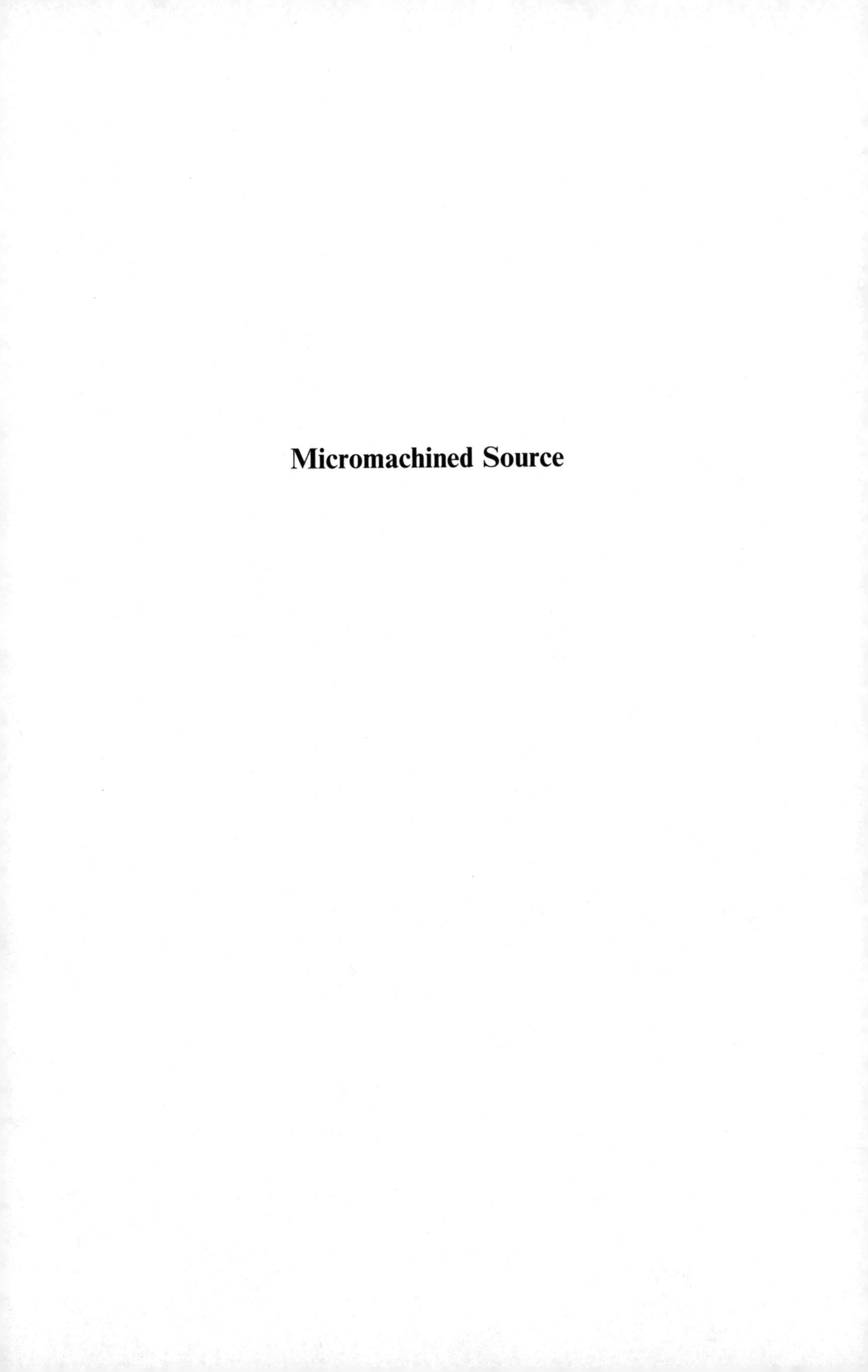

Micromachined Source

CHAPTER 3

A Silicon-based ESI Chip with Integrated Counter Electrode and its Applications Combined with Mass Spectrometry

GARY A. SCHULTZ

Advion BioSciences, Inc., Advion BioSystems, 19 Brown Road, Ithaca, NY 14850, USA

3.1 Introduction

In the mid-1990s, nanoelectrospray ionization (nanoESI) was introduced as an ionization method using tapered glass capillaries with 1 μm diameters that enabled mass spectrometry (MS) analysis of solution volumes of 1 μL or less.[1,2] Charged droplets with diameters less than 1 μm were formed at flow rates less than 40 nL min^{-1}.[2] These small, highly charged droplets provided efficient ionization of analytes contained within the droplets due to high surface-to-volume ratios and small radii through which analytes needed to diffuse to reach the charged surface of the droplets. Analytes on or near the surface of evaporating charged droplets will be preferentially ionized over analytes residing within the bulk of the droplets.[3] Using nanoESI, a 1 μL sample could be sprayed for 30–60 minutes and signal averaging improved signal-to-noise ratios and ion statistics enabling high mass accuracy and multiple MS-MS experiments on analytes.

The electric field generated on a sprayer causes the separation of positively and negatively charged ions in solution and pushes ions of one polarity to the solution surface. The higher the electric field the greater the surface charge

Miniaturization and Mass Spectrometry
Edited by Séverine Le Gac and Albert van den Berg

Published by the Royal Society of Chemistry, www.rsc.org

repulsion counteracting the fluid surface tension.[4] A Taylor cone forms on the tip of a sprayer when these two forces are balanced. If the electric field is higher than that necessary to offset the surface tension with surface charge repulsion, excess charge will escape off the tip of the Taylor cone in the form of a liquid jet that destabilizes to form charged droplets. A high electric field will replenish the surface charge maintaining an environment of continuous formation of charged droplets. If the electric field is significantly greater than that required to balance the surface tension, excess charge is emitted from the Taylor cone. With an electric field equal to that formed when producing a nanoelectrospray from a 1–2 μm diameter sprayer, droplets with a maximized surface charge form resulting in an excess of charge relative to analyte contained within the droplet which leads to less competition for the charge as the droplets evaporate to form gas-phase ions.[5]

Although nanoESI proved to be useful, it was also referred to as an 'art' and needed a person skilled in the art to make the technique routinely useful. NanoESI with pulled capillaries was not easily automated making it a low throughput technique. Two things made automating use of pulled capillaries for nanoESI challenging. One was making pulled capillaries with reproducible tip dimensions. Tip dimensional changes resulted in flow rate and spray stability variations. The second was recovering the sample from a capillary if plugging of the capillary tip occurred during a spray.

The goal of this work was to develop a robust, automated nanoESI device that enabled all researchers to benefit from use of nanoESI. The ESI Chip™ is a silicon-based microfabricated array of nozzles etched in the planar surface of a silicon wafer.[6] Silicon was chosen as the substrate due to its semiconductor properties, ability to etch features with high selectivity and uniformity across a wafer, and availability of deep reactive ion etching (DRIE) equipment capable of etching through-wafer features of less than 10 μm with a 20 : 1 aspect ratio.[7] The fabrication process enables the physical dimensions of each nozzle to be within a few hundred nanometers resulting in improved spray stability and nozzle-to-nozzle reproducibility.[6] The ESI Chip works within the NanoMate® system whereby a pipette tip is used to deliver a sample to the inlet of a nozzle.[8] The sample remains in the pipette tip except for that which is being sprayed enabling unused sample to be returned to a sample well or moved to a new nozzle. The system is designed for use of one pipette tip and a new nozzle for each sample and was the first commercially available nanoESI ion source that was fully automated for up to 384 samples.

3.2 Nozzle Fabrication and Description

3.2.1 Fabrication of a Silicon Nozzle

The ESI Chip is fabricated from a double-side polished silicon wafer of 500 μm thickness. Double-side polished silicon wafers enable an inlet feature to be etched opposite the nozzle making the delivery of liquids to the nozzle possible

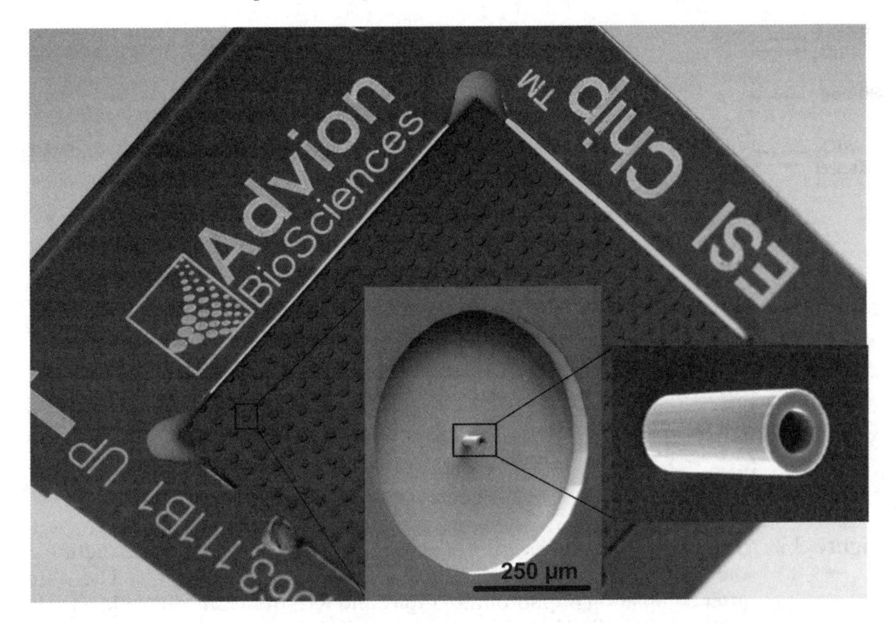

Figure 3.1 ESI Chip consisting of a 20 by 20 array of microfabricated nozzles etched from a silicon wafer. The inset shows scanning electron micrographs showing a nozzle with annular space and a nozzle prior to dielectric coating of the wafer. Nozzle dimensions in the inset are 10 µm i.d., 22 µm o.d. by 55 µm height.

by sealing of a pipette tip or capillary around the inlet and flow of liquid through-wafer exiting at the tip of the nozzle. Figure 3.1 shows the nozzle side of the high-density HD ESI Chip. This chip is a 20 by 20 array of nozzles etched every 1.125 mm. Scanning electron micrographs of a nozzle and surrounding annular space and a closer view of the nozzle are shown in the inset of Figure 3.1. The nozzle dimensions are controlled to within a micrometer and changes in the inner diameter of the nozzle are used to change the flow rate for nanoESI. There are currently three versions of the ESI Chip with inner diameters of 2.5 µm, 4.1 µm and 5.5 µm covering nominal flow rate ranges of 20–40 nL min^{-1}, 60–100 nL min^{-1} and 100–500 nL min^{-1}, respectively.

The fabrication sequence is outlined in Figure 3.2. A hard mask consisting of silicon oxide is grown to a thickness of 2 µm on each side of the silicon wafer. A resist is spin-coated onto side one of the wafer. Using non-contact profilometry, a mask is used to expose ultraviolet (UV) light to patterns defining nozzle inner (through-channel) and outer diameters as well as annular spaces surrounding the nozzles. The resist exposed to UV light is developed and removed exposing the underlying silicon oxide. The exposed silicon oxide is etched using reactive ion etching (RIE) to the underlying silicon. A resist is spin-coated onto side two of the silicon wafer. Using non-contact profilometry, a mask containing inlet features is aligned to the nozzles etched on side one of the wafer. The resist is exposed to UV light, developed and removed exposing

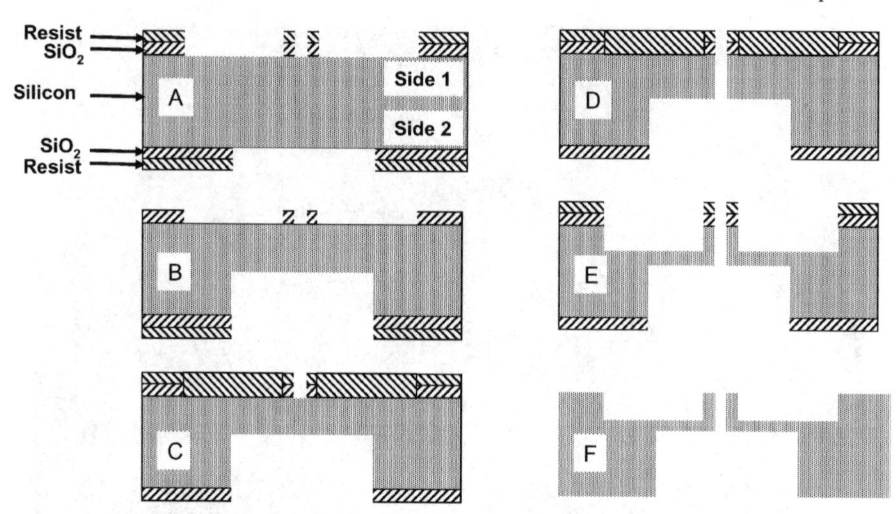

Figure 3.2 Schematic of the fabrication sequence used to etch the silicon structure of the ESI Chip using a double-side polished silicon wafer. (A) The wafer after completion of photolithography and RIE of the silicon oxide on sides one and two. (B) The wafer after etching of the inlet structure on side two (bottom side on the figure). (C) The wafer after spinning resist on side one (top side on the figure), photolithography and development to define the through-channel structure. (D) The wafer after DRIE of the through-channel structure to the inlet structure. (E) The wafer after DRIE of the annular space to define the nozzle. (F) The wafer after removal of the resist and silicon oxide from the wafer.

the underlying silicon oxide. The silicon oxide is etched using RIE exposing the underlying silicon (Figure 3.2A). Side two of the wafer is then etched to form inlets to a depth of approximately 350 μm (Figure 3.2B). Another layer of resist is spin-coated onto side one of the silicon wafer. A mask consisting of holes defining the through-channel structures of each nozzle is aligned to the previously etched features and exposed to UV light. The resist is developed and removed to expose the underlying silicon (Figure 3.2C). The silicon is then etched to form through-channels connecting each nozzle to an inlet feature on side two of the wafer (Figure 3.2D). Finally, the resist protecting the annular spaces is removed and the exposed silicon is etched using DRIE to define the nozzles (Figure 3.2E). The resist and silicon oxide are then removed (Figure 3.2F).

After etching of the silicon structure, a conformal dielectric coating covers all surfaces of the silicon wafer. The dielectric coating consists of three layers. The first is growth of silicon oxide by conversion of silicon in a high-temperature oxygen furnace. Then chemical vapor deposition is used to deposit conformal silicon nitride and silicon oxynitride to fill the silicon oxide pores.[9] The dielectric prevents ions in solution from creating a conductive path to the silicon in the presence of a high electric field necessary for formation of a nanoelectrospray. The dielectric coating enables the nanoelectrospray of salt solutions as high as 100 mM ammonium acetate with fluid voltages up to 2.5 kV while maintaining a silicon voltage of zero volts.

The silicon wafer consisting of an array of microfabricated nozzles is then cut into smaller arrays. The cutting process exposes the silicon underlying the dielectric coating on four sides of the chip. The chip is placed in an aluminium frame containing pads for alignment of the chip to points along the frame. A conductive path is created between the frame and silicon using conductive epoxy completing the ESI Chip. The frame serves as a means to handle the chip without touching the silicon as well as a means of applying a voltage to the silicon structure underlying the dielectric layer which is necessary for the optimal formation of a nanoelectrospray.

3.2.2 The Electrode Inside the ESI Chip

What makes the ESI Chip unique from all other ESI sprayers is the presence of a counter electrode integrated into the nozzle structure. This is accomplished by use of the semiconducting properties of silicon and the dielectric coating conformal on all surfaces of the silicon structure. The dielectric allows application of voltage to liquids delivered to the nozzle structure while also applying a voltage to the silicon underlying the dielectric. Application of a voltage difference between the applied fluid and silicon creates an electric field across the dielectric and surrounding the nozzle when a fluid exits the nozzle. If the electric field is high enough to enable the surface charge repulsion of ions contained in the solution to be greater than the solution surface tension, a Taylor cone will form on the nozzle and excess surface charge will escape from the Taylor cone forming a nanoelectrospray.

Equation (3.1) is a simplified formula containing the primary variables necessary to estimate the electric field (E_c) at the tip of a capillary sprayer based on the capillary radius (r_c), its distance from a counter electrode (d) and the solution voltage (V_c):[10,11]

$$E_c = \frac{2V_c}{r_c \ln(4d/r_c)} \tag{3.1}$$

Using Equation (3.1), the electric field at the tip of a 2 μm capillary is calculated to be 2.1×10^8 V m^{-1} with a solution voltage of 1 kV at a distance of 3 mm from a counter electrode. In the design of the ESI Chip, the distance between the solution and counter electrode voltages is a few micrometers and Equation (3.1) cannot be used to estimate the electric field.

3.2.3 Electric Field Modeling

SIMION® 6.0 was used to model the electric field of a pulled capillary, the ESI Chip and a chip without the integrated counter electrode to gain insights into the field surrounding each of these nozzles.[12] The models used a nozzle outside diameter of 28 μm (the diameter of the ESI Chip nozzle at the time of this writing) positioned 4.7 mm from a counter electrode held at ground potential. Electrodes were defined for the silicon and the liquid forming a Taylor cone on the nozzle. For the ESI Chip model the dielectric was defined as a non-conductor. In this way, the electric field would penetrate through the dielectric. A fluid voltage of 1.6 kV (voltage commonly used with the ESI Chip) was used in this model to estimate the electric field present after establishing a Taylor cone on the sprayer.

Figure 3.3 shows the equipotential field lines in 50 V increments for each model. Figure 3.3A shows that the electric field for the pulled capillary model is $8.5 \times 10^7\,\mathrm{V\,m^{-1}}$ at the tip of the Taylor cone. Figure 3.3B shows that the electric field for the ESI Chip model is $2.2 \times 10^8\,\mathrm{V\,m^{-1}}$ at the tip of the Taylor cone. For the ESI Chip model, the electric field is 74% greater at the outer diameter of the nozzle where the fluid is thin and close to the silicon counter electrode. When the electric field is larger than that required to overcome the surface charge repulsion, the excess charge emanates from the nozzle by forming multiple nanoelectrosprays on the nozzle edges.[12] Also, since the silicon is held to ground potential, there is a nearly field-free region between the ESI Chip and the counter electrode located 4.7 mm away.

Figure 3.3C shows that the electric field formed when the chip-based counter electrode is removed by applying the same voltage to the silicon and fluid is $4 \times 10^6\,\mathrm{V\,m^{-1}}$. Comparing the electric field values of Figure 3.3B and C, use of the integrated counter electrode increases the electric field by a factor of 53. This model helps to explain some of the reasons why it was difficult to sustain a nanoelectrospray from short nozzles in close proximity to the chip surface as well as channels exiting on the edge without the presence of a surface that defined the outer diameter of a nozzle.[13,14] This was evident in several of the early attempts to form a nanoelectrospray from microchips which required solution voltages of several kilovolts to form a stable spray.[13,14] For the model described in Figure 3.3C, the electric field can be increased to $1.4 \times 10^8\,\mathrm{V\,m^{-1}}$ by applying a fluid voltage of 6 kV and moving the chip within 0.5 mm from the counter electrode.[12] Using Equation (3.1), a pulled capillary of 2 μm diameter with an applied voltage of 1 kV and distance of 3 mm from the counter electrode leads to an electric field of $2.1 \times 10^8\,\mathrm{V\,m^{-1}}$. Therefore, even though the physical size of the chip nozzle is large when compared to a pulled capillary, the electric field is of similar magnitude to that of a 2 μm pulled capillary.

Data with the models were generated with decreasing separation between the sprayer and counter electrode to establish how closely these models compared with the calculations of Equation (3.1). Figure 3.4 shows the electric field dependence on nozzle distance from the counter electrode for Equation (3.1), for the SIMION Capillary and the SIMION Chip models. Equation (3.1) and the SIMION Capillary model generally agree with a slight overestimation of the electric field by the SIMION model. The SIMION Chip model shows little dependence on the separation of the nozzle from the counter electrode. This is due to the close proximity of the grounded silicon nozzle surrounding

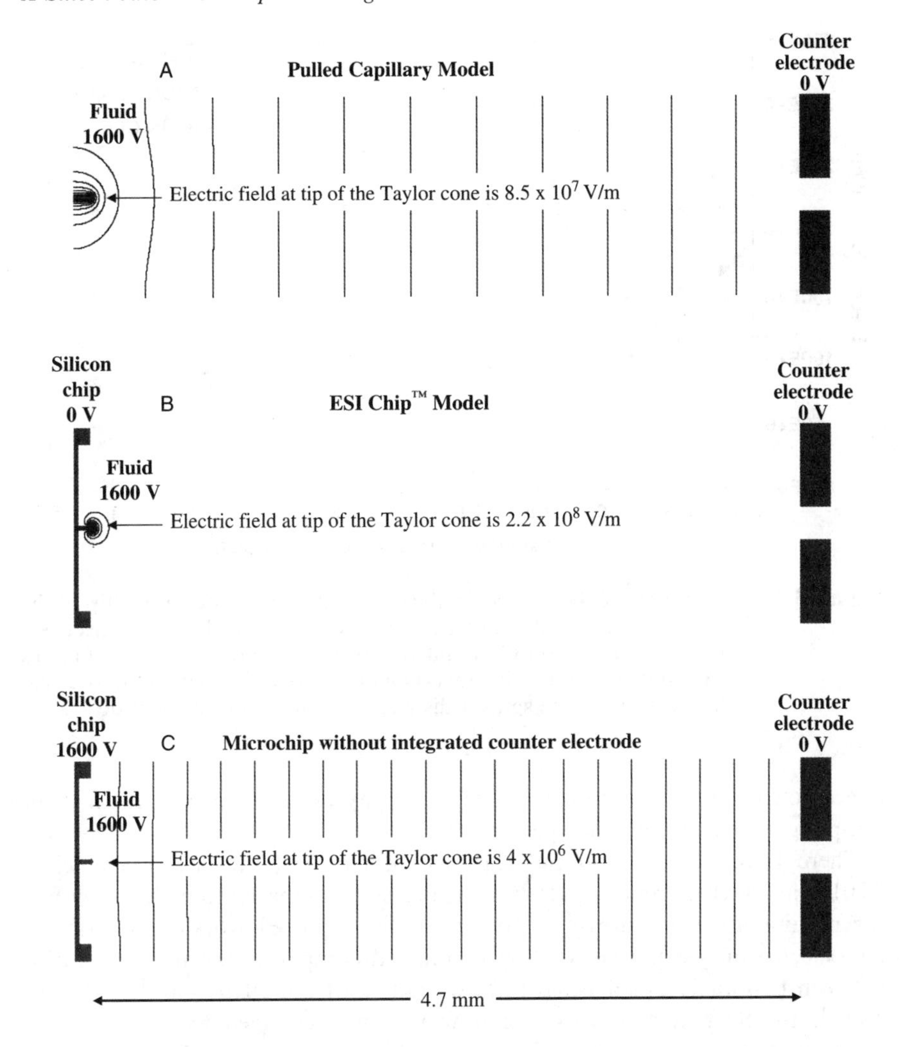

Figure 3.3 SIMION electric field models for 28 μm diameter sprayers located 4.7 mm from the ion orifice counter electrode and a spray (solution) voltage of 1.6 kV. Equipotential lines are shown every 50 V. Electric field calculated at the tip of the Taylor cone for each model is shown. (B, C) A comparison of the models shows a 50-fold increase in the electric field generated at the tip of the Taylor cone when the silicon underlying a dielectric film is held at ground potential rather than at the spray potential. Electric field shown in (B) is equivalent to that of a 2 μm diameter pulled capillary with a 1.0 kV spray voltage at a distance of 3 mm from a counter electrode.

the fluid as it exits the nozzle. One benefit of integrating a counter electrode into the nozzle structure is that formation of the electric field surrounding the nozzle is independent of the relationship of the nozzle position with respect to the mass spectrometer ion orifice. This enables independent formation of the

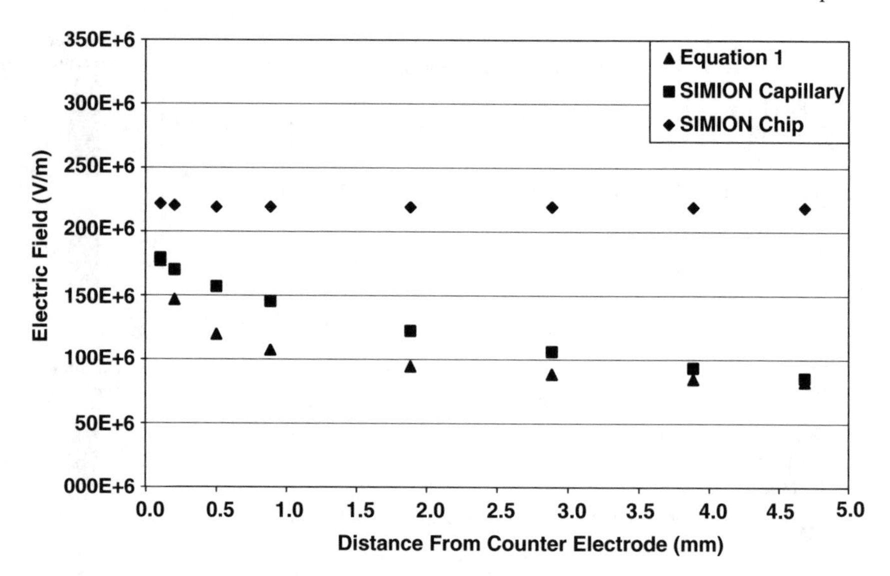

Figure 3.4 Graph plotting the changes in the electric field calculated at the tip of the Taylor cone as the distance between the sprayer and counter electrode varies using Equation (3.1) and the SIMION models shown in Figures 3.3A and B. The ESI Chip model (Figure 3.3B) shows that the electric field is independent of the sprayer distance from the counter electrode.

nanoelectrospray and optimization of the spray position relative to the ion-sampling orifice of the mass spectrometer.

There have been several publications on nanoESI (at flow rates below 25 nL min^{-1}) which reported less ion suppression compared to higher flow ESI for multi-analyte mixtures.[15–18] In one of the reported works,[16] a series of nanoelectrospray experiments were conducted comparing ion intensity ratios between two analytes subjected to flows ranging from 10 nL min^{-1} to 700 nL min^{-1}. In the reported work the flow rate was changed by changing the capillary diameters and applied voltages which, according to Equation (3.1), varies the electric field at the capillary tip. At 10–20 nL min^{-1}, the capillary diameters were 1 μm with an applied voltage of 800 V resulting in a calculated electric field of 3.3×10^8 V m^{-1}. Experiments conducted in the flow rate range of 500–700 nL min^{-1} used capillary outer diameters of 10–11 μm with an applied voltage of 1 kV, resulting in a calculated electric field of 5×10^7 V m^{-1}. The calculated electric fields show a 6-fold difference between the extremes of the experimental flow rate range and capillary dimensions used for this work. As an example, Figure 3.5A shows the results of one experiment that used a 10 molar excess of sodiated turanose (10^{-5} M) to *n*-octylglucopyranoside (10^{-6} M) sprayed from capillaries of different diameters to affect the flow rate. The results indicated a 6-fold change in the response ratios between the sodiated molecular ions at the low and high flow rates. The conclusions from this work suggested a dependency of ion intensity ratios on flow rate and that as the

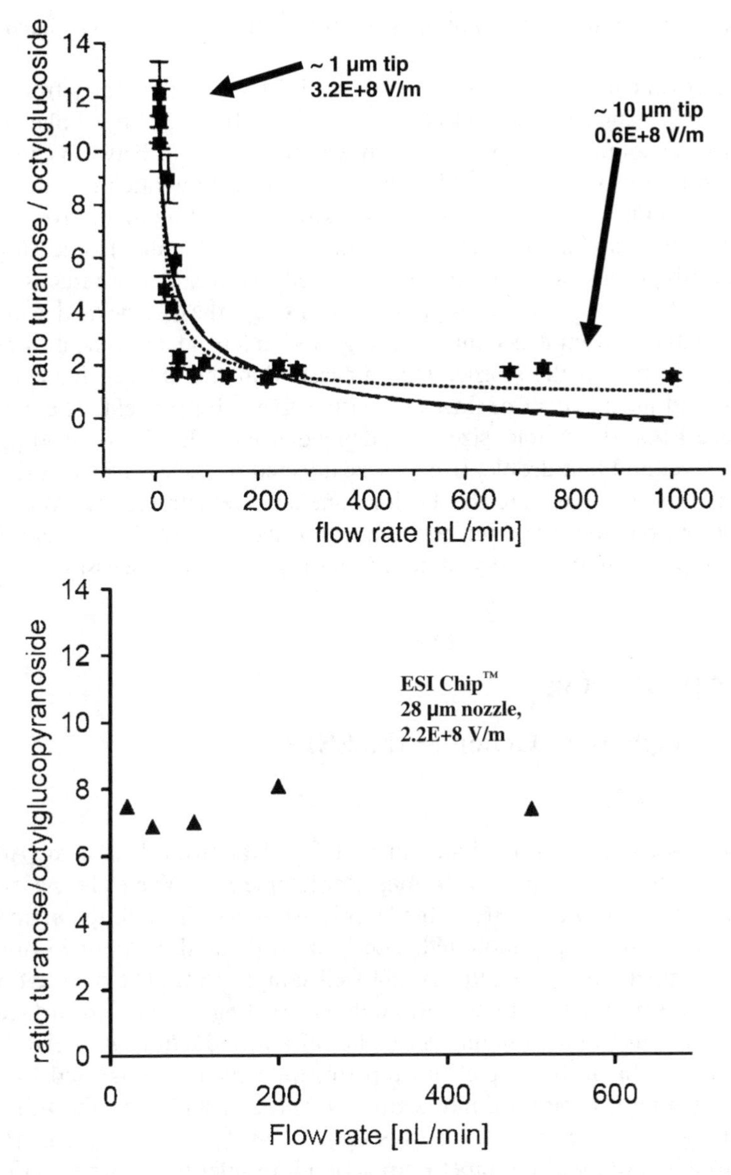

Figure 3.5 Plot of the signal intensity ratio for 10^{-5} M sodiated turanose and 10^{-6} M *n*-octylglucopyranoside using (A) pulled capillaries[16] and (B) the ESI Chip. The capillary graph contains labels of the diameter and electric field calculated using Equation (3.1) for the capillary dimensions identified in the text. The ESI Chip graph shows a ratio independent of flow rate. Data show a strong correlation between the electric field at the sprayer tip and the signal intensity ratios indicating that the electric field is an important variable in minimizing ion suppression.

flow rates were reduced to less than 20 nL min^{-1}, the ratio reflected the analyte molar ratio.

This experiment was duplicated using the ESI Chip at flow rates from 20 to 500 nL min^{-1} using a constant electric field of 2.2×10^8 V m^{-1} by application of 1.6 kV to the solution.[19] Figure 3.5B shows that the ratio between these two analytes was approximately 7 and independent of the flow rate in a range from 20 to 500 nL min^{-1}. A ratio of 7 was approximately that obtained from a 2 μm diameter pulled capillary (Figure 3.5A, flow rate ~50 nL min^{-1}) indicating that the ESI Chip produces a spray similar to that of a 2 μm capillary with a solution voltage of 1 kV. Based on the design of the ESI Chip, the electric field is linearly proportional to the applied solution voltage and independent of the flow rate of the solution through the nozzle. This suggests that the droplet diameters are constant and independent of flow rate with a fixed electric field. These results also suggest that the droplet size has a dependence on the electric field present at the sprayer and that droplet size may be affected by the electric field strength and its ability to repopulate the Taylor cone surface with charge over a wide variation of flow rate. Experiments designed to understand electric field effects on ion suppression are worthy of additional research and discussion.

3.3 Applications

3.3.1 Sample Introduction to the ESI Chip

3.3.1.1 Infusion

The ESI Chip is used in conjunction with the NanoMate® system (Advion BioSystems) when interfaced with mass spectrometry.[8] When placing an ESI Chip into the NanoMate system, the chip is grounded through the aluminium frame in which the chip is mounted. The NanoMate holds a 96- or 384-rack of conductive pipette tips and a 96- or 384-well sample plate. The pipette tips are coated with a proprietary material to enhance sealing to the chip surface surrounding the inlet and to minimize oxidation and reduction of analytes. The NanoMate has a probe to pick up pipette tips which is connected to a gas syringe for sample aspiration into a tip. To deliver a sample to the ESI Chip, the probe picks up a pipette tip, moves to aspirate sample from the plate then moves to align and seal the pipette tip around an inlet to a nozzle of the ESI Chip. This is shown in Figure 3.6. After the pipette tip is aligned, a voltage is applied to the probe, transferring the voltage to the tip and the sample. A pressure of 0 to 0.5 psi is applied to the headspace behind the sample contained in the tip. The voltage and pressure are optimized through software control to deliver the fluid sample to the ESI Chip nozzle to form a stable nanoelectrospray. The NanoMate system uses a new pipette tip and nozzle for each sample eliminating the chance for sample-to-sample carryover. Sample delivery is software controlled (ChipSoft, Advion BioSystems) enabling automated nanoelectrospray. The NanoMate was the first commercially available

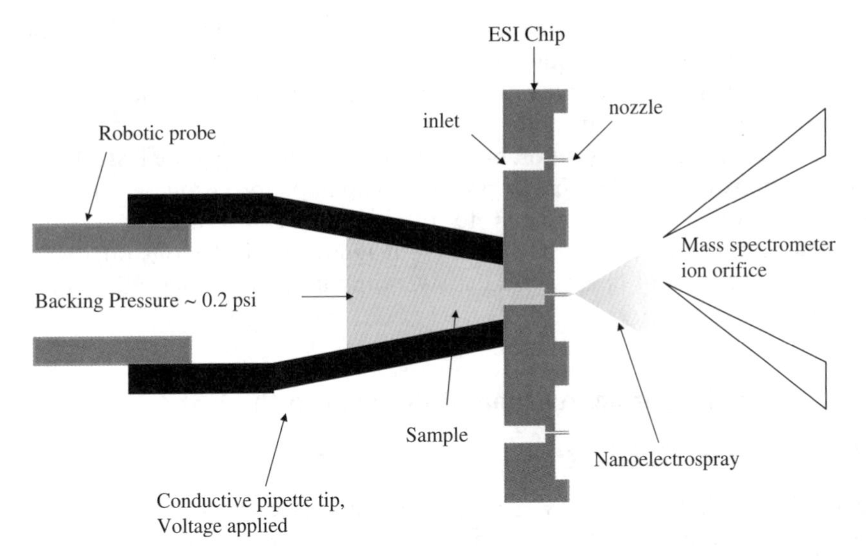

Figure 3.6 Pipette tip aligned and sealed around the inlet to a nozzle of the ESI Chip using the NanoMate system that automates nanoelectrospray infusion. Spray voltage and backing pressure are optimized based on the solution solvent composition to obtain a stable spray. The NanoMate system loads a new sample into a new pipette tip and automatically positions it to a new nozzle. The chip is automatically moved to the optimized spray position.

automated nanoelectrospray system that interfaced with mass spectrometry. It can be interfaced to over 85% of the existing atmospheric pressure ionization mass spectrometry systems worldwide.

Formation of a stable nanoelectrospray is dependent on many variables including the fluid surface tension, solvent composition, conductivity of the fluid and the applied voltage and pressure. Flow rate flowing through a nozzle is dependent on these variables as well as the inner diameter of the nozzle. As an example, a nozzle with a 5.5 μm inner diameter and 28 μm outer diameter will spray a solution of 50% methanol with 0.1% acetic acid at a flow rate of 100 nL min^{-1} with an applied voltage of 1.4 kV and pressure and 0.2 psi. A nozzle with a 2.5 μm inner diameter and 28 μm outer diameter will spray this same solution at a flow rate of 20 nL min^{-1} with an applied voltage of 1.2 kV and pressure of 0.3 psi.

Among the first uses of the NanoMate with ESI Chip system was rapid analysis of proteins separated on two-dimensional polyacrylamide gels (2D gel).[8] These 2D gels stained with colloidal coomassie blue demonstrated a limit of detection on a bovine serum albumin (BSA) standard of 75 fmol loaded on-gel using a 6 min infusion combined with data dependent acquisition (DDA) on an ion trap mass spectrometer. Zhang and Van Pelt wrote a comprehensive review of the use of the ESI Chip for protein characterization.[20] The development of top-down proteomics involving the MS analysis of intact proteins is emerging as an important application facilitated by nanoESI with the ESI Chip.[21–23]

Lipids analysis has benefited from nanoESI with the ESI Chip.[24–27] Lipids isolated from cells using liquid–liquid extraction can be quantified and separated using MS-MS scans such as multiple precursor and neutral loss scanning to identify head groups and side chain information.[26]

Small-molecule quantification by automated infusion using the ESI Chip has been demonstrated to be able to provide comparable precision, accuracy and linear dynamic range compared to traditional LC-MS-MS.[28–32] A linear dynamic range greater than five orders of magnitude was demonstrated which enabled validation of bioanalytical assays without the carryover limitations of LC-MS.[32]

3.3.2 Liquid Chromatography Interfaces to the ESI Chip

3.3.2.1 Nano Liquid Chromatography

Nano liquid chromatography (nanoLC) was first coupled to the ESI Chip as shown in Figure 3.7.[19] A fused silica capillary (20 μm i.d.) connected to the outlet of a nanoLC column was interfaced to the ESI Chip by positioning the proximate end of the capillary inside of a pipette tip such that when the tip was sealed around the chip inlet the end of the capillary was approximately 100 μm from the chip inlet. A liquid junction was used to apply a voltage to the LC effluent exiting the capillary prior to entering the inlet of the ESI Chip. A nanoLC separation of 100 fmol BSA digest on a 75 μm by 15 cm (LC Packings PepMap C18) column coupled to the ESI Chip is shown in Figure 3.8. A 20 μm i.d. by 30 cm length transfer capillary was used to couple the nanoLC

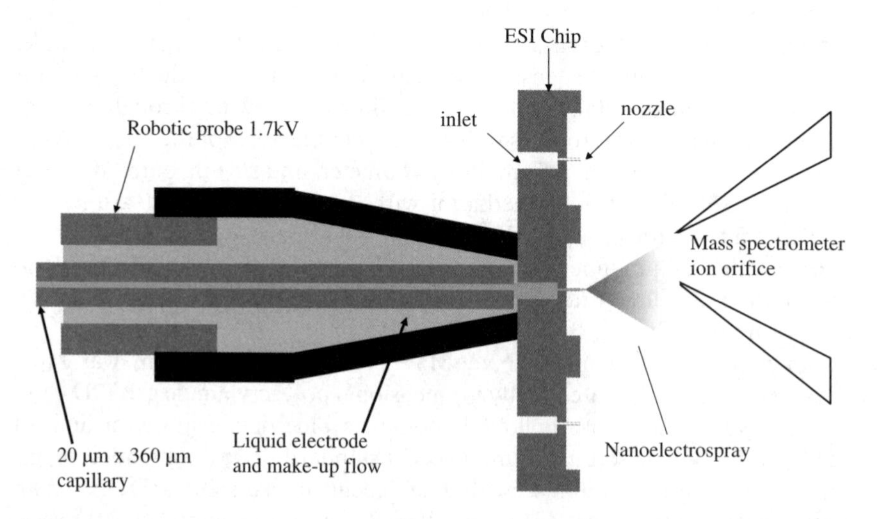

Figure 3.7 First coupling of a nanoLC column to the ESI Chip using a liquid electrode to apply spray voltage to the effluent exiting a capillary positioned at the inlet to the ESI Chip. Make-up flow was optionally applied to adjust solvent composition to enhance ionization.

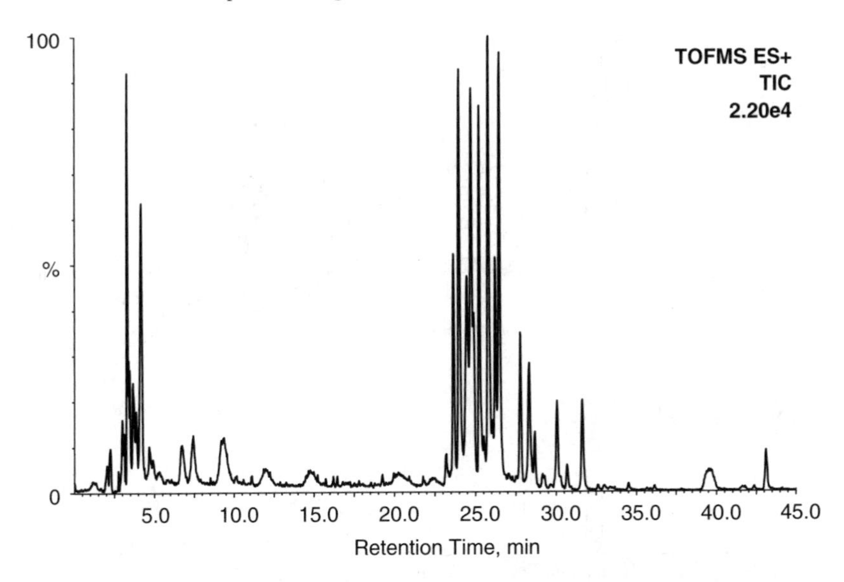

Figure 3.8 Total ion chromatogram (TIC) of a nanoLC-MS analysis of 100 fmol BSA digest using a 75 μm by 15 cm column coupled to the ESI Chip using the configuration shown in Figure 3.7. Mobile phase A: 10% acetonitrile, 90% water with 0.1% formic acid. Mobile phase B: 90% acetonitrile, 10% water with 0.1% formic acid. Gradient 0–4 min hold at 100% A, 4–24 min gradient to 60% A. 0.5 μL injection. Flow rate of 250 nL min^{-1} and spray voltage of 1.6 kV and no make-up flow.

column to the ESI Chip. The transfer capillary introduces an extracolumn variance of only 0.5 s due to its small inner diameter. The BSA digest was prepared by digestion with trypsin at 37 °C for 16 h and quenched with formic acid. The digest was then reduced with dithiothreitol and alkylated with iodoacetamide. An LC Packings Micro Pump and FAMOS™ autosampler were used in combination with a NanoMate 100 on a Waters Q-TOF mass spectrometer. Benefits from this approach included the ability to automatically move the column to a new nozzle and the ability to adjust the solvent composition being sprayed by addition of a makeup solvent to, for example, increase the organic composition to enable negative ionization of high aqueous effluents. Coupling of nanoLC columns to the ESI Chip is now done using a 15 μm diameter fused silica capillary with conductive polymer injection molded at one end of the capillary. The conductive polymer tip provides a leak-free seal around the chip inlet and a means of applying the spray voltage.

3.3.2.2 Micro and Analytical Liquid Chromatography

The benefits of nanoESI using the ESI Chip were recently extended to micro and analytical HPLC with the release of the TriVersa™ NanoMate system (Advion BioSystems) incorporating a post-column splitter that delivers 50 to 400 nL

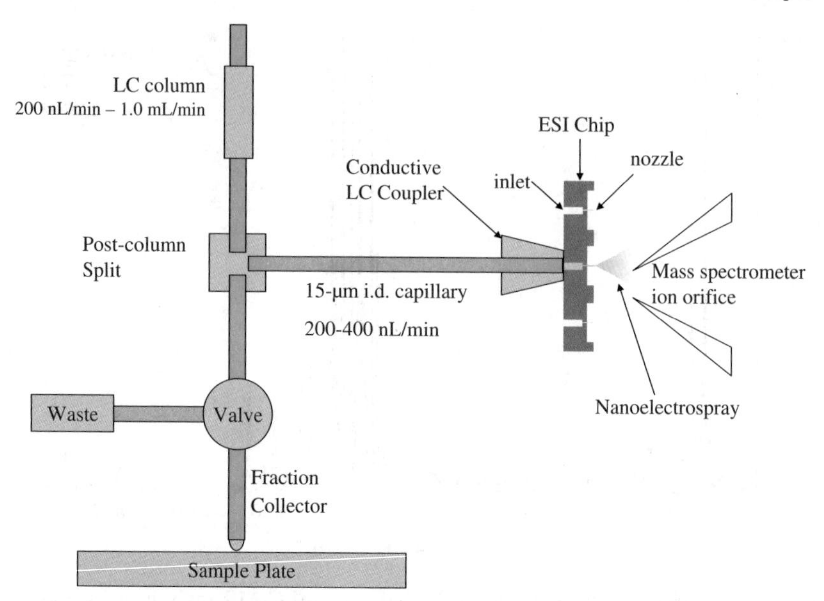

Figure 3.9 Schematic of the TriVersa NanoMate enabling the coupling of LC columns with flow rates up to 1.0 mL min^{-1} to the ESI Chip through use of a post-column splitter adjusted to deliver 200 to 400 nL min^{-1} of the post-column effluent to the ESI Chip while simultaneously collecting fractions into sample plates.

min^{-1} of the effluent to the ESI Chip with the remaining effluent directed through a six-port valve.[33] Figure 3.9 shows a schematic of the TriVersa system. The valve is software controlled and provides the option for collecting LC fractions into a 96- or 384-well sample plate. In this configuration, the TriVersa enables simultaneous acquisition of LC-MS data while collecting fractions. The system is being used in applications that would otherwise require repeated LC-MS experiments on a sample to collect sufficient MS-MS data for identification of unknowns within the chromatogram. Two main benefits of TriVersa to higher flow rate LC separations are the extension of nanoelectrospray and time for nanoelectrospray infusion of LC fractions for signal averaging and acquisition of multiple MS-MS and MS^n experiments on analytes from a single injection.[34,35] For complex samples analyzed using LC-MS, TriVersa enables the collection of 99.9% of the injected sample into purified fractions.

Drug metabolism, which often involves the use of radiolabels and fraction collection for scintillation counting, benefits by being able to reconstitute and spray LC fractions containing radiolabeled metabolites.[35,36] For complex samples analyzed using LC-MS, TriVersa enables the LC-MS experiment to focus on quantification and infusion of LC fractions on acquisition of qualitative MS^n data. Drug metabolism often uses radiolabeling and LC fraction collection followed by scintillation counting to identify which LC fractions contain metabolites. For metabolite identification, TriVersa enables high quality MS^n spectral traces to be acquired on each metabolite for structural

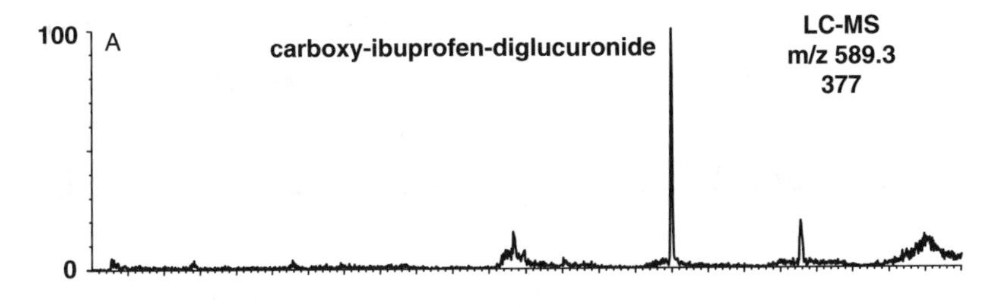

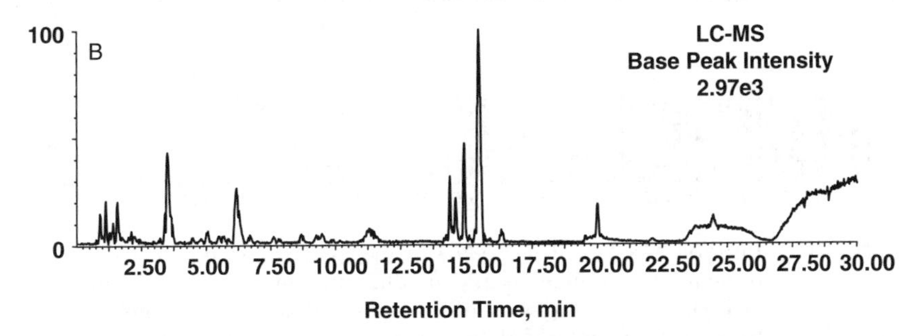

Figure 3.10 Gradient separation of ibuprofen metabolites from a human urine sample on a 2.1 by 100 mm reversed-phase C18 column using a TriVersa NanoMate coupled with a Waters Q-TOF micro mass spectrometer. (A) Extracted ion chromatogram for carboxy-ibuprofen-diglucuronide with MH^+ at m/z 589.3 and (B) base peak intensity chromatogram. LC fractions were collected simultaneously every 15 s into a 384-well plate.

elucidation. As an example for metabolite characterization, Figure 3.10 shows the LC-MS separation and analysis of a human urine sample containing ibuprofen metabolites. The effluent from a 2.1 mm by 100 mm C18 column was split 1000 :1 delivering 200 nL min^{-1} to the ESI Chip for positive ionization at 1.7 kV. LC fractions were collected simultaneously every 15 s into a 384-well sample plate.

Figure 3.10A shows the extraction ion chromatogram (XIC) for m/z 589.3 which corresponds to the MH^+ for carboxy-ibuprofen-diglucuronide. Figure 3.10B shows the base peak intensity (BPI) chromatogram. Infusion of the LC fraction corresponding to the carboxy-ibuprofen-diglucuronide peak at 20.2 min shown in Figure 3.11 enables the acquisition of a high-quality MS-MS spectrum by signal averaging the product ion mass spectrum of m/z 589.3. Infusion of LC fractions can easily improve the signal-to-noise ratio by a factor of 10 when compared to that obtained using LC-MS-MS and up to a 50-fold signal enhancement was reported.[35]

3.3.3 Spray Stability of the ESI Chip

Another benefit of nanoelectrospray from the ESI Chip is improved spray stability compared with Turbo IonSpray (TIS) when using selected reaction

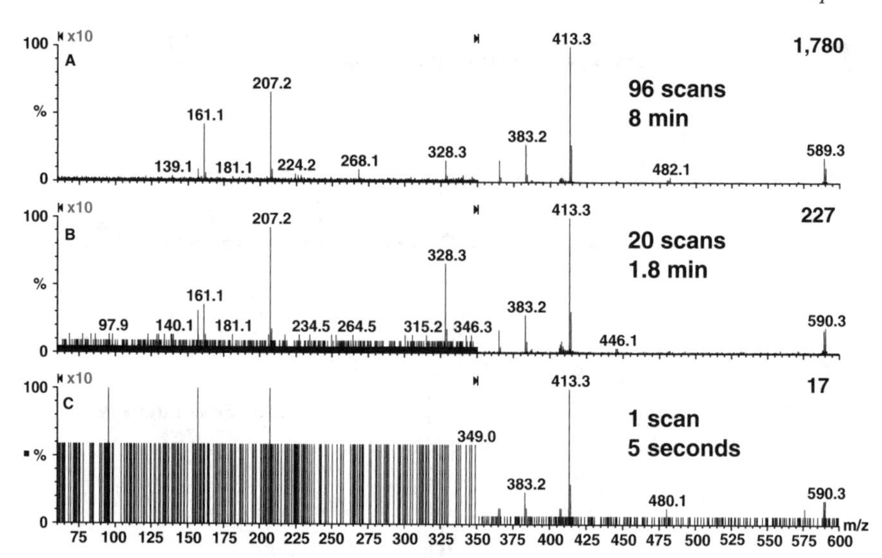

Figure 3.11 Infusion MS-MS of m/z 589.3 of the LC fraction for the carboxy-ibuprofen-diglucuronide peak at 20.2 min in the separation shown in Figure 3.10A: (A) sum of 96 scans; (B) sum of 20 scans; (C) single scan. Infusion increased the signal intensity 100-fold compared to a single 5 s MS-MS scan providing a 10-fold increase in the signal-to-noise ratio.

monitor (SRM) on a triple quadrupole mass spectrometer (4000 Q TRAP, AB/MDS Sciex, Toronto, Canada).[37] Figure 3.12 compares the ion current signal stability from infusion of TIS and the ESI Chip using a 10 ms SRM dwell time for venlafaxine (m/z 278.2 > m/z 58.1). The ion current signal stabilities measured from 1 min of spray for TIS and the ESI Chip were 15% and 3.4% relative standard deviation (RSD), respectively. This improvement in stability may be due to the difference in the droplet size formed between these two ionization techniques or the result of the relative calm of the gas surrounding the nanoelectrospray relative to the high flux of nebulization gas and turbo drying gas used to form gas-phase ions for high flow rate electrospray. The ESI Chip stability using a 10 ms SRM is better than TIS using a 200 ms SRM.[37] This improved stability provides improved quantification of chromatographic peaks as more data points can be acquired across the peaks without sacrificing precision. It also enables monitoring of more SRM transitions simultaneously for metabolomic applications that aim to quantify endogenous metabolites from biological samples. The spray stability was investigated with the LC-SRM-MS analysis of rat plasma extracts of alprazolam.[37] Figure 3.13 compares the TIS and ESI Chip SRM ion chromatograms for alprazolam and alprazolam-*d5* for a 10 ms SRM dwell time showing a significantly improved stability in the ion chromatograms with the ESI Chip without a loss in sensitivity.

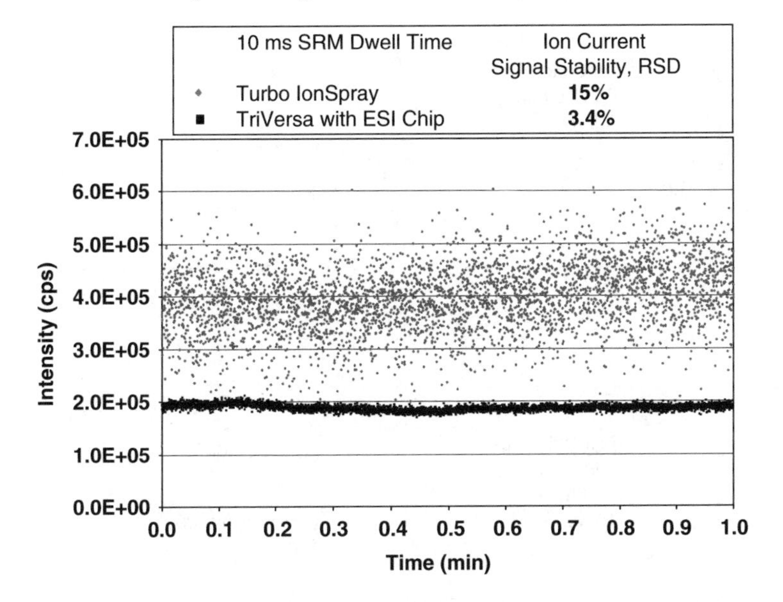

Figure 3.12 Plot of SRM data points collected every 10 ms using Turbo IonSpray (TIS) and the ESI Chip for venlafaxine (SRM m/z 278.2 > m/z 58.1). A 5 μM venlafaxine solution at 5 μL min^{-1} was mixed with 400 μL min^{-1} of 75% acetonitrile, 25% water and 0.2% formic acid. This solution was sprayed using TIS at 450 °C while monitoring the venlafaxine SRM from m/z 278.2 to m/z 58.1. A 5 μM venlafaxine solution at 5 μL min^{-1} was mixed with 600 μL min^{-1} of 75% acetonitrile, 25% water and 0.2% formic acid. This solution was split to deliver 200 nL min^{-1} to the ESI Chip while monitoring the venlafaxine SRM from m/z 278.2 to m/z 58.1. Data collected every 10 ms result in a 15% RSD with TIS and 3.4% RSD with the ESI Chip.

3.4 Concluding Remarks

The ESI Chip enabled nanoelectrospray to be routinely and widely used by researchers and its use with the NanoMate systems enabled the automation of nanoESI for the first time. The chip fabrication process produces nozzle features which are reproduced to within a micrometer from wafer to wafer and batch to batch. The integrated counter electrode of the ESI Chip forms a high electric field surrounding the nozzle and, when combined with the fabrication process, provides stable and consistent sprays between nozzles.

The coupling of LC to the ESI Chip expanded the applications for nanoelectrospray beyond static infusion and nanoLC. LC is rapidly evolving with the adoption of stationary phases and pumping technology that routinely results in chromatographic peak widths of a second or less. Quantification of these peaks using ESI-MS requires shorter measurement times for acquisition of the chromatographic data. The improved spray stability of the ESI Chip will provide for multiple analyte quantification with improved precision by enabling the use of acquisition times of less than 10 ms.

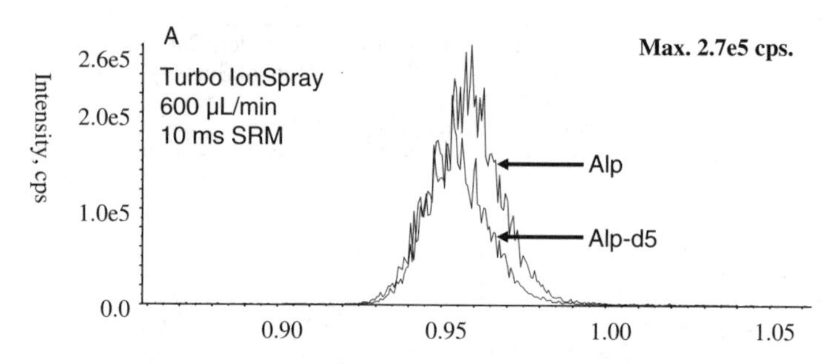

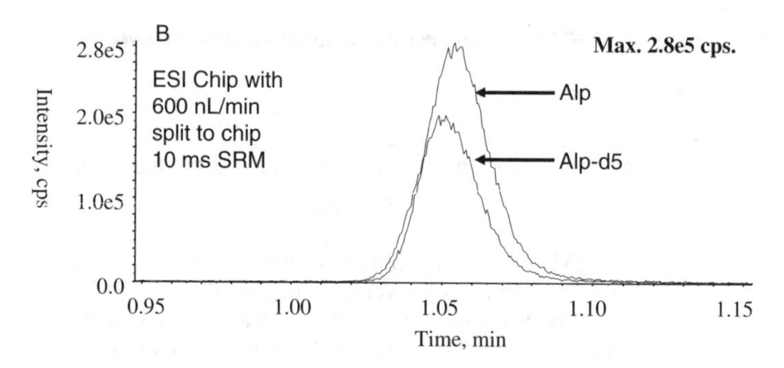

Figure 3.13 Comparison of the LC-MS-MS analysis of alprazolam and *d5*-alprazolam using (A) Turbo IonSpray and (B) the ESI Chip with a 10 ms SRM dwell time. TIS conditions were 600 μL min^{-1} at a turbo gas temperature of 450 °C. For the ESI Chip, a 1000 : 1 post-column split was used to deliver 600 nL min^{-1} to the chip.

Acknowledgements

This work was made possible due to the vision of Jack Henion and Tom Kurz, co-founders of Advion BioSciences, Inc. Their leadership enabled the innovation that culminated in the commercialization of the ESI Chip and associated products. I am grateful to them for the opportunity to work on such an exciting technology. I am also grateful to all of my colleagues from Advion who make it all possible.

References

1. M. Mann and M. Wilm, *Int. J. Mass Spectrom. Ion Processes*, 1994, **136**, 167.
2. M. Wilm and M. Mann, *Anal. Chem.*, 1996, **68**, 1.
3. L. Tang and P. Kebarle, *Anal. Chem.*, 1993, **65**, 3654.
4. M. Yamashita and J. B. Fenn, *J. Phys. Chem.*, 1984, **88**, 4451.
5. N. B. Cech and C. G. Enke, *Mass Spectrom. Rev.*, 2001, **20**, 362.

6. G. A. Schultz, T. N. Corso, S. P. Prosser and S. Zhang, *Anal. Chem*, 2000, **72**, 367.
7. F. Laermer and A. Schilp, *US Pat.*, 5, 501, 893, 1996.
8. S. Zhang, C. K. Van Pelt and J. D. Henion, *Electrophoresis*, 2003, **24**, 3620.
9. J. Li, T. N. Corso and G. A. Schultz, *US Pat.*, 6, 891, 155, 2005.
10. R. J. Pfeifer and C. D. Hendricks, *AIAA J.*, 1968, **6**, 496.
11. P. Kebarle and Y. Ho, *On the mechanism of electrospray mass spectrometry*, in *Electrospray Ionization Mass Spectrometry*, ed. R. B. Cole, Wiley, New York, 1997, p. 9.
12. G. A. Schultz, T. N. Corso and S. P. Prosser, *Proceedings of the 49th ASMS Conference on Mass Spectrometry and Allied Topics*, 27–31 May 2001.
13. Q. Xue, F. Foret, Y. M. Dunayevskiy, P. M. Zavracky, N. E. McGruer and B. L. Karger, *Anal. Chem.*, 1997, **69**, 426.
14. R. S. Ramsey and J. M. Ramsey, *Anal. Chem.*, 1997, **69**, 1174.
15. R. Juraschek, T. Dulcks and M. Karas, *J. Am. Soc. Mass Spectrom.*, 1999, **10**, 300.
16. A. Schmidt and M. Karas, *J. Am. Soc. Mass Spectrom.*, 2003, **14**, 492.
17. C. E. C. A. Hop, Y. Chen and L. J. Yu, *Rapid Commun. Mass Spectrom.*, 2005, **19**, 3139.
18. G. A. Valaskovic, L. Utley, M. S. Lee and J. T. Wu, *Rapid Commun. Mass Spectrom.*, 2006, **20**, 1087.
19. G. A. Schultz, R. E. Murphy and S. Zhang, *Proceedings of the 52nd ASMS Conference on Mass Spectrometry and Allied Topics*, 30 May – 3 June 2004.
20. S. Zhang and C. K. Van Pelt, *Expert Rev. Proteomics*, 2004, **1**, 449.
21. J. J. Coon, B. Ueberheide, J. E. Syka, D. D. Dryhurst, J. Ausio, J. Shabanowitz and D. F. Hunt, *Proc. Natl Acad. Sci. USA*, 2005, **102**, 9463.
22. Y. Du, B. A. Parks, S. Sohn, K. E. Kwast and N. L. Kelleher, *Int. J. Mass Spectrom.*, 2006, **234**, 686.
23. B. Macek, L. Waanders, J. V. Olsen and M. Mann, *Mol. Cell. Prot.*, 2006, **5**, 949.
24. A. Karin, B. Zemski and R. C. Murphy, *Anal. Biochem.*, 2006, **349**, 118.
25. D. Schwudke, J. Oegema, L. Burton, E. Entchev, J. T. Hannich, C. S. Ejsing, T. Kurzchalia and A. Shevchenko, *Anal. Chem.*, 2006, **78**, 585.
26. C. S. Ejsing, T. Moehring, U. Bahr, E. Duchoslav, M. Karas, K. Simons and J. Shevchenko, *J. Mass Spectrom.*, 2006, **41**, 372.
27. D. Linden, L. William-Olsson, A. Ahnmark, K. Ekroos, C. Hallberg, H. P. Sjogren, B. Becker, L. Svensson, J. C. Clapham, J. Oscarsson and S. Schreyer, *FASB J.*, 2006, **20**, 434.
28. J. M. Dethy, B. Ackerman, C. Delatour, J. Henion and G. Schultz, *Anal. Chem.*, 2003, **75**, 805.
29. J. Kapron, E. Pace, C. K. Van Pelt and J. Henion, *Rapid Commun. Mass Spectrom.*, 2003, **17**, 2019.
30. C. K. Van Pelt, S. Zhang, E. Fung, I. Chu, T. Liu, C. Li, W. Korfmacher and J. Henion, *Rapid Commun. Mass Spectrom.*, 2003, **17**, 1573.

31. L. Corkery, H. Pang, B. Schneider, T. Covey and K.W. Michaelsiu, *J. Am. Soc. Mass Spectrom.*, 2005, **16**, 363.
32. E. R. Wickremsinhe, B. L. Ackermann and A. K. Chaudhary, *Rapid Commun. Mass Spectrom.*, 2005, **19**, 47.
33. G. A. Schultz, R. Almeida, L. Klecha, V. Italiano, M. Lees, P. Weisz, J. Lesinski and S. J. Prosser, *Proceedings of the 53rd ASMS Conference on Mass Spectrometry and Allied Topics*, 30 May–3 June 2005.
34. E. T. Gangl, M. Annan, N. Spooner and P. Vouros, *Anal. Chem.*, 2001, **73**, 5635.
35. R. F. Staack, E. Varesio and G. Hopfgartner, *Rapid Commun. Mass Spectrom.*, 2005, **19**, 618.
36. K. O. Boernsen, J. M. Floeckher and G. J. Bruin, *Anal. Chem.*, 2000, **72**, 3956.
37. G. A. Schultz, E. Pace and J. D. Henion, *Proceedings of the 54th ASMS Conference on Mass Spectrometry and Allied Topics*, 28 May–1 June 2006.

CHAPTER 4

Microfabricated Multichannel Electrospray Ionization Emitters on Polydimethylsiloxane (PDMS) Microfluidic Devices

JIN-SUNG KIM,[†] AND DANIEL R. KNAPP

Medical University of South Carolina, Charleston, SC 29425, USA

4.1 Background

Microfluidic analytical systems, particularly for the analysis of biomolecules, are a subject of increasing interest,[1–4] and these analytical devices have been shown to be a powerful addition in the integration of a microscale total analysis system (μTAS)[5–7] with the potential for high-throughput mass spectrometric analysis. Microfluidic devices coupled to electrospray ionization mass spectrometry (ESI-MS) instruments have been reported by several groups. The first reports of microfluidic devices employing ESI-MS utilized electrospray from channels terminating at the edge of the device;[8,9] however, most subsequent reports have utilized conventional electrospray emitters (*e.g.* tapered fused silica capillary) attached to the device.[10–13] Spraying from the edge of a device was troubled by droplet formation resulting in a mixing volume that degraded the separation. Attachment of separate emitters can also introduce a dead volume in the connection in addition to the disadvantage of the need for an individual attachment step for each emitter in a multichannel device.

Use of multiple channels in a microfluidic device eliminates the problem of cross-contamination when analyzing a series of samples. Reduction of the size

[†] Present address: Analytical Research Laboratory, Korea Institute of Toxicology, Korea.

Miniaturization and Mass Spectrometry
Edited by Séverine Le Gac and Albert van den Berg

Published by the Royal Society of Chemistry, www.rsc.org

of emitter tips has been shown to reduce the amount of sample required thereby yielding increased sensitivity in microspray (nanospray) ESI-MS analysis.[14,15] Therefore, microfabrication of an array of ESI emitters with narrow channels and small tips can provide an attractive method for high sensitivity ESI-MS analysis of multicomponent samples such as those encountered in proteomics studies and for the miniaturization of LC-MS analytical instrumentation.

Many approaches and substrate materials have been used to fabricate microfluidic devices; there have been earlier reports of microfabricated ESI emitters made from silicon nitride,[16] parylene,[17] polycarbonate[18] and monolithic silicon.[19] Microfabrication processes for these are time-consuming and laborious, and require relatively sophisticated facilities. The micromachining process[18] is not well suited to production of multiple emitters since each emitter must be individually fabricated. The use of polydimethylsiloxane (PDMS) for microfluidic devices has many advantages in terms of cost, ease of fabrication and the inherent chemical properties of PDMS (*e.g.* the hydrophobic and uncharged surface, and relatively low tendency to adsorb biological samples). The devices are produced by casting PDMS against a photoresist-patterned master on a silicon wafer with the "soft lithography" technique.[20]

In this chapter, four methods (referred to as trimming,[21,22] two-layer photoresist,[22] resin casting[22] and three-layer photoresist with self-alignment features[23] methods) are described using soft lithography for microfabricating multichannel ESI emitters as an integral part of PDMS microfluidic devices, which were interfaced to an ESI time-of-flight mass spectrometry (ESI-TOF-MS) instrument. These methods obviate the need for manual attachment of separate components, which is labor intensive and can cause dead volume that degrades the separation. Soft lithography provides a relatively simple microfabrication process for PDMS devices compared to that of other microfabrication methods. PDMS devices made by these methods are also readily transferable to injection-molding processes for larger scale production of the devices. Four-channel PDMS emitters produced the by two-layer and resin casting methods are more reproducible and more robust than those produced by the trimming method. A further improvement in the three-layer photoresist process for producing 16-channel ESI emitters as an integral part of PDMS microfluidic devices facilitates not only the fabrication of the emitters, but provides a new approach to self-alignment of the top and bottom parts in the assembly of the device. This microfabrication method can be readily extended to larger multichannel devices, and should also be generally applicable to developing other PDMS microfluidic systems.

4.2 Soft Lithography Method for Fabricating Microfluidic Systems

4.2.1 Concept

The soft lithography method was developed for rapidly and inexpensively fabricating microfluidic devices with channels >20 μm width using high-resolution image setting and photolithography to produce masters for molding of PDMS, followed by contact sealing of oxidized PDMS surfaces.[24] The process, from the design idea to the finished device, can be completed in less than 24 h, significantly reducing the time to obtain prototype microfluidic systems. The steps of the method are illustrated in Figure 4.1 and described below.

1. A transparency containing the channel design produced from a computer drawing is used as a photomask to produce a positive pattern of the channels in photoresist on a silicon wafer.
2. Glass posts are attached to the wafer to define reservoirs.
3. PDMS prepolymer is cast on the photoresist master and heat cured.
4. The PDMS cast replica of the master pattern containing the channels is peeled off from the patterned wafer, and the glass posts are removed.
5. The channels in the PDMS replica are sealed by bonding a flat piece of PDMS following surface oxidization in air plasma. The oxidized PDMS surfaces bond when brought into conformal contact forming an irreversible

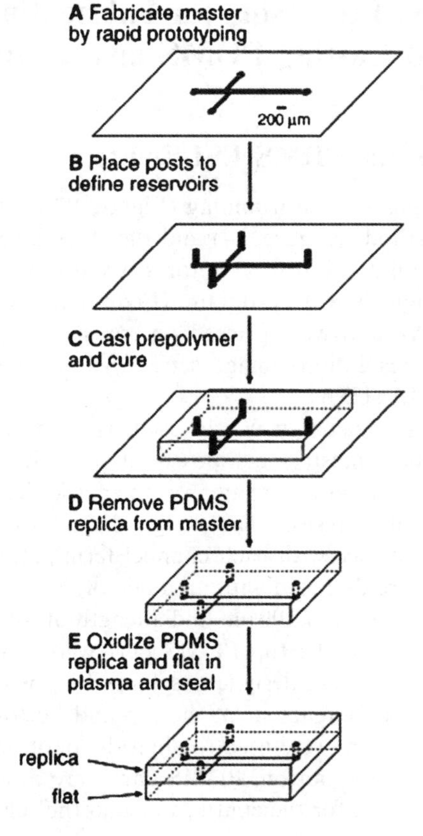

Figure 4.1 Schematic describing the fabrication of enclosed microscopic channels in oxidized PDMS. (Reprinted with permission from ref. 24.)

seal. Surface silanol (SiOH) groups on the channel walls produced by the oxidation can ionize in neutral or basic solutions and support electroosmotic flow (EOF) in the channels (although the "self-healing" properties of PDMS can limit the durability of this surface modification).

4.2.2 Advantages of Soft Lithography

Using soft lithography for creating microfluidic systems has several advantages over other methods. Microfluidic devices can be designed directly by the end user, and the devices can be produced quickly and simply from low-cost polymer using inexpensive transparencies as photomasks to create positive relief masters for PDMS molding. Multiple devices can be made from each master compared to glass or Si/SiO_2 devices that require individual photolithography and fabrication. The sealing of the channels is an easy process requiring only surface oxidation of the components to effect bonding. Finally, the resulting devices are durable and resist damage from dropping or bending.

4.3 Photoresist Processing for Fabricating the Master Wafer and Casting PDMS against the Patterned Master

4.3.1 Designs for the PDMS ESI Emitters

The photomask designs for the trimming (Figure 4.2), two-layer (Figure 4.3) and resin casting methods (Figure 4.4) and the three-layer method with self-alignment features (Figure 4.5) for photoresist patterns were created with a computer-aided design (CAD) program (Free hand 8; Macromedia, San Francisco, CA, USA), and were printed on transparency films with a high-resolution (3356 dpi resolution) image setter (Scitex Dolev 450; CreoScitex America, Bedford, MA, USA).

In Figure 4.2, this transparency was used as a mask in contact photolithography to produce masters composed of a positive relief of negative photoresist on a silicon wafer (white areas of the mask yield raised areas on the wafer with negative photoresist). The design defines the channels and edge profile of the PDMS device with each channel terminating at a point on the edge. The channels were designed with an angle of 30 ° or 60 ° for the sharply pointed emitter tips, a width of 100 μm and a length of 4 cm. In Figure 4.5, the images of the channels for the top PDMS part were defined as 30 μm wide channels, the emitter tips were designed as 1 mm long with a 60 ° point angle and the hexagons for self-alignment of the top and bottom parts were 6 mm. The 0.5 mm square markers were used to align the photomasks with the mask aligner. The images for the bottom PDMS part were designed with the same dimensions as the top part for the emitter tips and the self-alignment features.

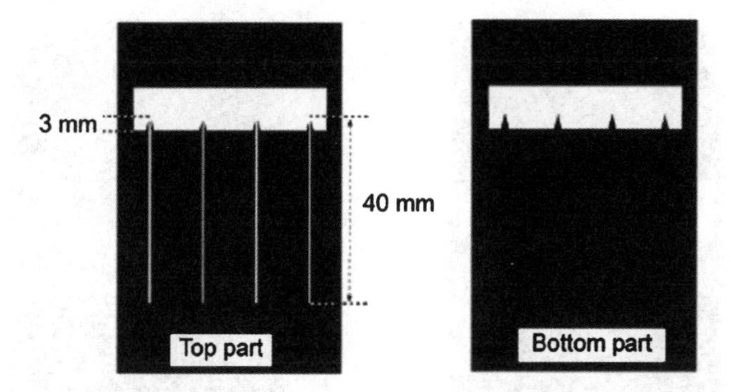

Figure 4.2 Photomask images for the photoresist patterns of the PDMS emitter device produced by the trimming method. (Reprinted with permission from ref. 22.)

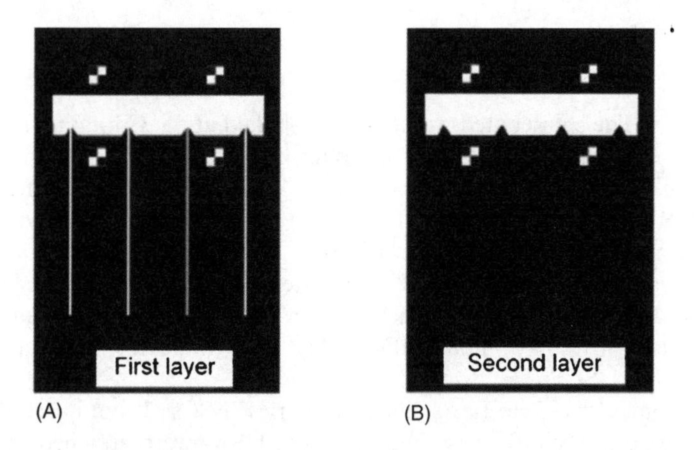

Figure 4.3 Photomask images for the photoresist patterns of the PDMS emitter device produced by the two-layer photoresist method. (A) First photoresist layer; (B) second photoresist layer. (Reprinted with permission from ref. 22).

4.3.2 Trimming Method

4.3.2.1 Photoresist Processing for Fabricating the Master

In order to make the thickness of the photoresist pattern 30 μm for the depth of the channel which is 4 cm long and 100 μm wide using the photomask design (Figure 4.2) for the trimming method, 3.5 mL of a solution of negative photoresist SU8-50 (PRS; Micorchemical, Newton, MA, USA) was dropped onto a 100 mm diameter silicon wafer (Silicon, Boise, ID, USA) chucked in a spin coater (model WS-200-4NPP; Laurell Technologies, North Wales, PA, USA) that was spun at 500 rpm followed by acceleration to 1000 rpm for a total spin

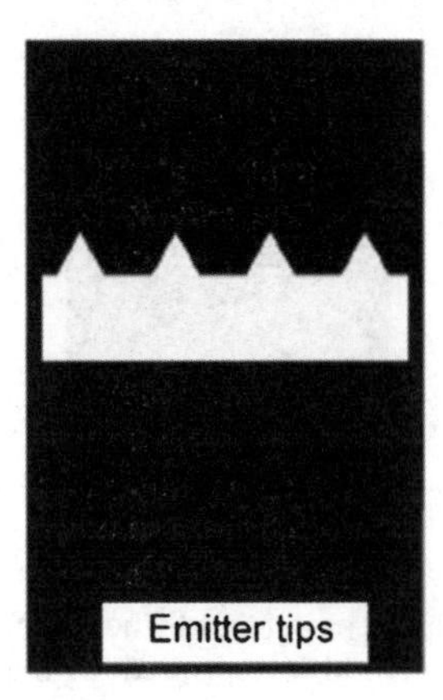

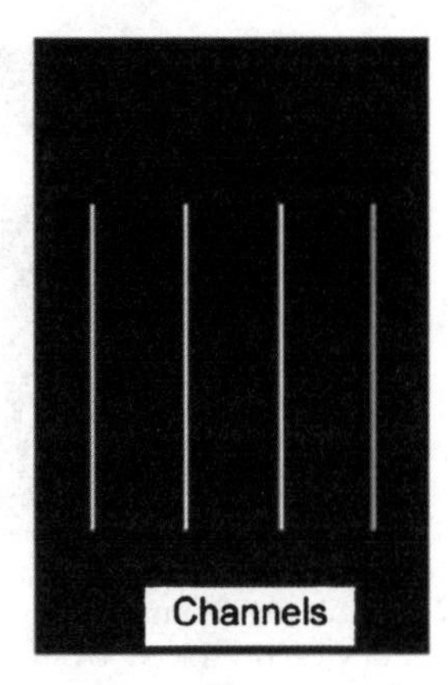

Figure 4.4 Photomask images for the photoresist patterns of the PDMS emitter device produced by the resin casting method. (Reprinted with permission from ref. 22).

time of 25 s. The spin-coated wafer was pre-baked at 55 °C for 3 min, and then at 95 °C for 25 min. After baking, the photomask image was patterned on the negative photoresist using an ultraviolet (UV) lamp in a mask aligner (Cobilt CA 800; BTS, Atco, NJ, USA) for 30 s. The wafer exposed with UV was post-baked at 95 °C for 15 min and developed in 2-(methoxy)propyl acetate (MPA; Acros Organics, Pittsburgh, PA, USA). After developing the wafer, the height of photoresist pattern on the master wafer was measured as 30 μm using a profilometer (Mitutoyo model ID-C112C Dial Indicator; McMaster-Carr, Atlanta, GA, USA).

Four segments of 5 cm fused silica capillaries (FSCs; 75 μm i.d. and 360 μm o.d.; Polymicro Technologies, Phoenix, AZ, USA) were attached at the end points of the channels of photoresist pattern as shown on Figure 4.6A on the master wafer using the PDMS premixture (Silgard 184, Dow Corning) as glue in order to provide openings for connecting longer FSCs to the cured PDMS device. PDMS glue was used 5 h after the PDMS premixture was prepared and kept at room temperature. Prior to attaching the FSCs on the photoresist, 2 mm long pieces of 360 μm i.d. polytetrafluoroethylene (PTFE) tubing were first attached to the photoresist, the FSC was inserted into the tubes and secured using PDMS glue and then the assembly was heated for 1 h at 70 °C to cure the PDMS.

To make the concave shape edge for the PDMS emitters, the ends of the tips on the photoresist pattern as seen on the right side of Figure 4.6A and B were covered with a curved piece of transparency film (3M PP2500 for laser printers; 3M Visual Systems, Austin, TX, USA) which was held in place by plastic tape (Scotch Magic Tape; 3M, Minneapolis, MN, USA) and pressurized by a brass block (0.5 in × 0.25 in), and the rectangular photoresist pattern on the master wafer was surrounded with three brass blocks of the same size.

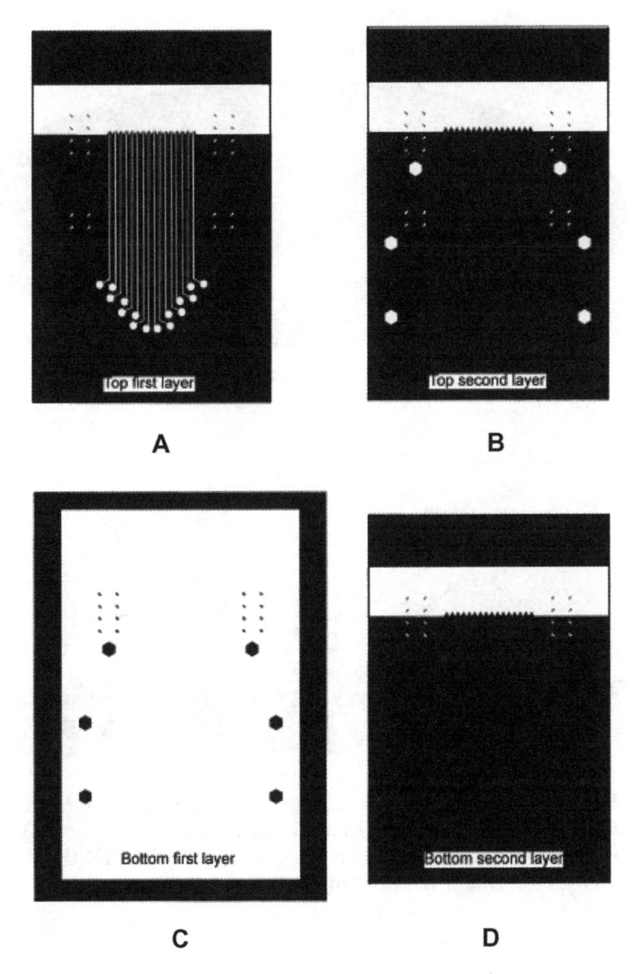

Figure 4.5 Photomask images for photoresist patterns of the PDMS emitter device with (A) top first layer, (B) top second layer, (C) bottom first layer, and (D) bottom second layer produced by the three-layer method with self-alignment features. (Reprinted with permission from ref. 23).

4.3.2.2 Casting PDMS against the Patterned Master

Before casting the top PDMS part against the master wafer, the master had been vapor-phase silanized with a (tridecafluoro-1,1,2,2-tetrahydrooctyl)-1-tricholorosilane (TCS; United Chemical Technologies, Bristol, PA, USA) releasing agent in a vacuum desiccator for 8 h. PDMS prepolymer cast on the master wafer was cured at 70 °C for 3 h in an oven, and the cured top PDMS part was peeled from the master wafer. After curing, the 5 cm FSCs in the top PDMS part were removed and replaced with new 20 cm segments of same-dimension FSCs to make fluid connections to the channels. The new FSCs were attached while blowing nitrogen gas through the capillary to prevent clogging,

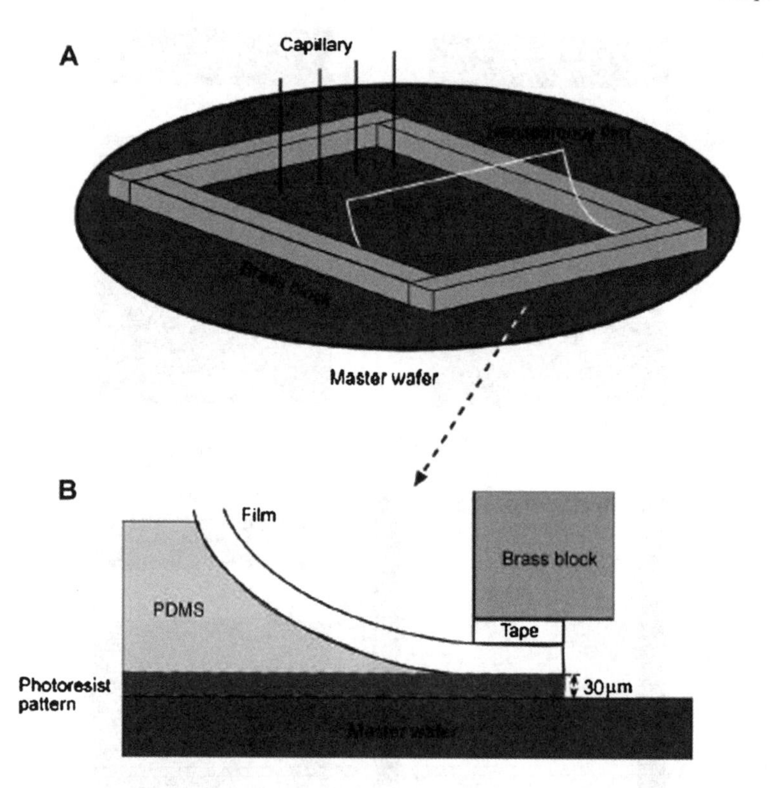

Figure 4.6 Schematic of (A) the molding of the PDMS replica on the master wafer with photoresist pattern by the trimming method, and (B) detail of the concave shape of the device edge leading to the PDMS emitter tip (side view). (Reprinted with permission from ref. 22).

and the FSCs were secured in place with PDMS premixture. The bottom PDMS part, which was symmetrical to the top PDMS part of the device, was cast on a silicon wafer with the photoresist pattern identical to that of the top part without the channels (Figure 4.6B).

The cured top and bottom PDMS parts were surface-oxidized at the same time in a plasma cleaner (model PDC-32G; Harrick Scientific, Ossing, NY, USA) at medium power setting for 1 min at 2 torr air pressure. After oxidation in air plasma, the top PDMS part was aligned to the bottom PDMS part using a thin layer of methanol between the parts, and then bonded by heating at 70 °C for 4 h to evaporate the methanol. The membrane edges for the emitter tips in the bonded PDMS device were trimmed to shape using iris scissors (Roboz Surgical Instruments, Rockville, MD, USA) and a scalpel blade under a stereomicroscope (model AO 569; American Optical, Southbridge, MA, USA) along the photoresist pattern in the cast PDMS device as a guide for the emitter shape. The angle of the emitter tip and the channel shape and size are shown in Figure 4.7.

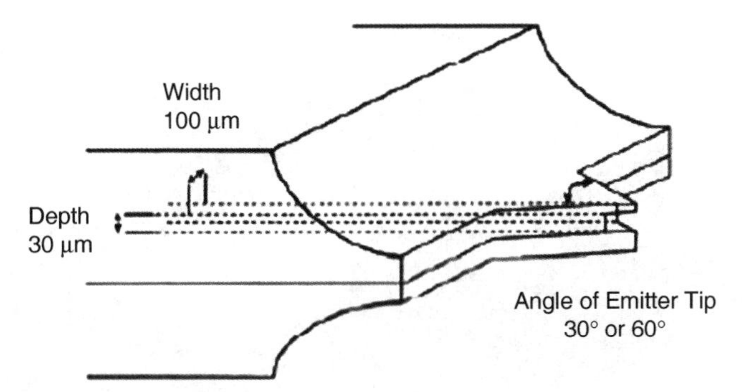

Figure 4.7 Detail of trimmed emitter showing the angles of the emitter tip and the dimensions of the channel. (Reprinted with permission from ref. 21).

4.3.3 Two-Layer Method

4.3.3.1 Photoresist Processing for Fabricating the Master

The photomask design for the two-layer photoresist pattern and the process for the first photoresist pattern (see Figure 4.3A) were made following the same procedure as the design for the trimming method described above. In order to make the pattern of the second layer of photoresist as ~150 μm high, without post-baking the exposed first layer, 3.5 mL of photoresist solution was dropped on the first photoresist, and spun initially at 2500 rpm followed by spinning at 1000 rpm for a total spin time of 20 s. The spin-coated wafer with two-layer photoresist was baked at 55 °C for 3 min, and then at 95 °C for 25 min. After baking, the transparency with the pattern of the second layer (Figure 4.3B) was aligned onto the exposed first layer photoresist using the mask aligner, and was exposed for 100 s. The silicon wafer with exposed first and second photoresist layer was hard-baked at 95 °C for 15 min, and developed in MPA. The heights of the photoresist pattern in each layer on the master wafer (Figure 4.8A) were measured as 30 μm and 150 μm, respectively, using a profilometer.

4.3.3.2 Casting PDMS against the Patterned Master

After connecting the FSCs as described for the trimming method, the master wafer with the two-layer photoresist pattern was surrounded with three PTFE blocks (0.5 in × 0.25 in). The ends of the reference points in the two-layer photoresist pattern on the master wafer were covered with a convex profile of an epoxy resin block (made from EpoFix; EMS, Ft. Washington, PA, USA) for producing the emitter tips in the top PDMS part (Figure 4.8B). The PTFE and resin blocks were pressurized with two aluminium plates (5 in × 5 in) held by four binder clips (Acco 72100, Large; Acco Brands, Lincolnshire, IL, USA). The degassed premixture for the top PDMS part was cast against the whole system on the master wafer which had been silanized with TCS under vacuum

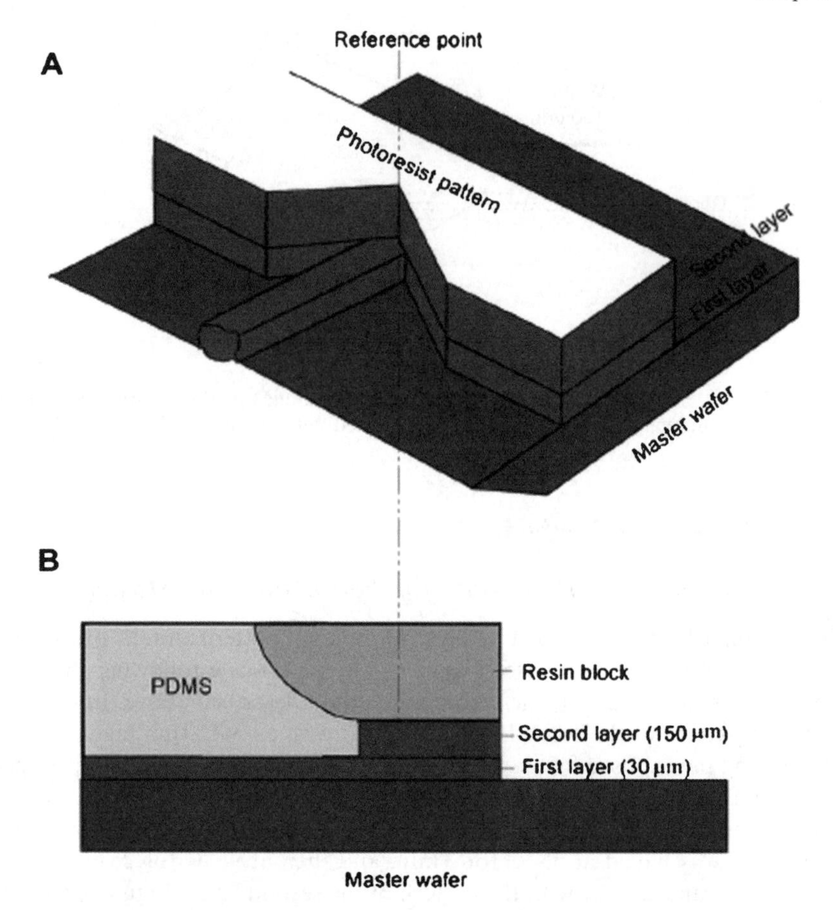

Figure 4.8 Schematic of (A) the molding of the PDMS replica on the master wafer with the pattern for the two-layer photoresist method, and (B) detail of the concave shape of the device edge leading to the membrane PDMS emitter tip (side view). (Reprinted with permission from ref. 22).

for 4 h. After casting the top part, the processes for curing, casting the bottom PDMS part and bonding the top and bottom parts were followed in the same way as for the processes in the trimming method. However, the two-layer method avoided the need for trimming the emitter tips.

4.3.4 Resin Casting Method

4.3.4.1 Photoresist Processing for Fabricating the Master

The production of the photomask designs for the photoresist patterns (Figure 4.4) and the processes for producing the photoresist patterns were the same as for the trimming method.

To make a mold for a more robust emitter tip in the PDMS device, ~1 μL of the epoxy resin (EpoFix) was dropped on each emitter tip position in the photoresist pattern (Figure 4.9A) of the master wafer held at a slope of 20 ° to allow the resin to flow to the end of the photoresist pattern for the emitter tips, and then allowed to cure at room temperature overnight. As a second step in the process, 10 μL of resin was placed immediately adjacent to the cured resin at each tip position of the photoresist pattern (Figure 4.9B), and then cured at room temperature for 5 h. In the third step, a concave-shaped PDMS support block (the shape produced on the PDMS device edge in the trimming method), which had been fabricated to have the same size as the molding size (Figure 4.4A), was aligned onto the cured resin pattern (Figure 4.9B and C), and then more resin was filled between the PDMS support and the master wafer (silanized with TCS under vacuum for 2 h) which had been surrounded with three PTFE blocks (0.5 in × 0.25 in). After curing overnight at 40 °C, the resin imprint was easily peeled off from the PDMS and PTFE blocks. The resin premixture had been mixed in a 16 : 1 ratio of resin and hardener, and was degassed to remove bubbles under vacuum for 30 min prior to the casting.

4.3.4.2 Casting PDMS against the Patterned Master

After connecting the FSCs in the same way as in the process for the trimming method, the resin imprint was aligned to the end points of the four channels in the photoresist pattern on the master (Figure 4.4B), and PDMS premixture was cast against this imprint and the master wafer (Figure 4.8D) with curing overnight in a 40 °C oven. The processes for replacing the FSCs, preparing the bottom PDMS part and bonding the two parts were the same as for the trimming method. The shape of the emitter tips and the channel shape and size are shown in Figure 4.9E.

4.3.5 Three-Layer Method with Self-Alignment Features

4.3.5.1 Photoresist Processing for Fabricating the Master

A SU8-100 photoresist solution (PRS) was used in all of the procedures. In order to make the three-layer photoresist patterns for the top master, the first layer was coated as 50 μm thick on a 100 mm diameter silicon wafer; 2 mL PRS was spun at 500 rpm for 10 s following at 5000 rpm for 30 s with a spin coater. The spin-coated wafer was baked at 60 °C for 10 min, and then at 95 °C for 30 min. After baking, the photomask image (Figure 4.5A) was patterned on the first layer using a UV lamp (370 nm) in a mask aligner for 30 s. On the exposed first layer without a developing step, the second layer for the profile of the ESI emitter was coated as 200 μm thick; 4 mL PRS was spun at 500 rpm for 20 s followed by 2000 rpm for 30 s. The wafer coated with two layers was baked at 60 °C for 20 min, and then at 95 °C for 60 min. The baked wafer was exposed with a second

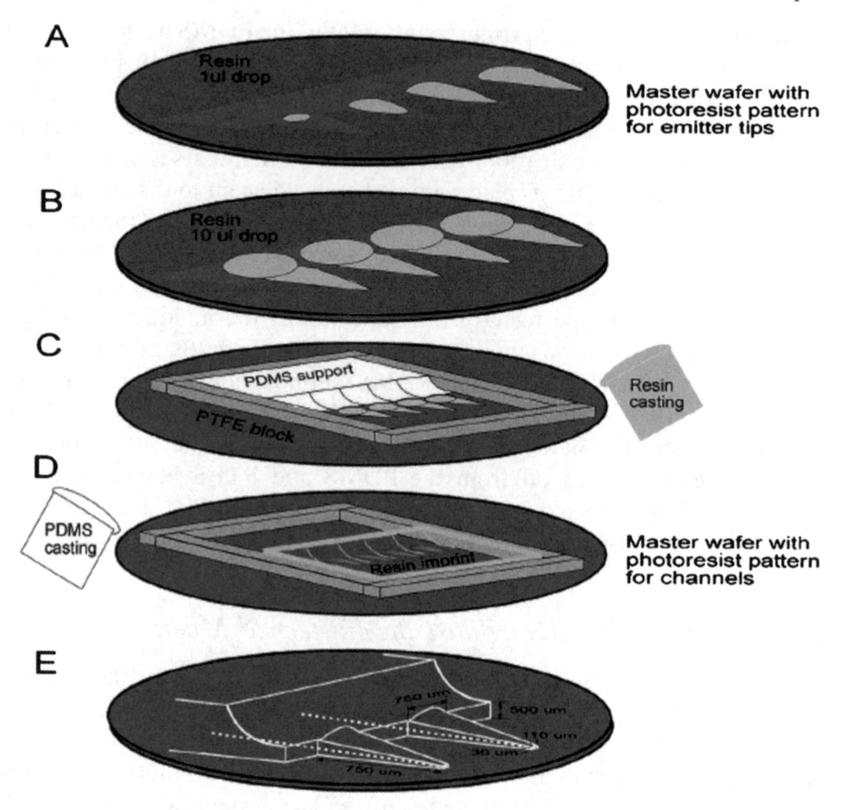

Figure 4.9 Schematic of (A–C) the fabrication of the mold for making the PDMS replica using the resin casting method, (D) the casting of the device and (E) the convex shape of the resulting emitter tip. (Reprinted with permission from ref. 22).

photomask image (Figure 4.5B) for 90 s. Next, without a developing step, the third layer was coated as 200 μm thick on the double-layer wafer, and baked using the same conditions as for the second layer. The wafer coated with three layers was exposed with the photomask image containing the self-alignment features (Figure 4.5B) (of which the emitter tip area was covered with black tape), and then post-baked at 60 °C for 15 min following by 95 °C for 40 min.

For the bottom master, the first layer of 340 μm thickness was coated using 4 mL PRS that was spun at 500 rpm for 20 s following at 1000 rpm for 20 s. The coated wafer was baked at lower temperature (60 °C) for ~24 h, and then was exposed for 120 s using the photomask image presented in Figure 4.5C. Next, the second layer of 250 μm thickness was coated on the first layer using 4 mL PRS that was spun at 500 rpm for 10 s followed by 2000 rpm for 20 s, and then was exposed for 90 s with the photomask image shown in Figure 4.5D. The post-baked wafers were developed in MPA. The thickness of the photoresist patterns for each layer (see Figure 4.10A and B) was measured using a profilometer.

4.3.5.2 Casting PDMS against the Patterned Master

The casting process for the top and bottom PDMS parts is shown in Figure 4.10. For the top PDMS part, the three-layer photoresist pattern on the master wafer, which had been silanized with the releasing agent (TCS) under vacuum for 4 h, was surrounded with three PTFE blocks (1 cm high × 1 cm wide × 5 cm long). The ends of the reference points (Figure 4.10C) for the emitter tips on the wafer were covered with an epoxy resin block (EpoFix) of convex profile[22] for producing the emitter tips (Figure 4.10D). The PTFE and resin blocks were clamped onto the wafer with two aluminium plates and screws. On this

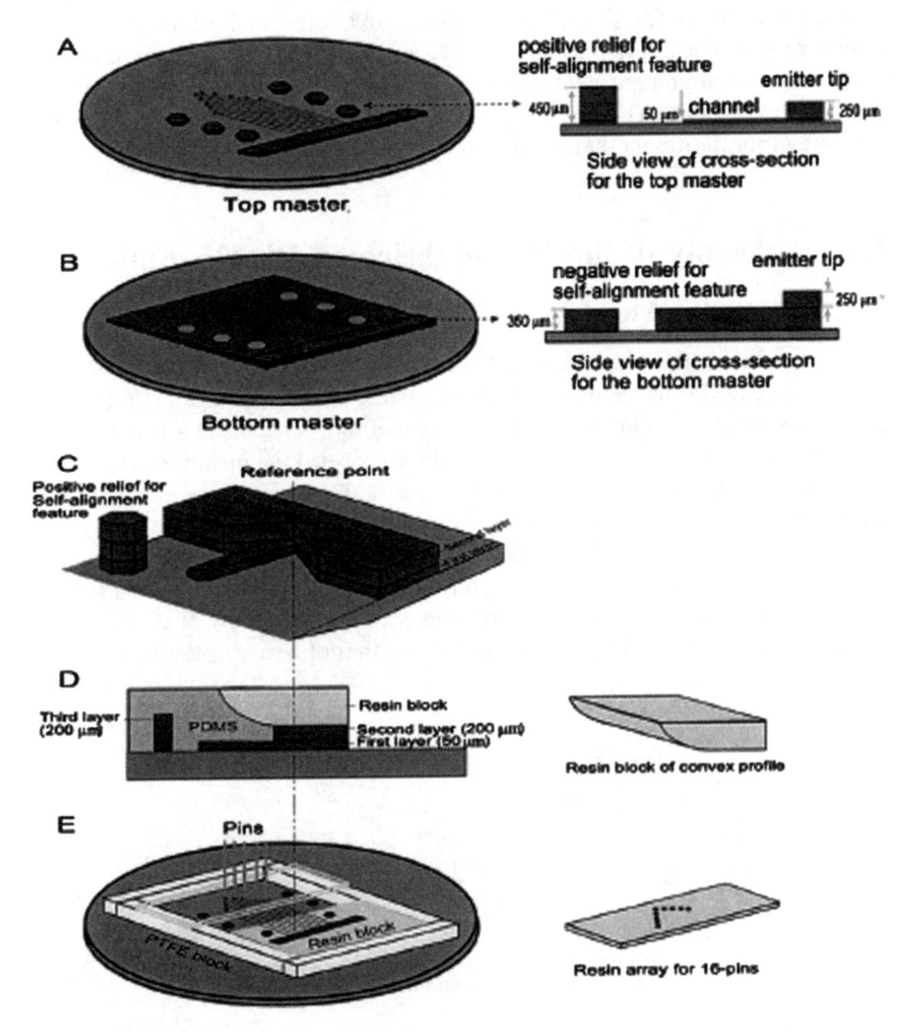

Figure 4.10 Schematic of the fabrication of the mold for making the PDMS replica on the master wafer with three-layer photoresist pattern. (Reprinted with permission from ref. 23).

assembled system, a PDMS plate with an array of 16 pins (stainless steel of 1.6 mm diameter) was placed over the photoresist pattern for the sample reservoirs in the master wafer and "glued" into place with viscous, partially polymerized PDMS (Figure 4.10E). This array was made by casting epoxy resin against another master wafer patterned with the photomask image coinciding with the positions of the 16 reservoirs for the top part (photoresist image not shown). The PDMS prepolymer for the top part was cast against the master wafer with the attached pin array, and then cured at 40 °C for 24 h in an oven. The cured top PDMS part was peeled off from the master wafer and the PTFE blocks, and the pins were removed from the top part. The peeled-off top part was heated again at 70 °C for 2 h. The bottom PDMS part was cured using the same conditions as for the top part. The cured top and bottom parts were surface-oxidized using a plasma cleaner, and then aligned by matching the hexagonal self-alignment features, a thin layer of methanol being placed between the two parts (Figure 4.11). The aligned parts were heated at 70 °C for 72 h to evaporate the methanol and allow bonding of the two parts.

4.4 Assembly of the Microfabricated PDMS Emitters

4.4.1 Trimming Method

Figure 4.6 shows how the concave shape ending in a thin membrane along the edge of the PDMS device was formed by casting the PDMS against a piece of curved transparency film in the trimming method. The 0.5 in × 0.25 in brass block provided sufficient pressure to the film to control the membrane thickness at the channel openings of the emitters to less than 100 μm using the tape to fix the film on the end of the channel openings on the master wafer.

When making the bond between the top and the bottom PDMS parts of the device by plasma oxidation, the edge surface of the membrane between the top and bottom parts was aligned along the profile formed by the photoresist pattern (Figure 4.6A) using a thin layer of methanol between the two parts.[25]

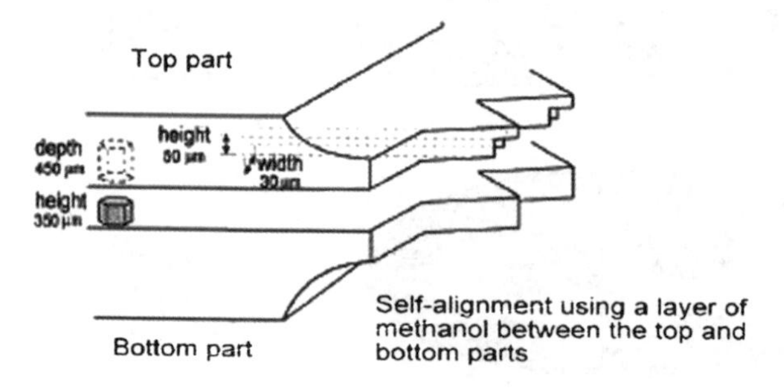

Figure 4.11 Schematic of the self-alignment features in the three-layer method. (Reprinted with permission from ref. 23).

The methanol allows precise alignment of the two parts. The emitter tips were trimmed to points at each channel opening following along the photoresist pattern profile as a guide using iris scissors. An emitter tip length of 3 mm was found to be optimal for easy trimming and for stability of the electrospray. After trimming, the PDMS device was heated at 70 °C to remove the pre-polymer residue and to evaporate the methanol in the device. The 3 mm long pointed emitter tips at each channel opening are depicted with the channel dimensions of 100 μm width and 30 μm depth in Figure 4.7.

4.4.2 Two-Layer Method

In the two-layer photoresist method of producing PDMS emitters, the first layer of photoresist was spin-coated onto the master wafer and exposed to UV with the transparency of the photomask image for the first layer (Figure 4.3A). The second layer of photoresist was spin-coated on top of the undeveloped first layer. After spin-coating the second layer, the photomask for the second layer was aligned to the cross-linked photoresist pattern in the first layer using the alignment marks, and then exposed for 100 s (more that the 30 s in the first layer because the thickness of the second layer was 150 μm). Figure 4.8A shows the resulting profile of the two layers of photoresist that were developed at the same time. Although the developing time for two layers (total 180 μm) was longer than that for a single layer, the developed first and second patterns on the master wafer were stable because the negative photoresist (SU8) has a high mechanical strength.

After development, the master wafer photoresist pattern was surrounded with three PTFE blocks and a resin block to make the replica molding for top PDMS part (Figure 4.8B). In order to make the resin block, resin was cast against a concave-shaped PDMS block like that made in the process for the bottom PDMS part in the trimming method. The resin block does not adhere to the PDMS surface even without treatment with the releasing reagent (TCS). The use of EpoFix resin cast against PDMS has been previously reported.[13] The convex-shaped resin block was fixed on the reference points (Figure 4.8). Tape was used to cover the ends of the two-layer photoresist pattern on the master wafer for the top PDMS part; then the PTFE and the resin blocks were pressurized by four binder clips on two aluminium plates, which have a hole in the area for connecting FSCs. After casting the PDMS premixture against this assembled molding system, the PDMS replica peeled off easily, and the membrane of the emitter tips was robust and very reproducible. The optimal emitter tip length was found to be 0.75 mm. The thickness of emitter tips in the bonded device with top and bottom PDMS parts was ~400 μm. If the thickness of the emitter tip membrane was made less than 180 μm in either the top or the bottom PDMS parts, the emitter tip membranes were prone to distortion when peeling off the PDMS replicas.

4.4.3 Resin Casting Method

To form the convex emitter tip master to produce the mold for the fabrication of the emitters using the resin casting method, 1 μL of EpoFix resin was dropped on the positions of each emitter tip (Figure 4.9A). The master wafer was held at a slope of 20 ° for 10 min, and then the wafer was placed on a horizontal surface in order to allow the resin to flow to the end of photoresist pattern. The resin was cured overnight at room temperature because the resin overflows from the photoresist pattern when curing at high temperature for fast curing. Next, 10 μL of resin was dropped immediately adjacent to the cured resin to form a convex shape for the emitter tip and serve as a barrier to prevent the overflowing of additional resin which was filled between the PDMS support and the master wafer held at a slope of ~50° (Figure 4.9B). The cured resin imprint was easily peeled off from the silanized molding system after overnight curing at 40 °C (Figure 4.9C). The resulting resin imprint mold was aligned to the photoresist patterns on another master wafer that had been made using the photomask image for channels (Figure 4.4B). Efforts to make the imprint mold using PDMS instead of EpoFix resin were unsuccessful because it was difficult to align a PDMS imprint mold along the patterns for the channels due to the elastomeric properties of PDMS and the slight shrinkage that occurs upon heating. Emitter tips from 0.5 to 2.5 mm long were tested; all of the sizes were robust because of the convex shape in spite of the 70 μm thickness of the emitter tips.

4.4.4 Three-Layer Method with Self-Alignment Features

As shown in Figure 4.10A, the first (50 μm), the second (200 μm) and the third layers (200 μm) for the top master were coated consecutively with PRS on the silicon master as a three-layer photoresist pattern. In the two-layer pattern for the bottom master, the first and the second layers were coated as 350 μm and 250 μ, respectively (Figure 4.10B). In the side views of the master wafer of the top part and the master wafer of the bottom part, respectively, the thickness of the photoresist pattern for the emitter tips in each master was 250 μm with a point angle of 60 °. It was required to carefully control the spin speed and time modified from the recommended conditions[26] of the manufacturer due to the high viscosity of SU8-100 PRS. Because the first layer was coated with a smaller amount of PRS (2 mL) than normally used for this PRS in order to make a thin coating of 50 μm, the perimeter of the first layer of photoresist sometimes became thicker than the center area after baking. (Generally SU8-100 PRS is used for photoresist layers over 100 μm thick with an amount of 4 mL for a 100 mm wafer.) This edge effect makes it difficult to align the photomask image onto the exposed wafer because a vacuum is created between the bowed wafer when the wafer is pressed against the photomask glass carrier. This effect, which restricts lateral movement of the glass carrier, was prevented by controlling the distance between them using a small piece of thin tape attached under the glass carrier.

When the first photoresist layer was exposed over 70% of the surface area on the bottom master wafer with UV radiation (370 nm), the wafer exhibited bowing after baking. The stress causing wafer bowing can result from both thermal and intrinsic stresses.[27] With a large exposed area and depending on the thickness of the photoresist, wafer bowing can be very high after baking. This bowing phenomenon makes it difficult to align the photomask for the next layer to the cross-linked photoresist pattern because the mask aligner cannot chuck a bowed wafer. In order to solve this problem, we used a PTFE gasket on the perimeter of the chuck to facilitate holding the bowed wafer.

PDMS shrinks if PDMS prepolymer is cured at high temperature.[28] In order to prevent the shrinkage, the curing condition of the prepolymer was changed from the normal 70 °C for 2 h to lower temperature (40 °C) for 24 h. The cured PDMS was peeled off from the PTFE block, and then heated again at 70 °C for 2 h to complete the curing. When peeling-off the PDMS parts from master wafers, it was necessary that the distance between emitter tips and the original shape be carefully preserved to enable matching of the top and bottom parts. To maintain the shape, the thickness of the emitter tips was made as 250 μm even though thinner tips are better for optimum electrospray sensitivity. In the fabrication of the top master, although the reservoirs could be made with the pins attached individually on the photoresist patterns for them in the top master, the 16-pin array resin block was used to facilitate the placement of multiple pins (Figure 4.10E). Such an array will be even more useful in making larger devices (*e.g.* 96 channels) or for producing a number of disposable devices without any laborious steps.

For the self-alignment features, the thickness (450 μm) of the positive relief on the top master (which forms hexagonal holes in the top PDMS part) was made greater than the depth (350 μm) of the negative relief on the bottom master (which forms hexagonal posts on the bottom PDMS part to fit into the corresponding holes on the top part) (Figure 4.11) in order to make the coated photoresist for each layer on the silicon uniform (the thickness was measured to have a deviation of 2–3%). The position of the coated wafer was maintained horizontal and at the same level during the process. In order to assemble the PDMS device, after both surfaces of the cured parts were exposed to air plasma,[29] the top and bottom PDMS parts were matched at the exact point for the channel and emitter tip, and then a layer of methanol was used to separate them during the alignment.[25] The methanol layer allowed an automatic match of the self-alignment features between the hexagonal holes in the top part and the hexagonal posts in the bottom part (Figure 4.11) with easy handling. This alignment system is similar to the alignment track system[28] previously reported for making the alignment between a PDMS master and a wafer master to produce the thin PDMS layers of three-dimensional microfluidic devices. Our system, however, eliminates the need for mechanical alignment stage in the assembly of PDMS devices as used in previous work.[28] The self-aligned parts were heated at 70 °C for 72 h in order to remove the residue in the cured PDMS during the bonding process and to reduce the chemical background from PDMS.[30] The resulting thickness of emitter tips in the bonded PDMS device was 500 μm.

4.5 Electrospray Mass Spectrometry and Data Acquisition

The channels of the microfabricated PDMS device were washed with methanol and water using a syringe pump (model 11; Harvard Apparatus, South Natick, MA, USA), and then filled with standard solutions of a mixture of angiotensin I and bradykinin, or a mixture of adrenocorticotropic hormone fragment 1-17 (ACTH 1-17), agiotensin I and angiotensin III. Standard solutions were prepared by dissolving 1 mg mL^{-1} of peptides in a methanol–water (1 : 1 v/v) solution, to which 0.1% acetic acid was added, and then by diluting them to 10, 1, 0.1 and 0.01 μM with the same solvent to determine the limit of detection (LOD).

For four-channel PDMS devices (Figure 4.12) prepared by the methods of trimming, two-layer and resin casting, a standard solution of angiotensin I and bradykinin was injected into the channel of the devices using a syringe pump and a flow rate of 1–20 μL min^{-1}. The standard solution of ACTH 1-17, angiotensin I and angiotensin III was filled into a reservoir of the 16-channel PDMS device (Figure 4.13) prepared by the three-layer method, and then injected into the corresponding channel by pressurizing with nitrogen gas via a 1.6 mm PTFE tube. The tube was fixed into a disposable pipette tip using epoxy glue. To compare the electrospray MS LODs of the PDMS emitters with that of a standard fused silica emitter, FSC tubing (75 μm i.d. and 360 μm o.d.) was pulled with a laser-based micropipet puller (model P-2000; Sutter Instrument, Novato, CA, USA) to a 25 μm i.d. tip and trimmed to a 20 cm length. The ESI performance of the FSC emitter was examined with sample concentrations of 10, 1, 0.1 and 0.01 μM and flow rates of 0.1–20 μL min^{-1}.

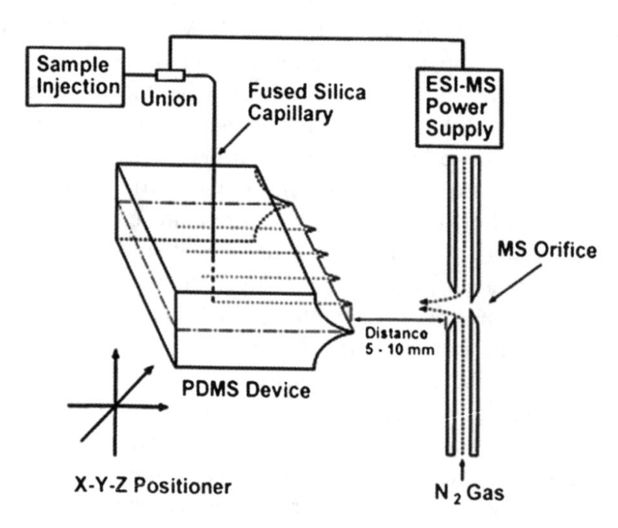

Figure 4.12 Schematic of the PDMS multichannel device interfaced to ESI-TOF-MS with sample injection system and ESI voltage connection. (Reprinted with permission from ref. 21).

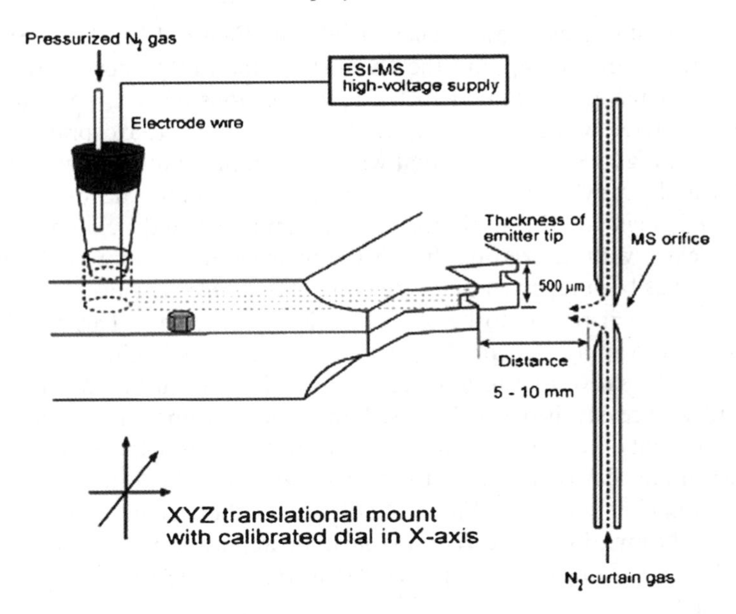

Figure 4.13 Schematic of self-aligned PDMS 16-channel device interfaced to ESI-TOF-MS with sample injection system and ESI voltage connection. (Reprinted with permission from ref. 23).

A Mariner ESI-TOF-MS instrument (Applied Biosystems Inc., Framingham, MA, USA) was used to acquire MS data. The instrument was modified by adding a *z*-axis adjustment made from an acrylic plate that was attached edgewise to a microscope mechanical stage mechanism (Fisher Scientific) mounted to the existing *xy*-adjustable ESI mount. The distance of the emitter tips was varied from 5 to 10 mm in front of the orifice of the ESI-TOF-MS system using the *xyz* translational stage. The tip position of the multichannel emitters was controlled by a calibrated dial movement to shift different emitters in front of the inlet. The flow rate of nitrogen curtain gas was between 300 and 2000 mL min^{-1}, and the interface was heated to 120 °C. Images of PDMS emitters and Taylor cones were captured using a CCD video camera with a 10 cm extension tube and a ×10 microscope objective as lens and a video capture interface and software (Snappy Video Snapshot; Play, Rancho Cordova, CA, USA).

4.6 PDMS Emitters' Electrospray Performance

4.6.1 Four-Channel PDMS Emitter Prepared by the Trimming Method

The PDMS emitter tips prepared by the trimming, two-layer and resin casting methods were positioned from 5 to 10 mm in front of the MS orifice by the *xyz* translational stage, as shown schematically in Figure 4.12. The FSC lines between the syringe pump and the channels of the PDMS devices were connected with a metal union (model ZU.5T; Valco, Houston, TX, USA) to which ESI high voltage was applied in order to generate the electrospray from the emitter tips. Figure 4.14A shows that when the sample solution of angiotensin I (10 μM) and bradykinin (10 μM) was injected at a flow rate of 3 μL min^{-1} without ESI high voltage, a solution drop accumulated on the emitter tip of the PDMS device without wetting due to the hydrophobic nature of the PDMS surface. It has been reported that when a PDMS device is oxidized, it becomes hydrophilic but it reverts to be hydrophobic in ~30 min.[31] This droplet formation prior to applying high voltage is consistent with the initially hydrophilic oxidized surface reverting to a hydrophobic character after PDMS curing for 72 h. It has previously been discussed that the hydrophobic surface of the emitter prevents the sample solution from spreading over the edge surface of the microfluidic device and helps to focus the electric field at the surface of the liquid exiting the channel.[8] Figure 4.14B shows the change in shape of the droplet and formation of the Taylor cone upon application of ESI high voltage (2.7 kV) on a 30° angle emitter placed 8 mm from the orifice, using a flow rate of 3 μL min^{-1}.

In order to observe the Taylor cone of the electrospray, a single-channel device (100 μm wide × 30 μm high) was used for facilitating the position of the CCD camera. When the ESI high voltage was applied to single-channel and four-channel devices, it was observed that a range of 1.8–2.8 kV produced a good Taylor cone at a 5 mm distance from the mass spectrometric orifice, while a range of 2.3–3.1 kV was suitable for a 10 mm distance. As shown in Figure 4.15,

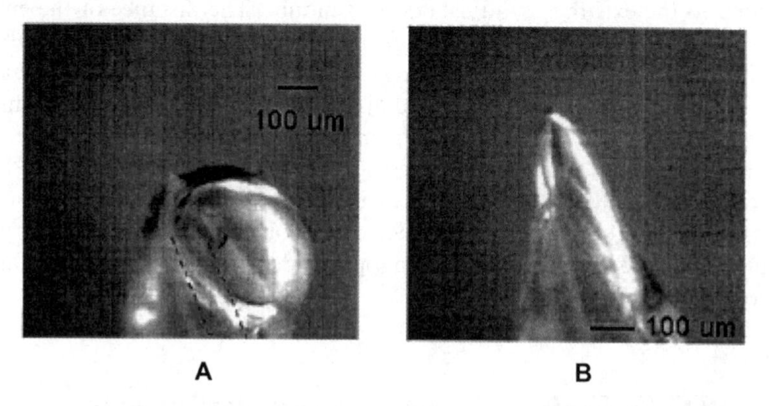

Figure 4.14 (A) Photomicrograph showing the sample solution (10 μM angiotensin I and 10 μM bradykinin in 1:1 methanol–water containing 0.1% acetic acid) droplet on the hydrophobic surface of PDMS emitter without ESI voltage (the dashed lines show 100 μm wide channel). (B) Photomicrograph of the electrospray of the same solution at a flow rate of 3 μL min^{-1} with ESI voltage of 2.7 kV at a distance of 10 mm from the mass spectrometric orifice. (Reprinted with permission from ref. 21).

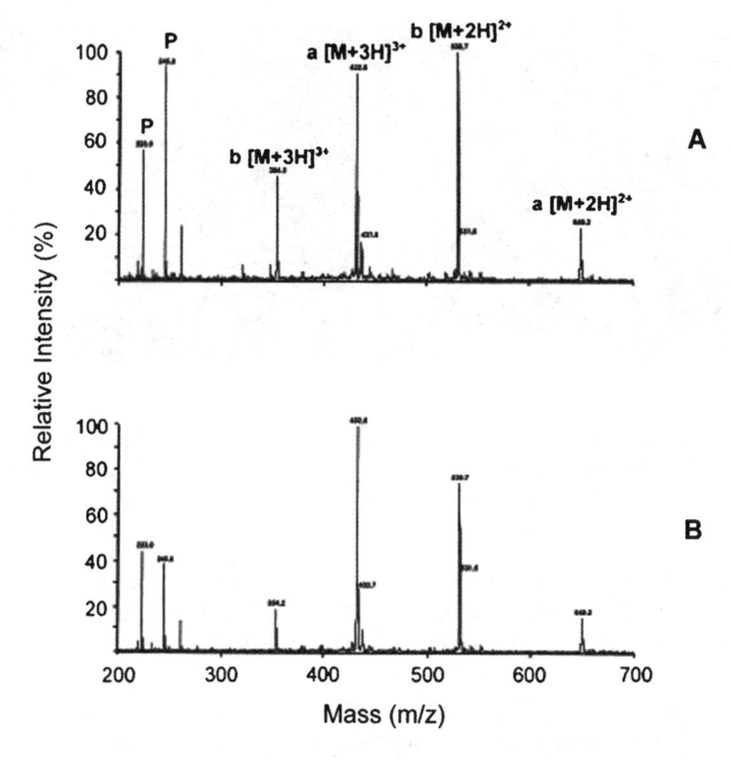

Figure 4.15 ESI mass spectra for (A) angiotensin I and (B) bradykinin analyzed as a mixture (10 μM each) with a flow rate of 3 μL min^{-1}, emitter–orifice distance of 10 mm and acquisition time of 0.5 s using an emitter prepared by the trimming method (P indicates PDMS background signal of 223 and 245 *m*/*z*). (Reprinted with permission from ref. 21).

when the emitter of a single-channel device was positioned 5 mm from the orifice with a 2.7 kV high voltage using a 10 μM angiotensin I (a) and bradykinin (b) solution and a flow rate of 3 μL min^{-1}, the signal intensities (Figure 4.15A) of the PDMS background peaks (223 and 245 *m*/*z*) were higher than when the device was placed at 10 mm (Figure 4.15B). These results indicate that, in order to minimize lower mass background noise, it is preferable to locate the emitter tip at a greater distance from the orifice.

4.6.2 Sixteen-Channel PDMS Emitter Prepared by the Three-Layer Method with Self-Alignment Features

In order to generate the electrospray from PDMS emitters (Figure 4.13) fabricated by the three-layer method, high voltage was applied to a platinum wire electrode inserted in the sample reservoir via a pipette tip, which was also fixed into a tip using epoxy glue. The tight connection was made as a result of the

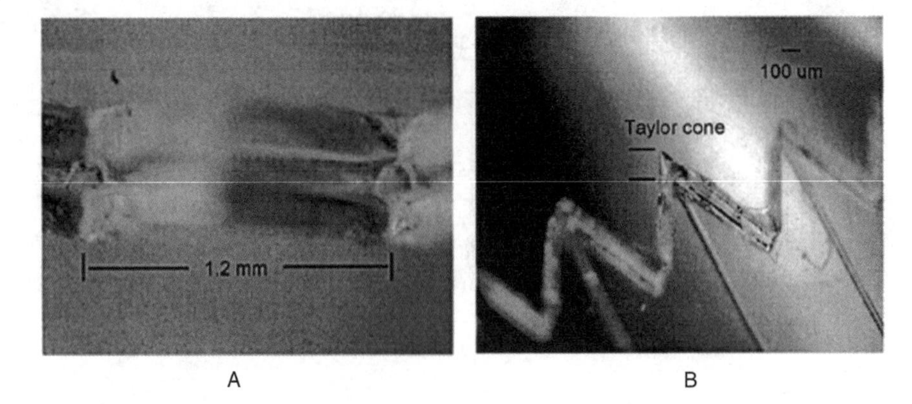

Figure 4.16 Images of (A) 16-channel emitter tips, and (B) electrospray of 10 μM ACTH 1-17 in 1:1 methanol–water containing 0.1% acetic acid. (Reprinted with permission from ref. 23).

elasticity of the PDMS around the tip in the reservoirs. Stable electrospray was obtained from the 16-channel emitter (30 μm wide × 50 μm high) as shown in Figure 4.16A. Figure 4.16B shows a typical Taylor cone obtained using a sample solution of 10 μM ACTH 1-17. In this test of ESI performance of the 16-channel emitters, the flow rate of 300 nL min^{-1} from each emitter was ten times lower than that with the single-channel emitter (100 μm wide × 30 μm high). To maintain this flow rate, the pressure of nitrogen gas was maintained in the range 5–20 mm Hg applied to the sample reservoir. When the pressure of nitrogen gas was increased to 400 mm Hg, the channel structure remained stable without any leakage to adjacent channels or reservoirs.

The electrospray performance of the PDMS emitter was durable for more than 30 h. In the previous reports of electrospray from the edge of glass microfluidic devices,[8,9] ESI voltages of ~4 kV were applied to generate the electrospray directly from the edge opening at 5 mm in front of the orifice. The electrospray from the hydrophobic PDMS emitter device with a thin point could be achieved using lower ESI voltages.

4.6.3 Signal Stabilities of the Total Ion Current (TIC)

In order to evaluate the signal stability in ESI-TOF-MS for two standard solutions, standard solutions at 10, 1, 0.1 and 0.01 μM were injected into a PDMS device at flow rates of 100 nL min^{-1}–20 μL min^{-1}. The signal stability of the TIC was evaluated by measuring the relative standard deviation (RSD) of the signal intensity. Figure 4.17 shows the signal intensity as a function of time for electrospray data acquired with various spectrum acquisition times (spectral summing times for TOF spectra) from a typical emitter of a four-channel PDMS device prepared by the trimming method. The average of the RSDs for

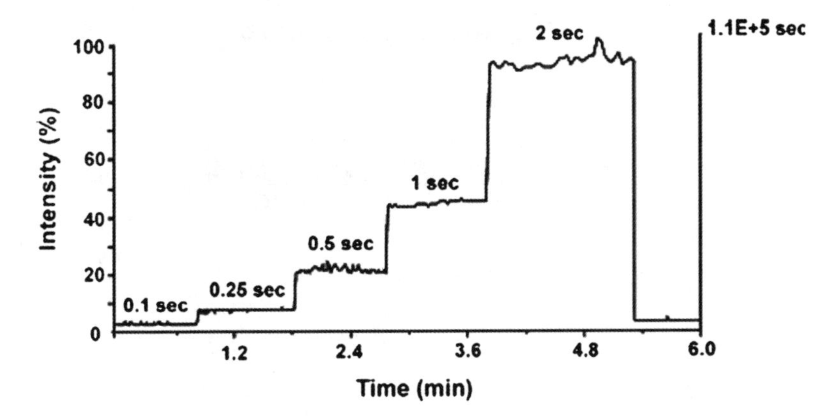

Figure 4.17 Signal stabilities for the total ion current observed using a 10 μM peptide mixture solution with spectral acquisition times of 0.1 to 2 s. The mass range of the TIC was from 200 to 700 *m*/*z* with a flow rate of 3 μL min^{-1} and emitter–orifice distance of 10 mm. (Reprinted with permission from ref. 22).

the different acquisition times of 0.1 to 2 s was 3.2% in the mass range 200–700 *m*/*z* with a flow rate of 3 μL min^{-1} at a 10 mm distance. These data demonstrate that the emitters are capable of producing a stable electrospray.

Figure 4.18 shows the TIC stabilities and the signal intensities monitored with an acquisition time of 1 s per spectrum for 20 min using similar conditions as for Figure 4.17 (flow rate of 3 μL min^{-1}; 2.7 kV, 10 mm distance from the spectrometer orifice) for concentrations of 10 and 1 μM as well as PDMS background in the mass range between 200 and 700 *m*/*z*. The PDMS background (Figure 4.18A) had a 12.5% RSD. When the concentration was increased from 1 μM (Figure 4.18B) to 10 μM (Figure 4.18C), the RSD dropped from 6.8% to 2.8% and the S/N (signal-to-noise) ratio improved from 181 to 532. The S/N ratio was calculated as the ratio between the signal intensities of the $[M+3H]^{3+}$ ion of angiotensin I and background.

The absolute signal intensity (in arbitrary units) for the sample of 10 μM (61 809) was ~2 times higher than that of 1 μM (32 697). In the case of the 10 μM sample, the average (2.95%) of signal stabilities and the average (538) of S/N ratios for the four channels of the four-channel PDMS device measured individually had standard deviations of 0.2% and 32, respectively. The measured molecular masses of angiotensin I and bradykinin were within 0.01% of the calculated values. These results of electrospray for the four-channel device demonstrate good long-term stability with a signal intensity correlated to the sample concentration.

Sweedler's laboratory has also reported the use of a PDMS ESI emitter.[32] In this work, flow rate and electrospray high voltage were found to be optimal at 0.2 μL min^{-1} and 4.9 kV, respectively, with steady ion signal for periods of at least 20 min (Figure 4.19). The total ion chromatogram (TIC) in Figure 4.19A

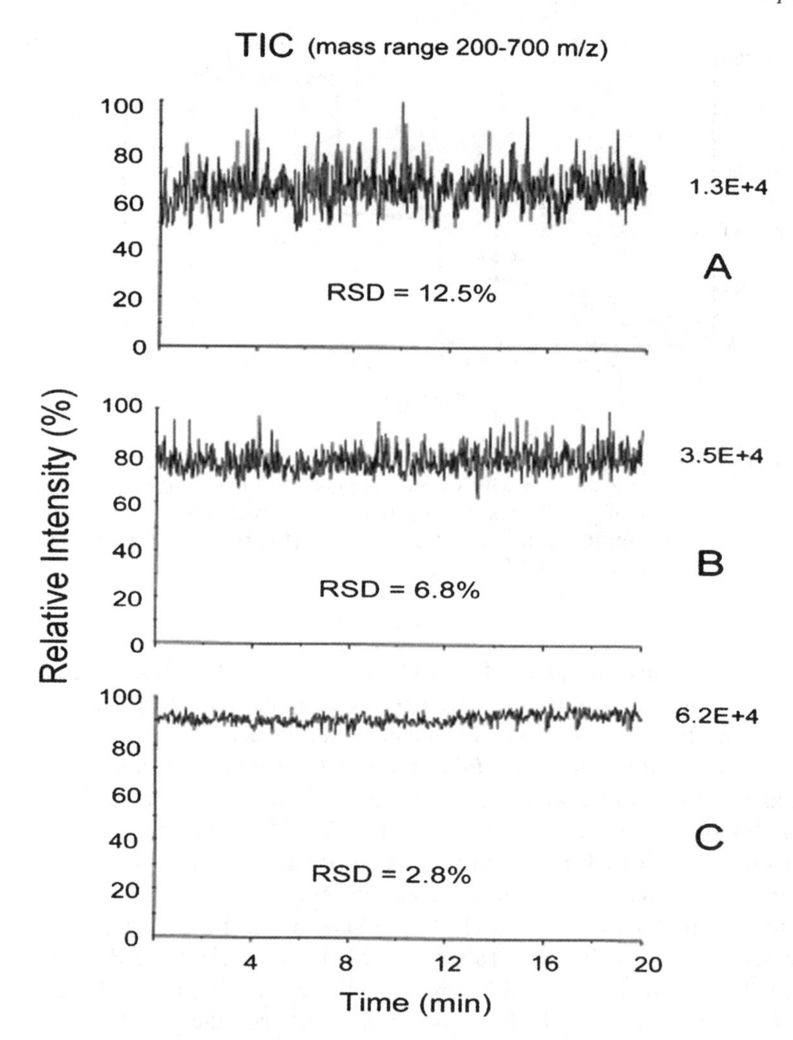

Figure 4.18 Signal stabilities obtained from PDMS emitter by the trimming method: (A) TIC of PDMS background signal when using a solution of 1:1 methanol–water containing 0.1% acetic acid; (B) TIC of the sample concentration of 1 μM peptide mixture; (C) TIC of 10 μM peptide mixture solution. The mass range of all TICs was between 200 and 700 *m/z* using a flow rate of 3 μL min^{-1} with an emitter–orifice distance of 10 mm. (Reprinted with permission from ref. 21).

was obtained with a 10 μM bovine insulin sample in 50 : 50 methanol–deionized water containing 1.6% formic acid. During the 10 min analysis, the ion signal for a mass range of 150–2000 *m/z* remained steady with an RSD of 10%. The mass spectrum of insulin in Figure 4.19B observed for a 30 s time period with S/N = 23, which was spanned by the sample plug beginning at 2.9 min, shows mass accuracy of 0.04% compared to the calculated molecular mass of bovine

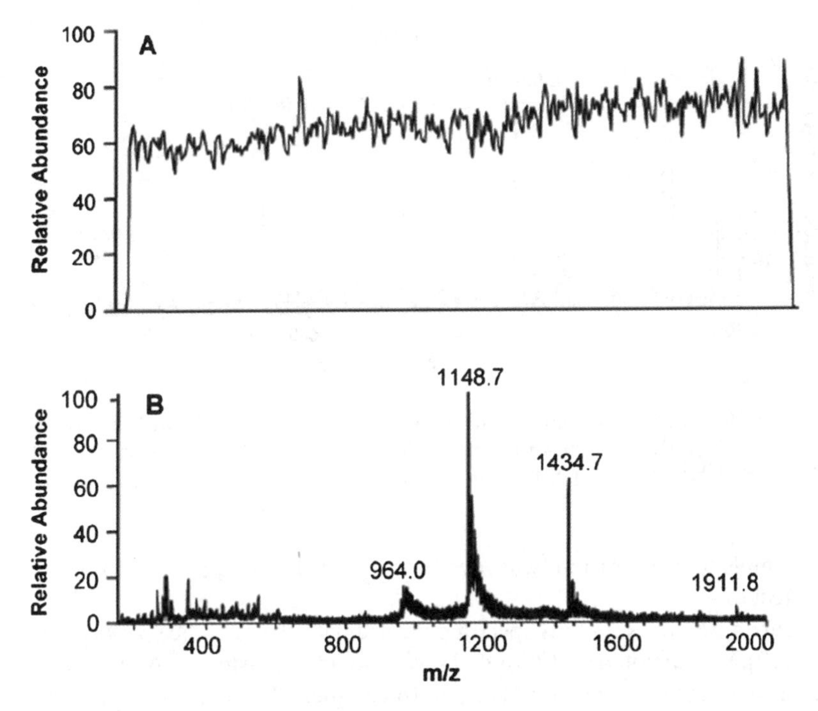

Figure 4.19 (A) Signal from a PDMS ESI emitter in direct spray mode showing stability of total ion chromatogram (10 μM bovine insulin sample in 50:50 methanol–deionized water + 1.6% formic acid; mass range, 150–2000 *m/z*; RSD, 10%). (B) Mass spectrum recorded during a 30 s time period spanned by the insulin plug. (Reprinted with permission from ref. 32).

insulin (5733.5 Da). The ion chromatograms and mass spectra were found to be reproducible between successive injections (integrated peak area RSD $< 3\%$).

4.6.4 Limit of Detection and Examination of Cross-Contamination

Figure 4.20 shows the ESI spectrum for a 1 μM angiotensin I sample, which was sprayed through a PDMS emitter prepared by the trimming method, with a flow rate of 1 μL min^{-1} at a distance of 10 mm from the orifice using an acquisition time of 0.1 s per spectrum. The LOD was observed as 1 μM at a 10 mm position with a S/N ratio of 18 for the $[M + 3H]^{3+}$ peak of angiotensin I (a). To make the comparison with results obtained with a pulled FSC electrospray tip, the sample concentration and the parameters of ESI-TOF-MS were kept constant as well as the 10 mm distance from the mass spectrometric orifice. The LOD with the 25 μm i.d. FSC tip was 0.01 μM (S/N = 25). When

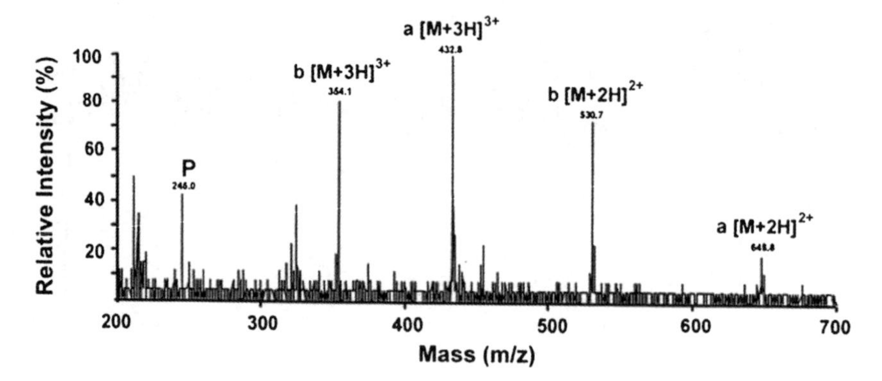

Figure 4.20 Mass spectrum for the 1 μM angiotensin I sample with a flow rate of 1 μL min^{-1} at a distance of 10 mm from the spectrometer orifice. (Reprinted with permission from ref. 21).

the sample concentration was decreased to 0.1 μM, angiotensin I became undetectable.

Figure 4.21 shows the ESI spectra obtained with the reference peptides ACTH 1-17 (10 μM), angiotensin I (AngI, 1 μM) and angiotensin III (AngIII, 1 μM) in the mass range 50–1000 *m/z* using a 16-channel PDMS emitter. To examine cross-contamination from adjacent emitter tips, sequential electrospray analysis of ACTH 1-17 (Figure 4.21A) and AngI (Figure 4.21B) was performed through channels 7 and 8 in the middle of the device, respectively, and then the spectral data were collected. No mass spectral signals were observed indicative of cross-contamination between the channels. Even if contamination occurred due to spattering as the sample was exhausted, it was not detected. The third sample of AngIII was injected into channel 9 (Figure 4.21C). There was no signal of AngIII in the spectrum for the fourth run of AngI in channel 8 (Figure 4.21D). These results showed that the electrospray analysis using different channels does not suffer from any contamination from adjacent emitter tips.

Although intense peaks were obtained for AngI and AngIII at 1 μM using a PDMS emitter prepared by the three-layer method with self-alignment features, these peptidic species were undetectable at an order of magnitude lower concentration. The LOD was therefore determined to be something lower than 1 μM but greater than 0.1 μM. Comparing with previous results, the LOD, flow rate and channel size of this PDMS emitter device were greater than those of the previous results; micromachined nozzles (3 μm orifice) of silicon nitride[16] and parylene emitters (5 μm wide × 10 μm high rectangular opening)[17] had LODs of 0.4 μM at 50 nL min^{-1} and 5 μM at 50 nL min^{-1}, respectively, and etched nozzles (10 μm i.d. and 20 μm o.d.) of monolithic silicon substrate[19] had a 0.01 μM LOD at a flow rate of 100 nL min^{-1}. To improve the LOD of the PDMS emitter, the dimensions and shape must be modified, but fabricating PDMS emitters as small as those reported in the other materials will be difficult due to the resolution limit of soft lithography.

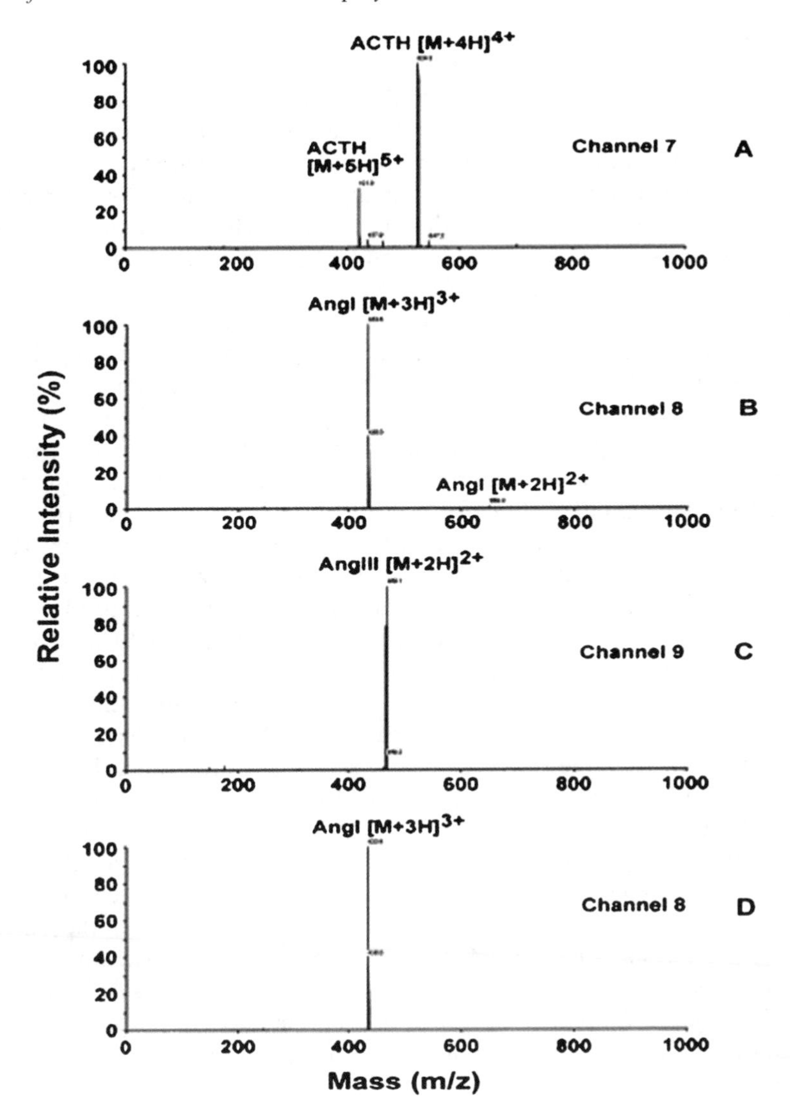

Figure 4.21 Spectra for the samples of (A) 10 µM ACTH 1-17, (B, C) 1 µM using a flow rate of 0.3 µL min^{-1} at 8 mm from the spectrometer orifice. (Reprinted with permission from ref. 23).

4.7 Comparison of the Four Microfabricated PDMS Emitter Devices

Although a more sharply pointed emitter tip was expected to yield better electrospray performance, no difference was observed between point angles of 30 ° and 60 (Figure 4.7) for PDMS emitters prepared by the trimming method. However, the 60 ° angle was found to be preferable for tips prepared by other methods as it facilitates the bonding between the top and bottom parts of the device. Also, when the flow rate of nitrogen curtain gas was varied from 300 to 2000 mL min^{-1}, the curved shape of the PDMS emitter device produced a smooth flow of the curtain gas with no disturbance of the electrospray.

The thickness of the emitter tips was the lowest with the resin casting method (~70 μm,) compared to the trimming method (~100 μm), the two-layer method (~400 μm) and the three-layer method (~500 μm). The microfabrication process was simplest for the trimming method (although a separate labor-intensive manual trimming was required) compared to the more complex resin casting and two-layer methods as well as to the most complex three-layer method with self-alignment features. The other methods except the trimming method eliminated the manual trimming step, but required more equipment in that a mask aligner is needed to align the second layer on the first photoresist layer. The accuracy and reproducibility of the tips produced by the trimming method were limited by the manual trimming step, while the resin casting method was limited by the reproducibility to form each tip in the mold. The two-layer method provided the most accurate and reproducible tips in this work. However, the three-layer method gave the easiest assembly process when the top part was bonded to the bottom part using the self-alignment features.

The 16-channel PDMS emitter tips (30 μm wide × 50 μm high) showed that the electrospray from the 10 μM sample was normally stable for 30 min at a flow rate of 300 nL min^{-1}. However, the electrospray sometimes became unstable due to small gas bubbles and partial clogging in the channel. The gas bubbles moved from the spray channel to emitter tip and distorted the Taylor cone. The bubbling was presumably caused by the ESI high voltage on the metal union for electrospray. Water hydrolysis is likely to occur at the surface of the metal, producing gas bubbles that exit through the spray channel.[33] Electrospray resumed in a matter of seconds following the release of these gas bubbles, and was not significantly affected.[32] The clogging of the narrow channel may result from PDMS particles abraded by the insertion into the reservoirs of either FSC segments connected on the spray channel in the two-layer method or the pipette tip which is connected also to the spray channel made by the three-layer method, for connecting nitrogen gas.

4.8 Concluding Remarks

The methods described here permitted the microfabrication of ESI emitters as an integral part of a PDMS microfluidic device using soft lithography. The trimming method provided a simple means to demonstrate the basic approach. The subsequently developed two-layer and resin casting methods provided more efficient processes for producing emitters without any manual trimming step for the four-channel emitter tip device. The 16-channel ESI emitter device fabricated using a three-layer photoresist process was easily aligned and bonded using self-alignment features (posts and holes) between the top and bottom PDMS parts.

The PDMS emitter devices resulting from these methods exhibited durable and stable performance of electrospray with adequate detection sensitivity. Although the relatively large channel size emitters demonstrated in the prototype devices described here do not yield ESI sensitivity as good as pulled fused silica "nanospray" emitters, it is expected that the design of smaller channels for the PDMS emitters could exhibit sensitivity comparable to standard fused silica emitters. The PDMS emitter design for a 16-channel device located within 2 cm will also be applicable to larger arrays such as 96-emitter tips to interface to a standard 96-well plate for making a miniaturized multichannel LC-MS system, and for developing μTAS for mass spectrometric analysis of protein and peptide mixtures. The self-alignment method used here should also be helpful for implementing other types of more complex microfluidic systems. These approaches to microfabricating ESI emitters will facilitate production of microfluidic analytical devices that utilize mass spectrometric sample detection.

Acknowledgements

Current work at MUSC on microfluidic device development is supported in part by NIH grant CA 86285 and the NIH NHLBI Proteomics Initiative via contract N01-HV-28181.

References

1. J. M. Ramsey, S. M. Jacobson and R. M. Knapp, *Nat. Med.*, 1995, **1**, 1093.
2. F. E. Regnier, B. He, S. Lin and J. Busse, *Trends Biotechnol.*, 1999, **17**, 101.
3. R. D. Oleschuk and D. J. Harrison, *Trends Anal. Chem.*, 2000, **19**, 379.
4. S. C. Jakeway, A. J. De Mello and E. L. Russel, *Fresenius' J. Anal. Chem.*, 2000, **366**, 525.
5. P. Gravesen, J. Branebjerg and O. J. Sondergard Jensen, *J. Micromech. Microeng.*, 1993, **3**, 168.
6. D. J. Harrison, K. Fluri, K. Seiler, Z. Fan, C. S. Effenhauser and A. Manz, *Science*, 1993, **261**, 895.
7. C. S. Effenhauser, G. J. M. Bruin and A. Paulus, *Electrophoresis*, 1997, **18**, 2203.
8. Q. Xue, F. Foret, Y. M. Dunayevskiy, P. M. Zavracky, N. E. McGruer and B. L. Karger, *Anal. Chem.*, 1997, **69**, 426.
9. R. S. Ramsey and J. M. Ramsey, *Anal. Chem.*, 1997, **69**, 1174.
10. D. Figeys and R. Aebersold, *Anal. Chem.*, 1998, **70**, 3721.
11. J. Li, P. Thibault, N. H. Bings, C. D. Skinner, C. Wang, C. Colyer and D. J. Harrison, *Anal. Chem.*, 1999, **71**, 3036.
12. I. M. Lazar, R. S. Ramsey, S. Sundberg and J. M. Ramsey, *Anal. Chem.*, 1999, **71**, 3627.
13. H. Liu, C. Felten, Q. Sue, B. Zhang, P. Jedrzejewski, B. L. Karger and F. Foret, *Anal. Chem.*, 2000, **72**, 3303.

14. R. M. Emmett and R. M. Caprioli, *J. Am. Soc. Mass Spectrom.*, 1994, **5**, 605.
15. G. A. Valascovic, N. L. Kelleher, D. P. Little, D. J. Aaserud and F. W. McLafferty, *Anal. Chem.*, 1995, **67**, 3802.
16. A. Desai, Y. C. Tai, M. T. Davis and T. D. Lee, *International Conference on Solid State Sensors and Actuators*, (Transducers'97), Piscataway, NJ, 1997, p. 927.
17. L. Licklider, X. Q. Wang, A. Desai, Y. C. Tai and T. D. Lee, *Anal. Chem.*, 2000, **72**, 367.
18. J. Wen, Y. Lin, F. Xiang, D. W. Matson, H. R. Udseth and R. D. Smith, *Electrophoresis*, 2000, **21**, 191.
19. G. A. Schultz, T. N. Corso, S. J. Prosser and S. Zhang, *Anal. Chem.*, 2000, **72**, 4058.
20. Y. Xia and G. M. Whitesides, *Angew. Chem. Int. Ed.*, 1998, **37**, 551.
21. J. S. Kim and D. R. Knapp, *J. Am. Soc. Mass Spectrom.*, 2001, **12**, 463.
22. J. S. Kim and D. R. Knapp, *J. Chromatogr. A*, 2001, **924**, 137.
23. J. S. Kim and D. R. Knapp, *Electrophoresis*, 2001, **22**, 3993.
24. D. C. Duffy, C. McDonald, O. J. A. Schueller and G. M. Whitesides, *Anal. Chem.*, 1998, **70**, 4974.
25. B. H. Jo, L. M. VanLerberghe, K. M. Motsegood and D. J. Beebe, *J. Microelectromech. Syst.*, 2000, **9**, 76.
26. SU-8 technical data sheet, MicroChem Corp., Newton, MA.
27. M. T. A. Saif and N. C. MacDonald, *J. Microelectromech. Syst.*, 1996 **5**, 79.
28. J. R. Anderson, D. T. Chiu, R. J. Jackman, O. Cherniavskaya, C. McDonald, H. Wu, S. H. Whitesides and G. M. Whitesides, *Anal. Chem.*, 2000, **72**, 3158.
29. M. K. Chaudhury and G. M. Whitesides, *Langmuir*, 1991, **7**, 1013.
30. J. H. Chan, A. T. Timperman, D. Qin and R. Aebersold, *Anal. Chem.*, 1999, **71**, 4437.
31. J. C. McDonald, D. C. Duffy, J. R. Anderson, D. T. Chiu, H. K. Wu, O. J. A. Schueller and G. M. Whitesides, *Electrophoresis*, 2000, **21**, 27.
32. J. M. Iannacone, J. A. Jakubowski, P. W. Bohn and J. V. Sweedler, *Electrophoresis*, 2005, **26**, 4684.
33. G. E. Yue, M. G. Roper, E. D. Jeffery, C. J. Easley, C. Balchunas, J. P. Landers and J. P. Ferrance, *Lab Chip*, 2005, **5**, 619.

CHAPTER 5

Microfabricated Nanoelectrospray Emitter Tips based on a Microfluidic Capillary Slot

SÉVERINE LE GAC,[a,†] STEVE ARSCOTT[b] AND CHRISTIAN ROLANDO[a]

[a] Université des Sciences et Technologies de Lille (Lille 1), UFR de Chimie, Bâtiment C4, UMR CNRS 8009, Chimie Organique et Macromoléculaire, 59655 Villeneuve d'Ascq Cedex, France; [b] Institut d'Electronique, de Microélectronique et de Nanotechnologies (IEMN), UMR CNRS 8520, Avenue Poincaré, Cité Scientifique, Villeneuve d'Ascq Cedex, France

5.1 Introduction

An impetus in the field of mass spectrometry (MS) analysis occurred in the early 1990s with the invention of two novel and "soft" ionization methods, electrospray ionization (ESI) by John Fenn and matrix-assisted laser desorption ionization (MALDI) by Koichi Tanaka, who both shared the Nobel Prize in Chemistry in 2002. A second impetus, which is more diffuse, is currently occurring and consists of miniaturization. Whereas the intrinsic sensitivity of mass spectrometers has roughly remained the same for a couple of decades, the amount of material required for recording one spectrum has

† Present address: BIOS the Lab-on-a-Chip Group, University of Twente, PO Box 217, 7500 AE Enschede, The Netherlands.

Miniaturization and Mass Spectrometry
Edited by Séverine Le Gac and Albert van den Berg

Published by the Royal Society of Chemistry, www.rsc.org

dramatically decreased by scaling down the ionization. A typical ESI analysis uses one nanomole of sample whereas a miniaturized nanoESI experiment routinely only requires one picomole and can go down to one femtomole of material. This decrease can be accounted for by two factors. First, the ionization yield of nanoESI is three orders of magnitude higher than for ESI; and second, the flow rate is two orders of magnitude lower. Consequently, ESI and nanoESI both present the same sensitivity in terms of concentration, but a thousand times less material is needed for nanoESI. This is obviously crucial for most biological applications. For instance, in human plasma the most abundant protein, namely albumin, is at a millimolar concentration whereas the concentration of the least abundant signaling proteins, *e.g.* interleukins, is 12 orders of magnitude lower, and in the femtomolar range. By using 1 mL of plasma, the sensitivity of ESI enables one to detect the so-called classical plasma proteins down to the micromolar range, whereas nanoESI allows for detecting proteins in the nano- to picomolar range, these proteins bringing more information about disease as they result from tissue leakage. One other example is proteomics: 10^4 cells are required at least for performing a full proteomics analysis; this represents a huge amount of work when cells must be extracted from tissues using microdissection techniques in order to obtain a homogeneous set of cells in a given physiological state. As to other aspects, standard proteomics analysis only applies to proteins with codon factor values higher than 0.1, whatever the species is. It is of no help for the detection of proteins which are expressed at a lower level. Therefore novel tools which afford enhanced analysis sensitivity are needed, and this stimulates the race toward analysis miniaturization.

In nanoESI-MS, the samples to be analyzed are introduced in capillary tube-based emitter tips composed of fused silica or glass whose fabrication results from heating and pulling techniques.[1] Samples are introduced manually into the capillary tip and air bubbles must be carefully removed. Both the ill-controlled capillary fabrication process and manipulation protocol lead to poor analysis reproducibility accounted for by the low reproducibility of the emitter tips themselves and preclude the use of tips with a diameter smaller than 1 μm. In addition, the way the analysis is carried out also prevents the implementation of much needed automation. Using microtechnology techniques for the fabrication of electrospray sources will help to control the fabrication process, the geometry and the size of devices. In this way, ionization sources are more reliable, more reproducible and will give analysis enhancement. Moreover, the geometry can be chosen to match the automation criterion. Finally, using microtechnology techniques also implies that batch production of sources can be implemented. In terms of microfabrication, a first idea would be to mimic existing capillary-based sources and subsequently to fabricate three-dimensional sources having the shape of a capillary tube.[2–4] This idea is not particularly compatible with the use of microtechnology techniques, as planar fabrication processes of such sources are difficult to realize. Another alternative reported in the literature is to fabricate two-dimensional (2D) sources which consist in the end of a microfabricated channel.[5–8] Therefore, we came to the idea of developing novel 2D ionization sources, whose geometry lends itself better to microtechnology-based fabrication. Such 2D sources would have the shape of nibs and borrow much from the principle of a common fountain pen (see Figure 5.1).[9–11] This idea was initially tested using commercial nibs for writing which were mounted on a holder placed at the inlet of the mass spectrometer; after application of a high voltage (HV) to the liquid, a spray was formed. This illustrated the potential of such nib-shaped geometry for electrospray ionization applications.

We describe here the development of nib-shaped sources for nanoESI applications. This development started with preliminary microfabricated devices that enabled us to validate the idea of microfabricated ionization nibs based on a microfluidic capillary slot. Thereafter, more sophisticated structures were fabricated using the negative photoresist SU-8, and, subsequently, more in-depth tests for nanoESI-MS applications were performed. Following this, we produced even more sophisticated structures based on polycrystalline silicon (polySi) using appropriate fabrication techniques. The last step consisted of on-line or continuous analysis, either by coupling a capillary to a SU-8 structure or though the production of covered polySi-based micronibs and the design of a dedicated holder. Through the fabrication of these different generations of nib tips, we were able to identify and to control critical parameters that govern the process of nib filling, the quality of the nanoESI-MS analysis and the enhancement of its sensitivity.

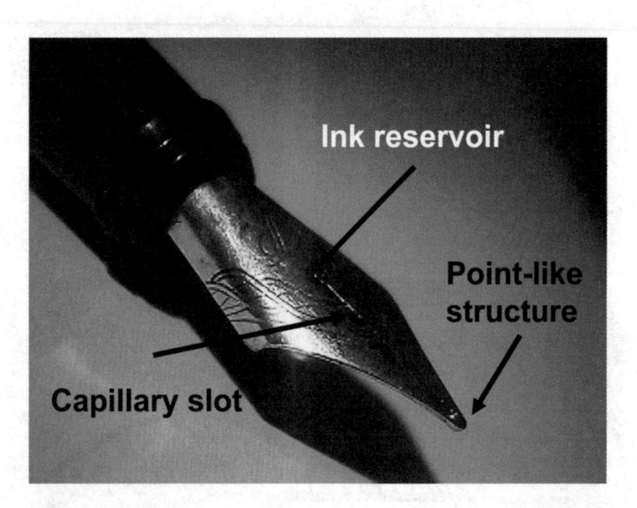

Figure 5.1 Photograph of a fountain pen. Arrows indicate the main components of the nib system, the ink reservoir, the capillary slot for guiding the fluid to the point-like structure where the ink is ejected. (Reprinted with permission from Ref. 9).

5.2 Proof-of-Concept: First Generation

5.2.1 Description

As a first step in our study, we decided to assess our idea of microfabricated nib tips with simple structures based on the negative photoresist SU-8. Structures having a 2½ D topology were fabricated on a silicon wafer support;[11] it should be noted that the nib feature is not completely planar as the tip of the nib tended to point upwards due to stress in the thick SU-8 polymer layer. The nib structure is composed of a reservoir feature, a capillary slot leading the liquid to the tip of the nib where electrospray occurs upon HV application. These first nib prototypes have a microfluidic capillary slot with a width of around 20 μm. Figure 5.2 shows a scanning electron micrograph of a microfabricated nib tip in SU-8 and supported on a silicon support.

5.2.2 Fabrication

The fabrication of the emitter tip mainly relies on photolithography techniques, as illustrated in Figure 5.3. Details of the fabrication process can be found elsewhere.[11]

The emitter tips were fabricated using the negative photoresist SU-8 (Microchem, VA, USA) on a 3-inch silicon wafer. The silicon wafer was first

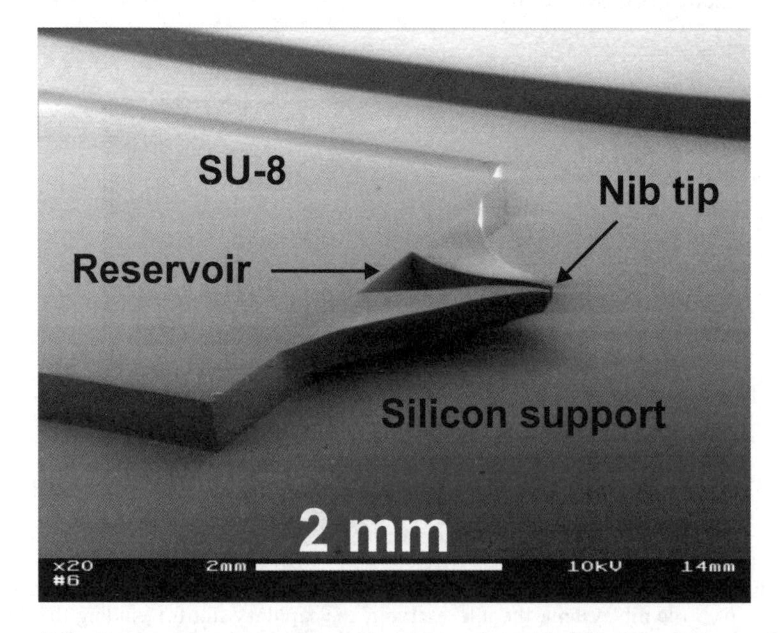

Figure 5.2 Image of a nib structure (first generation) fabricated in the epoxy-based negative photoresist SU-8 and supported on a silicon wafer (scanning electron microscopy). (Reprinted with permission from Ref. 11).

Figure 5.3 Novel fabrication process developed for manufacture of the first generation of micromachined electrospray emitter tips: silicon support wafer (blue), 400 μm thick negative photoresist SU-8 micro-nib layer (gold) and two-step UV photolithographic masking step (black) to form the reservoir structure and SU-8 support (using a long lithography exposure time) and the nib-like emitter tip containing the capillary slot (using a shorter lithography exposure time).

dehydrated at 170 °C for 30 min to ensure good adhesion of the SU-8 film to the silicon wafer. Following this, a 400 μm thick layer of SU-8 was deposited using spin-coating techniques. The photoresist layer was subsequently submitted to two steps of ultraviolet (UV) photolithography to pattern, successively, the main reservoir and the capillary slot, and the nib tip. The exposure time values for both steps were chosen carefully to achieve full polymerization of the SU-8 layer (step 1) or partial polymerization of the resist so as to form a free-standing structure (step 2). Finally, the silicon support was cleaved below the nib tip so as to create an overhanging structure. The depth of the membrane structure was seen to depend linearly on the exposure time and the exposure dose. Following wafer dicing, the emitter tips overhung the silicon substrate by approximately 500 μm. The resulting nib structure was of 30 μm thickness at its tip with a slot of approximately 20 μm. The thickness of the membrane structure is difficult to control during UV dosage and photoresist development, so that the critical dimensions of the nib tip were difficult to control. In addition to this, the high tensile stress incurred whilst employing thick films of SU-8 tends to reduce the planarity of cantilever structures.

5.2.3 Mass Spectrometry Testing

Mass spectrometry tests were carried out using a LCQ deca XP + ion trap mass spectrometer (Thermo Finnigan, San Jose, CA, USA). The electrospray emitter tip nib structure was placed on a dedicated holder that is introduced in the inlet of the mass spectrometer. A platinum wire was subsequently inserted in the reservoir feature of the nib, and was connected at its other end on a metallic part on the holder where HV is normally applied for ionization purposes. Solutions to be analyzed were dropped in the reservoir feature using a gel-loader tip; typically 5–10 μL of sample were deposited in the reservoir. Liquid was subsequently seen to fuse at the tip of the nib due to capillary action in the slot. HV was then applied, and a spray could be observed with HV values in a

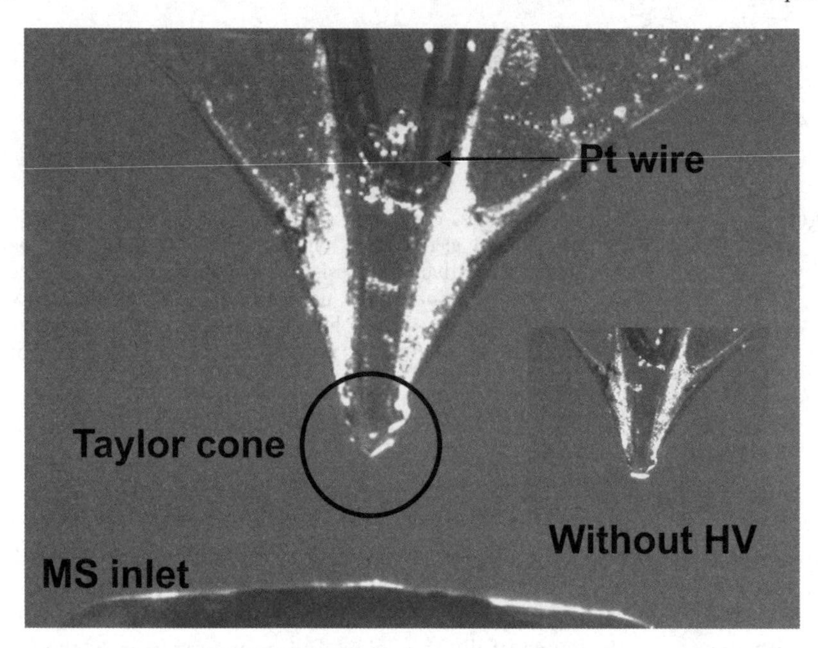

Figure 5.4 Photograph of the nib-like source in front of the mass spectrometer inlet; top view taken using a camera on the ion trap, with voltage and without voltage (inset). Under HV supply, one can clearly see the Taylor cone. Note the capillary slot of the nib filled in with liquid and the platinum wire inserted in the reservoir for HV supply. (Adapted with permission from Ref. 10).

1.2–2.5 kV range, as illustrated in Figure 5.4. This could be visualized using a camera mounted at the entry of the mass spectrometer that helps for positioning ionization sources in front of the heated capillary inlet.

Ionization performances of our nib tips were compared to the results routinely obtained with commercial borosilicate-based capillary sources having an inner diameter of around 1 μm (ES 380; Proxeon, Odense, Denmark).[10] In the latter case, HV was 1 kV and was directly applied on the conductive gold/palladium alloy coating the source. Mass spectra were recorded using Excalibur software (Thermo Finnigan, San Jose, CA, USA); typically, the ionization signal was recorded for 2 min and mass spectra were plotted as an average over this recording period.

Test samples consisted of peptide solution. Gramicidin S samples were prepared starting from a stock solution at 0.1 mM in a MeOH–H_2O 50:50 mixture acidified with 0.1% formic acid; the concentration range of Gramicidin S was of 50 down to 1 μM.

The MS tests of the first nib prototypes were conclusive; a spray was observed after application of an HV on the liquid and the signal was stable for the various tested conditions. Figure 5.5 shows the total ion current (TIC) spectrum recorded over a period of 2 min using a 5 μM Gramicidin S sample

5.3.2 Fabrication

The second generation nib tips were fabricated on standard 3-inch silicon n-type substrates orientated (100). The fabrication process is illustrated in Figure 5.9. Firstly, a 200 nm nickel etch-release layer was deposited onto the silicon wafer surface. This Ni layer was then patterned using a photomask to form localized etch-released pads. The silicon wafers were dehydrated at 170 °C for 30 min in order to ensure good adhesion of the SU-8 resist to the silicon surface. Subsequently, a very thin layer of SU-8 was spin-coated on the silicon wafer, so as to reach a thickness of 25 or 35 μm for the emitter tips. Care was taken during photoresist dispensing and spin-coating to achieve good planarity for the subsequent lithographic step. A single mask was used to form the emitter tips. After development of the resist, the Ni etch-release layer was etched using a nitric acid-based wet etch (HNO_3–H_2O). After drying, the individual devices were diced carefully so that emitter tips remain unaffected by the wafer cleaving step.

The main advantages of these second prototypes over the first ones are (i) improved control of critical tip dimensions due to the more 2D nature of the fabrication, (ii) smaller critical dimensions, (iii) improved cantilever planarity due to reduced stress of a thinner film of SU-8 and (iv) improved yield due to an improved fabrication process.

5.3.3 Mass Spectrometry Testing

5.3.3.1 First Series of Tests; Use of a Platinum Wire for HV Application

Tests were carried out in a similar way as before, on an ion trap mass spectrometer, firstly using the same setup wherein HV is applied on the liquid using a platinum wire.

The first tests aimed at clarifying the influence of the slot width on the nib performance and functioning for nanoESI applications. Therefore several nibs

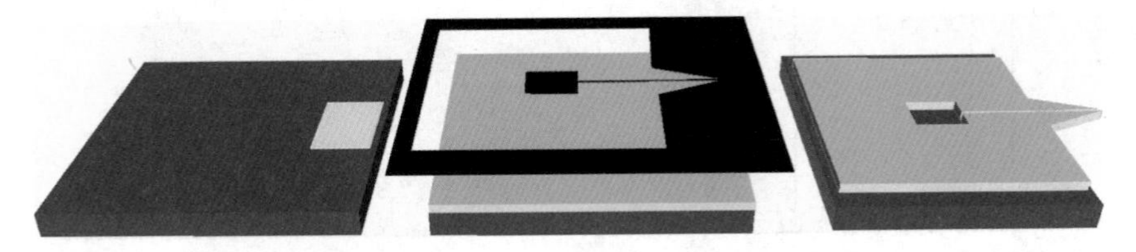

Figure 5.9 Novel fabrication process for the second generation of micromachined electrospray emitter tips: silicon support wafer (blue), 200 nm thick nickel etch-release layer (white) which is patterned using a HNO_3-based wet etch, negative photoresist SU-8 which forms the micro-nib support layer and tip which hosts the capillary slot (gold) and single photolithographic masking layer which defines the reservoir and tip (black).

having different slot widths were successively used for the analysis of a peptide solution, other test conditions being kept identical (HV value, peptide concentration, *etc.*). Figure 5.10 presents the results obtained when the comparison test was carried out on a Gramicidin S solution at 5 μM and using ionization HV of 1.2 kV. This figure clearly shows how the mass spectrum pattern was changed when moving from 16 to 8 μm; the relative intensities of the peaks corresponding to the mono- and the di-charged species were changed in favor

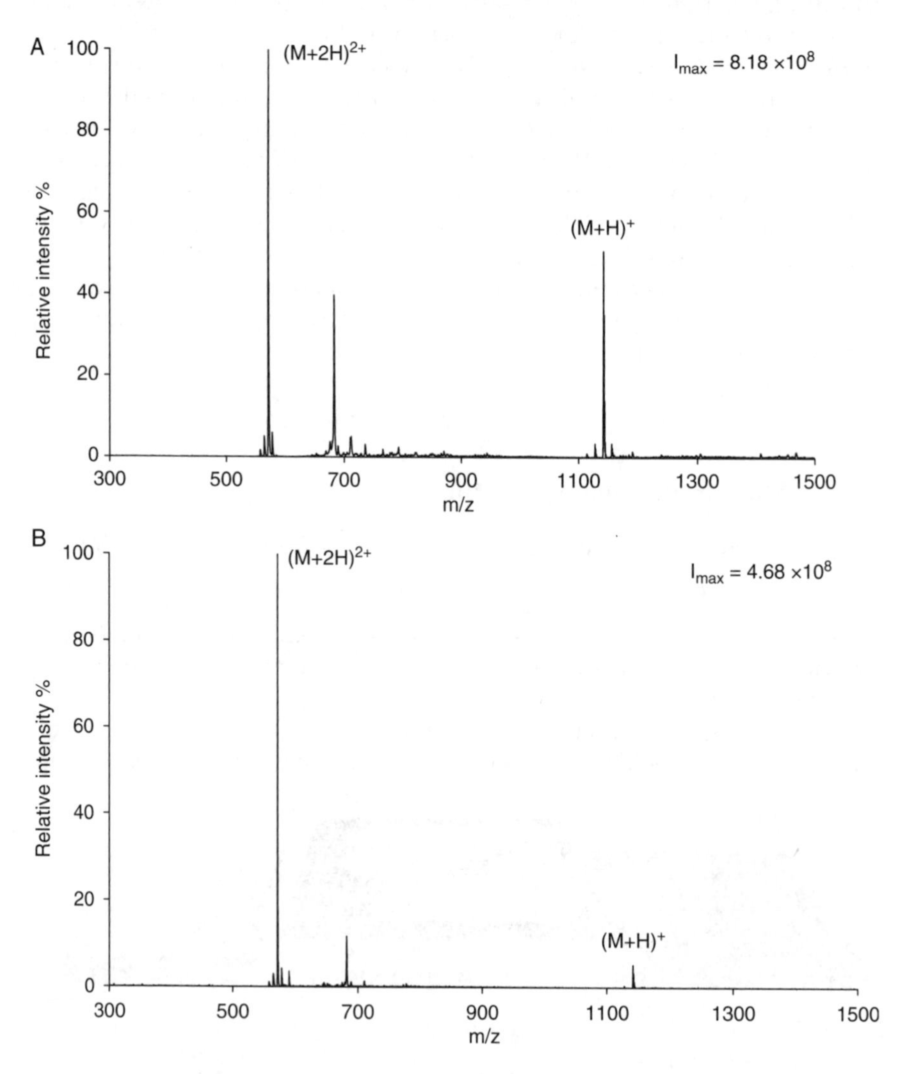

Figure 5.10 Mass spectra obtained using second-generation nib emitter tips. (A) Nib with a 16 μm slot width, HV of 1.2 kV and Gramicidin S sample at 1 μM ($I_{max} = 9.91 \times 10^6$). (B) Nib with an 8 μm slot width, HV of 1 kV and Gramicidin S sample at 5 μM ($I_{max} = 3.72 \times 10^8$). (Adapted with permission from Ref. 13).

of the formation of multi-charged species. This illustrates the enhanced ionization conditions brought by smaller nib tips and a smaller aperture. Another obvious difference between the two spectra is the signal-to-noise ratio (S/N) which is higher in the case of the 8 μm nib tip.

Typically the functioning range of 8 μm nib tips was in accordance with nanospray conditions, *i.e.* solution at ~1 μM and HV of ~1 kV; peptide concentration was in the micromolar range (1–10 μM) and the ionization voltage was successfully decreased to 0.8 kV for a Gramicidin S solution at 1–10 μM.[9,13]

Further experiments described in this chapter were done with the smallest nibs having an 8 μm wide slot as the latter lead to enhanced analysis and ionization conditions.

5.3.3.2 Change of Setup; Direct Application of HV onto the Silicon Support

Further tests on these nib tips were performed using a different experimental setup. Until now, the HV was applied to the liquid using a platinum wire that enables an electrical connection with the liquid and the metallic zone of the tip holder. However, the resulting experimental setup was cumbersome as the platinum wire had to be inserted in the reservoir and maintained in place between the metallic zone and the liquid. Another idea is to benefit from the semiconducting properties of the silicon wafer that supports the nib structure. A 3-inch diameter conducting n-type doped silicon substrate has been employed here; thus if voltage is applied at the reverse side of the wafer, it will be present at the reservoir feature and thus applied to the test liquid. Therefore, the silicon support of the micro-nib emitter tip was put in intimate contact with the metallic zone of the tip holder in order to establish a good electrical contact. Teflon tape was carefully wrapped around the silicon wafer and the moving part so as to fix the nib tip on the tip holder when inserted in the mass spectrometer inlet. This setup was seen to be robust and, as Teflon is inert chemically, there is a low risk of contaminating the test solution in the event of close contact with the Teflon tape.

This novel setup was firstly validated on two standard peptides, Gramicidin S as before and Glu-Fibrinopeptide B which is mostly detected at *m/z* 786.67 as a $(M + 2H)^{2+}$ species and also appears to a lesser extent at *m/z* 1570.60 as a $(M + H)^{+}$ species. For both species, the peptide sample concentration was 1–5 μM and the ionization voltage in the range 1.1–1.4 kV.[15] The mass spectrum pattern was consistent with what is observed in typical nanospray experiments, *i.e.* a very intense peak corresponding to the $(M + 2H)^{2+}$ species and a smaller peak corresponding to the $(M + H)^{+}$ species (Figure 5.11). It should be noted that by applying the ionization voltage via a platinum wire, we were not able to obtain such spectra with a 1 μM sample and voltage values of 1–1.2 kV; the resulting mass spectrum presented two peaks having comparable intensities (Figure 5.11). This illustrates the enhancement of ionization performances using this novel setup due to smoother ionization conditions as discussed later.

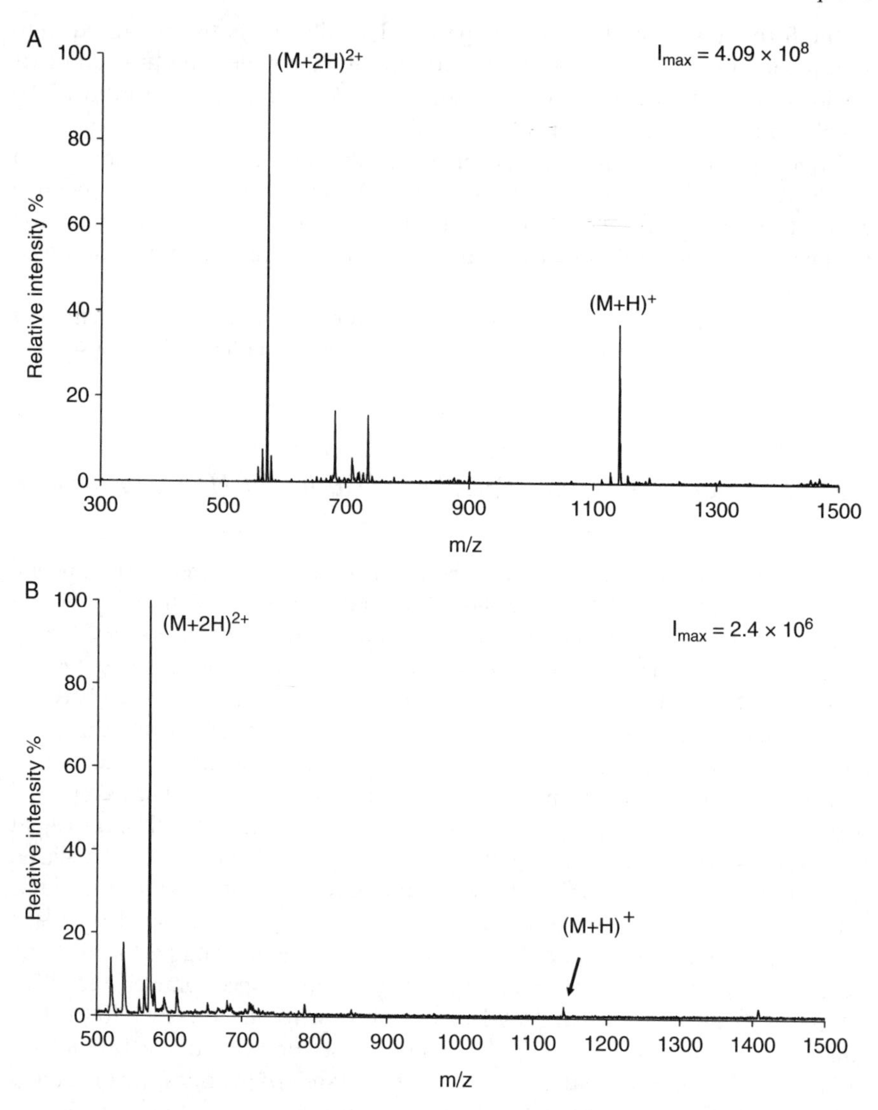

Figure 5.11 Mass spectra obtained using second-generation nib emitter tips (8 μm slot width) using a Gramicidin S sample at 1 μM and HV of 1.2 kV. (A) Former setup whereby HV is applied through a platinum wire. (B) Novel setup whereby HV is applied directly through the silicon support of the nib.

5.3.3.3 Limit of Detection of 8 μm Wide Nibs

A second series of tests was aimed at determining the limit of detection (LOD) of the nib tips having a dimension of 8 μm. We endeavored to decrease the concentration of peptide so as to reach "proteomics" conditions, where samples are at a concentration down to 1 nM which corresponds to the LOD of the

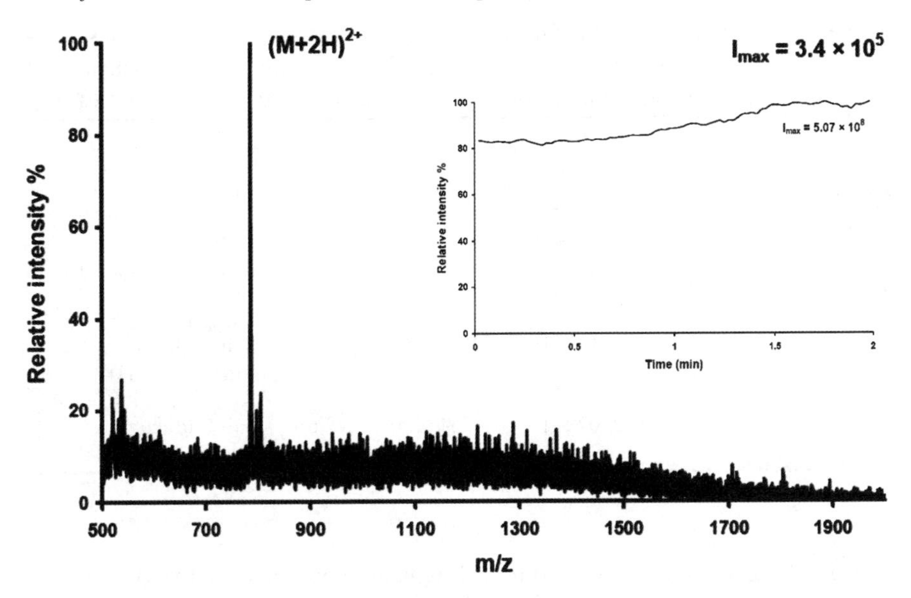

Figure 5.12 Mass spectrum averaged on a 2 min acquisition of the signal; analysis of a 10^{-2} μM Glu Fibrinopeptide B sample under 1.2 kV ionization voltage using the micro-nib source with an 8 μm slot dimension. The inset shows the corresponding TIC signal recorded for 2 min (RSD of 7.2%). (Adapted with permission from Ref. 13).

mass spectrometer. Diluted solutions of Glu-Fibrinopeptide B at 10 nM and 100 nM were prepared and tested in MS using a micro-nib emitter tip under the same conditions as before. These tests were successful without any increase of the voltage supply or any marked decrease in the spectrum intensity. The spray was stable in any of the tested conditions (Figure 5.12). For instance, the least concentrated solution was analyzed using a 1.1 kV HV. The resulting mass spectra present high intensity and signal-to-noise ratio values, even for the sample at 10 nM, but also abundant salt adducts on the doubly charged species. Moreover, the mass spectrum pattern remains unchanged and only presents the $(M + 2H)^{2+}$ species, as shown in Figure 5.12. Finally, the data from this series of experiments (see Table 5.1) with decreasing concentrations in peptide showed that our nib tips function according to the model proposed by Kebarle for a linear dependence of the analyte signal intensity as a function of its concentration with a unitary slope value as far as this latter is lower than 10 μM.

5.3.3.4 Proteomic Samples

As the results were conclusive using standard peptide samples, we then moved to a biological sample, closer to those that are routinely analyzed in the laboratory. A commercial cytochrome *c* digest was tested at 1 μM using ionization voltages

Table 5.1 MS analysis of Glu-Fibrinopeptide B samples at various concentrations. Ionization voltage values, maximal intensities measured on TIC signal and mass spectra and signal-to-noise ratios (S/N).

Sample concentration (μM)	*HV value (kV)*	I_{TIC}	I_{MS}	*S/N*	*Comments*
5	1.2	1.16×10^{10}	1.18×10^{8}	250	Little $(M + H)^+$ $I_{Mono}/I_{Di} = 4.3 \times 10^{-2}$
1	1.2	9.7×10^{9}	2.3×10^{7}	167	Little $(M + H)^+$ $I_{Mono}/I_{Di} = 1.5 \times 10^{-2}$
0.1	1.1	1.19×10^{9}	1.43×10^{6}	25	Salt adducts Little $(M + H)^+$ $I_{Mono}/I_{Di} = 1.46 \times 10^{-2}$
0.01	1.1	5.07×10^{8}	3.4×10^{5}	6	Salt adducts No $(M+H)^+$

Table 5.2 List of the cytochrome *c* fragments contained in the commercial digest sample with their sequence, mono-molecular weight, isoelectric point (pI) and hydrophobic character (log *P*, Kyte–Doolittle scale) calculated using Expasy software.

Fragment	*Fragment sequence*	*Mass* $(M + H)^+$	*Isoelectric point, pI*	*Hydrophobicity, log P*
56–73	GITWGEETLMEYLENPKK	2138.05	4.49	−0.961
56–72	GITWGEETLMEYLENPK	2009.95	4.09	−0.788
9–22	IFVQKCAQCHTVEK	1633.82	8.06	0.021
39–53	KTGQAPGFSYTDANK	1584.77	8.5	−1.187
40–53	TGQAPGFSYTDANK	1456.67	5.5	−0.993
26–38	HKTGPNLHGLFGR	1433.78	11	−0.877
89–99	GEREDLIAYLK	1306.7	4.68	−0.609
28–38	TGPNLHGLFGR	1168.62	9.44	−0.391
92–99	EDLIAYLK	964.53	4.13	0.212
80–86	MIFAGIK	779.45	8.5	1.6
74–79	YIPGTK	678.36	8.59	−0.567
9–13	IFVQK	634.39	8.75	0.82

of 1–1.2 kV. The interest in this sample lies in the diversity of the peptidic fragments that it contains; these fragments differ from each other by their molecular weight (MW), their hydrophobicity (log *P*) and also their isoelectric point (pI), as shown in Table 5.2. Due to the diversity of the fragment properties, this sample allowed us to study the surface properties of the micro-nib source and its influence on the response in nanoESI, *i.e.* if there is any class of peptides that was adsorbed or retained on the micro-nib surface through interaction, the respective intensity of the resulting mass spectrum could be thus reduced. Figure 5.13 presents the mass spectrum obtained using an ion trap mass spectrometer for

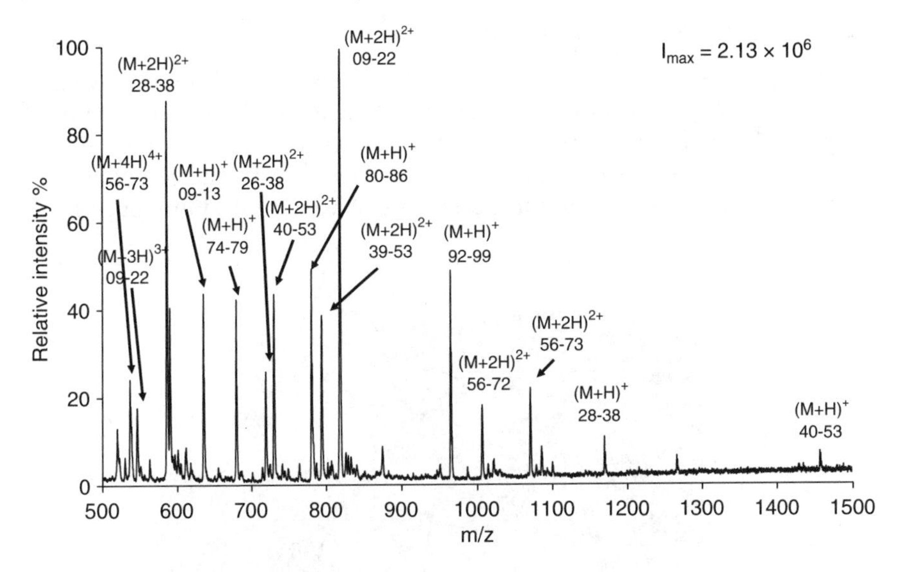

Figure 5.13 Mass spectrum averaged on a 2 min acquisition of the signal; analysis of a protein digest sample, cytochrome *c* at 1 μM under 1.2 kV ionization voltage using the micro-nib source with an 8 μm slot dimension. Peaks are labeled with the sequence of the peptidic fragments as well as the charge state.

the analysis of the cytochrome *c* digest at 1 μM using a micro-nib emitter tip with an ionization voltage of 1.1 kV. All the peptides were detected using a micro-nib source as the ionization emitter tip: no influence of the hydrophobic character or the basicity of peptides was observed. The SU-8 resist composing the emitter tip, which is more hydrophobic than glass or fused silica, does not influence the analysis conditions: no peptide was adsorbed, even partially, on the surface.

We compared these results with those that were obtained with a standard capillary tube-based emitter tip using a 1 μM cytochrome *c* digest sample and under a 1 kV ionization voltage. Firstly, the signal-to-noise ratio was much higher using a nib tip, even if the maximal intensity was lower than with a standard tip; these S/N and maximal intensity values were respectively 20 and 1.89×10^7 for the standard tip against 35 and 2.13×10^6 for a nib tip. Secondly, using a standard tip, the most basic peptide (26–38 fragment; pI = 11) could not be detected; it could be detected using a nib tip. This is in agreement with a recent paper that discusses the influence of the tip material on the adsorption of analytes, and especially the strong adsorption of basic peptides on glass-based capillary sources which results in their non-detection in MS.[16] Thirdly, the nib tips appeared to favor the formation of multi-charged species, compared to the standard tip. This last point can be accounted for by the smoother ionization conditions brought by this novel setup, as discussed later.

5.3.3.5 Reduced Nib Dimensions

We finally validated the functioning of nib emitter tips with reduced section area ($h \times w = 25\,\mu m \times 6\,\mu m$) (see Figure 5.14A) using a Glu-Fibrinopeptide B sample at 1 µM. The required ionization voltage here was lower than 1 kV, as illustrated by the mass spectrum presented in Figure 5.14B and resulting from 1 min acquisition of the signal for the analysis of a 1 µM Glu-Fibrinopeptide B

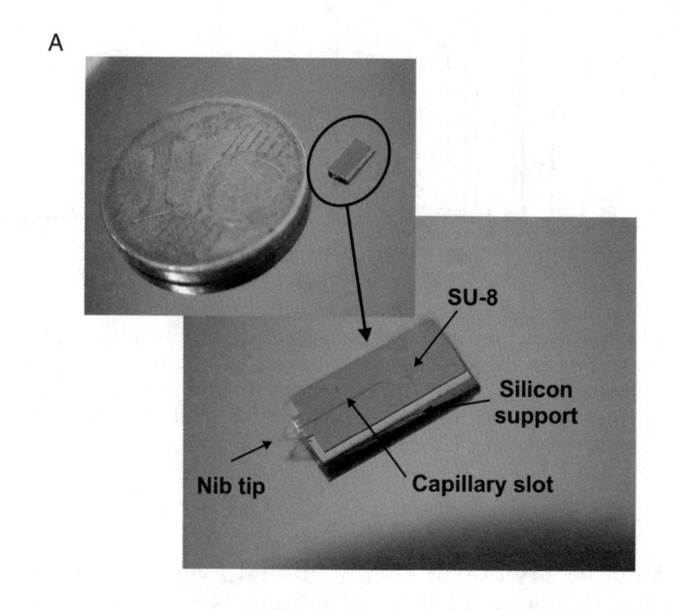

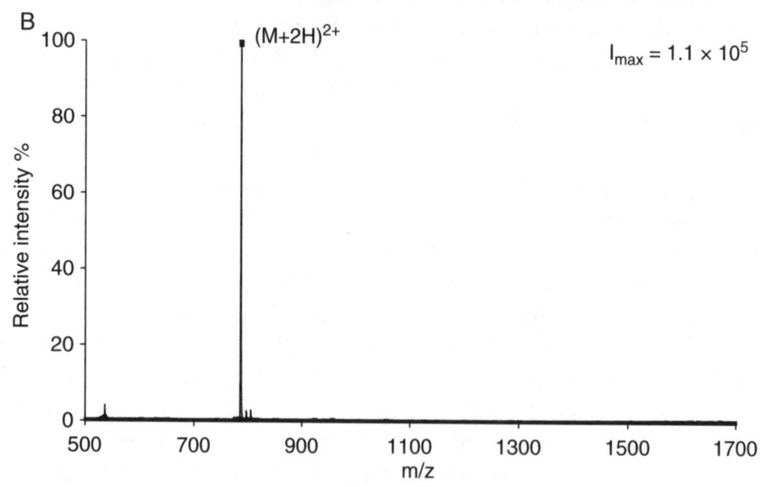

Figure 5.14 (A) Photograph of a smaller nib from the second generation and having an outlet section of $6\,\mu m \times 25\,\mu m$. (B) Mass spectrum (averaged over a period of 2 min) obtained using the nib, for a Glu-Fibrinopeptide sample at 1 µM and an ionization voltage of 1.2 kV.

sample under a 0.9 kV ionization voltage. The signal intensity was lower as a result of the lower ionization voltage. Nonetheless, it should be noted that the signal-to-noise ratio was much higher compared to the spectrum obtained with an 8 μm nib tip and under a 1.2 kV HV. It demonstrated a threefold increase, from 300 to 1000.

5.3.4 Discussion and Conclusions

5.3.4.1 Source Dimensions: Capillary Filling of a Slot

For this second series of nib prototypes, we tried to enhance the design of the nib tip and to study the influence of the aperture dimensions on the functioning of the nib.[14] Dimensions of the nib tip should be carefully chosen, so as (i) to observe the required capillary effect action in the slot so that the fluid can flow from the reservoir to the nib tip and (ii) to enhance ESI phenomena through the production of smaller droplets. Consequently, we endeavored to decrease the dimensions of the nib orifice, *i.e.* the surface area of the section at the tip, by decreasing the height h of SU-8 from 35 μm to 15 μm compared to the first generation. Capillarity was not observed for the nibs with a smaller height of resist ($h \times w = 15\,\mu m \times 20\,\mu m$) than width of slot. Consequently, we had two series of nib tips that work: a first series of nibs with an SU-8 height h of 35 μm and slot width w values of either 8 or 16 μm and a second series of nib tips with an SU-8 height h of 25 μm and slot width values w lower than 10 μm. These dimensions were appropriate for capillary action to occur in the slot and were fully compatible with electrospraying. In addition, the dimensions of nibs with either an 8 μm slot width w ($h = 35\,\mu m$) or a slot width w lower than 10 μm ($h = 25\,\mu m$) were comparable to those of standard emitter tips (< 10 μm) and so were their performances in nanoESI-MS experiments.

We decided to perform numerical simulations to get a better understanding of this phenomenon.[17] The main parameters to be taken into account are the slot width w, the slot height h, the contact angle of the liquid with the inside of the slot walls θ_w and the contact angle with the upper surface of the cantilever θ_s. In the case of our devices there was no differential surface treatment between the inside of the wall and of the cantilever surface, which is assumed to correspond to $\theta_w = \theta_s$. In that case, numerical simulations show that a stable regime is obtained for an aspect ratio w/h lower than the cosine of the wetting contact angle θ_w, *i.e.* $w/h < \cos\theta_w$. In the case of extremely wetting liquids ($\theta_w = \theta \approx 0°$) the aspect ratio of the slot (w/h) can be equal to unity. Decreasing the aspect ratio, *i.e.* lower w/h ratio values, results in a wider range of stability, or, in other words, a wider range of liquid types to be used. Aspect ratios w/h lower than 0.1 increases the stability area from $\theta_w = \theta \approx 0°$ up to near 90° which allows for using most of the common solvents in analytical experiments. The nibs used in our experiments with aspect ratios of w/h of $16/35 = 0.457$, $8/35 = 0.228$ and $10/25 = 0.400$ meet this stability criterion and work well. On the other hand, as pointed out previously, the liquid does not fill the slot of nibs with aspect ratio w/h of $20/15 = 1.333$, in agreement with the theoretical calculations.

5.3.4.2 Novel Setup: Smoother Ionization Conditions

The test setup was also changed for this second series of nibs. The ultimate goal of this work is to produce high-density automated arrays of microfabricated emitter tips for nanoESI applications. This goal could clearly not be reached with the first setup that implies a manual step to introduce the metallic wire onto the micro-nib emitter tip. Another alternative to facilitate analysis automation would be to use a conductive coating of the nib, but the latter was seen to deteriorate upon HV application. In addition, this also demands one more step in the fabrication of the nib tips. Moreover, further enhancement in ionization performances was observed; this is accounted for by the larger surface area for HV application in contact with the analyte solution. This leads to smoother ionization conditions explained by the lower current density. The surface area between the HV application zone and the test liquid corresponds here to the surface of the reservoir bottom plus that of the microchannel, whereas in the former case, this surface was reduced to that developed by the platinum wire, *i.e.* $0.4 \times 10^{-6}\,m^2$ (platinum wire) against $7 \times 10^{-6}\,m^2$ (silicon support). As a result, the current density is one order of magnitude lower in the latter case and the ionization conditions are improved.[18–20] Furthermore, the conducting microchannel between the reservoir and the nib is acting as a liquid junction bringing the HV very close to the tip. This design is clearly better than when the HV is applied in the reservoir contributing also to a smoother ionization.

Analysis enhancement has been clearly observed through the tests on this second series nib tips. When analyzing standard peptidic species, the formation of uniquely di-charged species $(M + 2H)^{2+}$ was seen with this new setup compared to the former one (see Figure 5.11). In the same way, the experiments with the cytochrome *c* digest revealed (i) the enhanced generation of multi-charged species when using a nib tip rather than a standard capillary tip and also (ii) that the signal-to-noise ratio obtained with a nib tip was higher than with a standard tip.

It is well known that ionization may be promoted in electrospray through protonation processes resulting from electrochemical phenomena. The latter occur in the source and consist of the decomposition of the solvent upon voltage application ($2H_2O \rightarrow 4H^+ + 4e^- + O_2$). We subsequently decided to probe the electrochemical behavior of our nib tips by using reserpine, an alkaloid compound which has been used as a probe for oxidation phenomena in an electrospray source. Reserpine is very prone to oxidation through its dehygrogenation[21] to give two products detected at m/z 607.3 and m/z 625.3 and corresponding respectively to its dehydrogenated product and the water adduct of the latter.[22,23] This investigation was carried out with a solution at

200 μM and under a voltage of 1.5 kV. The resulting mass spectrum only presents one peak at *m/z* 609.3 corresponding to the protonated molecular ion. No products resulting from the oxidation of reserpine could be detected over a wide range of concentrations (up to 2 mM). This showed that the enhancement of the ionization does not result from electrochemical phenomena as observed by grounding the ESI tip[24] but only from smoother ionization conditions.

5.3.4.3 Future Improvements

From these observations, one can easily conclude that further improvements can be expected from further miniaturization of the nib tips, as 8 μm nib tips lead to better results than 16 μm ones and as 6 μm × 25 μm dimensions gave even better analysis conditions than 8 μm × 35 μm ones. However, working with SU-8 the fabrication of smaller and reliable nib structures is difficult to conceive. Therefore, the production of smaller nibs implies a change in materials and fabrication techniques.

Another limitation of these nib sources comes from their opened structure; this opened configuration leads to higher sample consumption due to some in-source evaporation of the test liquid. To alleviate this problem, a cover plate can be included on the microchannel upstream of the capillary slot. Thereby, the contact surface area between the liquid sample and the air could be decreased to give reduced evaporation of the sample in the source; signal would thus be acquired for longer durations. It should be noted that this evaporation phenomenon did not result in an in-reservoir concentration of the peptide solution. For the experiments presented here, acquisition was done after extensive washing of the source and just after loading of the peptide sample.

5.4 Polycrystalline Silicon-Based Nano-Nib Devices for NanoESI-MS

5.4.1 Description

As mentioned above, we decided to investigate the fabrication and testing of reduced-dimension nibs for nanoESI applications. For this purpose, we targeted structures having a low micrometer dimension orifice at the tip. As small structures cannot be routinely fabricated using SU-8 we require the use of other materials and other fabrication routes. In order to do this, we chose to use polySi to produce the nib feature, *i.e.* to replace the former photoresist, this being still supported on an n-type silicon wafer. PolySi has a Young's modulus of 200 GPa, against 20 GPa for SU-8. In addition to this, heat annealing can result in the absence of stress in the material and enhances the planarity of the nib feature on its support. This material lends itself well to the fabrication of long, small, thin features as we aim at in this work.

Firstly, the polySi-based prototypes present an open configuration as did the SU-8-based structures. The polySi structures included a reservoir, a capillary

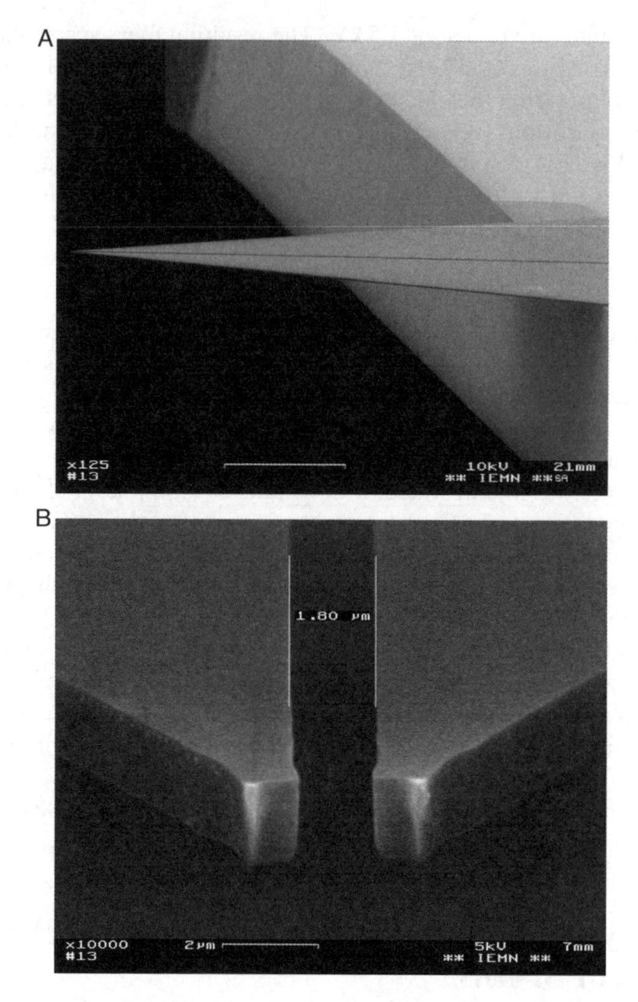

Figure 5.15 Scanning electron micrographs of polySi-based nanotips. (A) Lateral view of the tip of 800 µm length. (B) Enlarged front view of the slot at its outlet showing a cross-sectional area of 2 µm × 2 µm. (Adapted with permission from Ref. 25).

slot and a very sharp tip with a narrow slot so as to present a high aspect ratio. Typically, nib tips were of 800 µm in length for a slot of approximately 2 µm × 2 µm, giving thus a cantilever structure with aspect ratio of 400 (Figure 5.15).

5.4.2 Fabrication

The fabrication process is illustrated in Figure 5.16, the details of which can be found elsewhere.[25] As before, a 3-inch n-type doped silicon wafer was used for the fabrication of the cantilever structure. Firstly, a layer of silicon dioxide

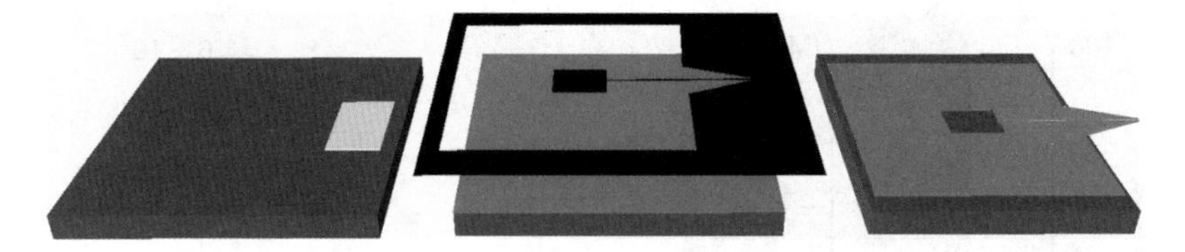

Figure 5.16 Novel fabrication process for the third generation of micromachined electrospray emitter tips: silicon support wafer (blue), 200 nm thick silicon dioxide etch-release layer (white) which is patterned using a HF-based wet-etch step, single polySi micro-nib layer (red) housing the reservoir and the nib tip which is defined using either a chlorine- or an SF_6-based dry etch and single UV photolithographic masking step (black).

(200 nm) is grown by dry oxidation. This oxide layer has a double purpose; it acts as (i) a masking layer during the wet-etch of silicon aiming at forming cleavage lines and (ii) a sacrificial layer for the release of the cantilevers from the silicon wafer surface. This layer is firstly patterned using photolithography and wet-etching techniques to form cleavage lines. Following this, a second step is used to form etch-release pads underneath the polySi cantilevers. A layer of polySi (2 and 5 μm) is deposited onto the wafer using low-pressure chemical vapor deposition (LPCVD). In order to minimize internal stresses in this layer a high-temperature annealing step at 1050 °C is performed for one hour. A low-temperature silicon dioxide layer is subsequently deposited on the polySi and serves as a mask-transfer layer. Following this, the slotted cantilever structures are defined in the polySi layer using chlorine-based (Cl_2/He) reactive ion etching (MDC-Trion, AZ, USA). The triangular cantilevers are released from the silicon surface by wet etching the sacrificial etch-release layer of thermal oxide; this is achieved with a 50% HF–H_2O solution for ensuring the selectivity between SiO_2 and polySi. Finally, the individual microfluidic devices were cleaved by making use of the predefined cleavage lines. Cantilevers are observed to be very planar; this results from minimal residual stress in the polySi via high-temperature annealing of the material.

5.4.3 Mass Spectrometry Tests

Mass spectrometry tests were conducted as described before, with an ion trap mass spectrometer using the second setup whereby HV is directly applied onto the supporting silicon wafer. The same standard peptide as before was used, Glu-Fibrinopeptide B, at a concentration of 1 μM. For the first series of tests, we decided to investigate the range of HV which would give an electrospray working with solutions of 50:50 MeOH–H_2O acidified with 0.1% HCOOH.

This time, experimental results were outstanding (Figure 5.17); typical HV to apply was lower than 1 kV to observe a stable spray and to enable peptide analysis. The aspect ratio h/w of the slot of $2/2 = 1.00$ equal to the maximum

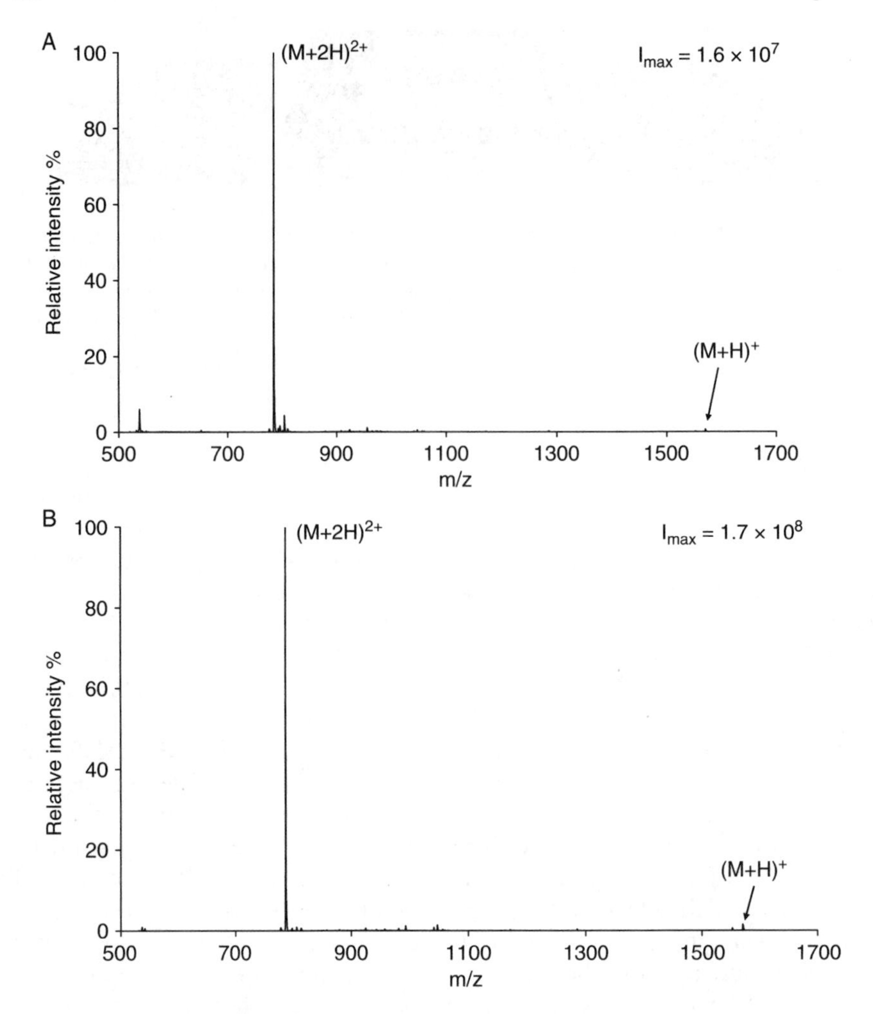

Figure 5.17 Mass spectra obtained using polySi-based nano-nibs for a Glu-Fibrinopeptide B sample at 1 μM. (A) 80:20 H_2O–MeOH, 0.1% HCOOH solvent composition and 1 kV HV; (B) 75:25 H_2O–MeOH, 0.1% HCOOH solvent composition and 0.7 kV HV. (Adapted with permission from Ref. 25).

allowed value did not preclude the nib to work perfectly. Moreover, this HV value could be decreased down to 0.5–0.7 kV without losing spray stability. Also, only the di-charged species of Glu-Fibrinopeptide B could be detected and the S/N ratio was very high, assessed to be higher than 2000.

Following this, another series of tests was aimed at varying the composition of the solvent used for preparing the peptide samples. We particularly decided to investigate the water content of the solvent and its influence on the spraying ability. Therefore, we prepared MeOH–H_2O, 0.1% HCOOH solvent mixtures varying from 50 to 95% H_2O so as to prepare peptide samples

(Glu-Fibrinopeptide B at 1 μM). These various solutions were tested in nanoESI as described before with an HV in the range 0.7–1 kV.It was observed that the higher the amount of water, the better the ionization and the higher quality the resulting mass spectra (*i.e.* in terms of high intensity, high signal-to-noise ratio).

These characteristics are of great interest for biological analysis, as biological compounds are prone to easily degrade in organic solvents. It should be noted that standard nanoESI tips work better with a high (*e.g.* 50%) amount of organic solvent; this may cause problems for nanoLC-nanoESI-MS analysis, whereby the composition of the solvent may vary from 100% aqueous to 100% organic within a separation run.

5.4.4 Discussion and Conclusions

Preliminary tests of these polySi-based nano-nibs have shown outstanding results in terms of analysis conditions and mass spectra quality (pattern and signal-to-noise ratio). We have demonstrated that HV can be decreased down to 0.5–0.7 kV, still keeping acceptable ionization conditions for peptide samples at 1 μM. These nibs are also compatible with analysis of solutions with a high amount of water without any degradation in analysis quality.

Another pure interest of these devices relies on the type of investigations they may enable. We have already discussed that the size of the ionization tip dictates the size of the generated droplets. Using a tip with a 2 μm aperture, droplets of 1 μm radius are created. Working with a peptide concentration of 1 μM, each droplet roughly contains 600 molecules but assuming a LOD lower than 10^{-2} μM as was the case for the 8 μm slot nib tips, analysis conditions whereby each droplet contains a single molecule can be investigated. In addition, the variations in the molecule number in such conditions have a relative importance which becomes greater, as illustrated in Table 5.3.

5.5 Towards an Integrated System Based on a Nib for on-line Analysis

All nib systems described in this chapter up to now are based on an open configuration: liquid is deposited in an open reservoir feature, the liquid then flows in an open capillary slot to reach the tip where electrospraying occurs upon application of the ionization voltage. We have already discussed that such an open configuration presents one major limitation: the evaporation of sample as the interface between air and the sample is quite high, this resulting in higher sample consumption. Therefore, we decided to investigate two routes to alleviate this, this enabling also an on-line connection of the nibs for continuous analysis of liquids.

Table 5.3 Dependence on the number of molecules per droplet and on the variations of this number on the size of the generated droplets, dictated by the size of the source.

Radius of the generated droplet (μm)	*Peptide concentration (μM)*	*ESI type (nib source)*	*Number of molecules per droplet*	*Fluctuation, $\sqrt{n}/n$ (%)*
100	1	ESI	2.4×10^9	2.04×10^{-3}
10	1	NanoESI (SU-8 slot)	2.4×10^6	0.06
10	10^{-2}	NanoESI (SU-8 slot)	2.4×10^4	0.6
1	1	NanoESI (polySi slot)	2.4×10^3	2.04
1	10^{-3}	NanoESI (polySi slot)	2.4	64.5

5.5.1 SU-8-Based Solution: Use of Capillary Tubing

The first solution we investigated consists of attaching standard capillary tubing (i.d. 10 μm; o.d. 280 μm) on a nib, the capillary being placed as close as possible from the tip of the nib (narrower area in the channel) so as to limit contact with air (Figure 5.18A). Gluing of the capillary was carefully done so as to avoid any possible contact of the liquid with glue; this may happen in the case of the liquid flow not being high enough and result in liquid flowing back along the capillary.

This nib system connected to a capillary was then placed at the outlet of a nanoLC column for on-line analysis after a separation step.[26] The column was monitored from a nanoLC setup (LC Packings-Dionex, The Netherlands). An amount of 1 μL of sample was injected (Famos injection system, LC Packings-Dionex) onto a commercial nanocolumn (PepMap, 75 μm i.d.) and eluted in gradient conditions at a flow-rate of 100 nL min^{-1} (Utlimate micro pump and accurate stream splitter, LC Packings-Dionex) using solvent mixtures based on AcCN and deionized water (solvent A: AcCN–H_2O 5:95; solvent B: AcCN–H_2O 95:5) acidified with 0.1% TFA. Ionization HV was 1.5 kV, as typically used for a standard emitter tip. The sample used for this test is a commercial cyctochrome *c* digest at a concentration of 800 nM. The spray was seen to be stable for several hours (several nanoLC runs) as expected with a standard emitter tip. The analysis of the solution eluted from the column was possible and the same as for routine conditions.

Figure 5.18B shows the base peak trace of one separation run, and the mass spectrum of the 92–99 fragment which is eluted at 23.8 min is shown in the inset. This preliminary test for on-line analysis using a nib tip showed that these kinds of tips are entirely reliable for on-line analysis, the sample being continuously fed into the capillary slot. This experiment enables us subsequently to validate the possible use of nib tips for continuous analysis, as is the case at the outlet of a nanoLC column. Following this, the integration of the nib tip on an SU-8-based microsystem was also investigated.

A

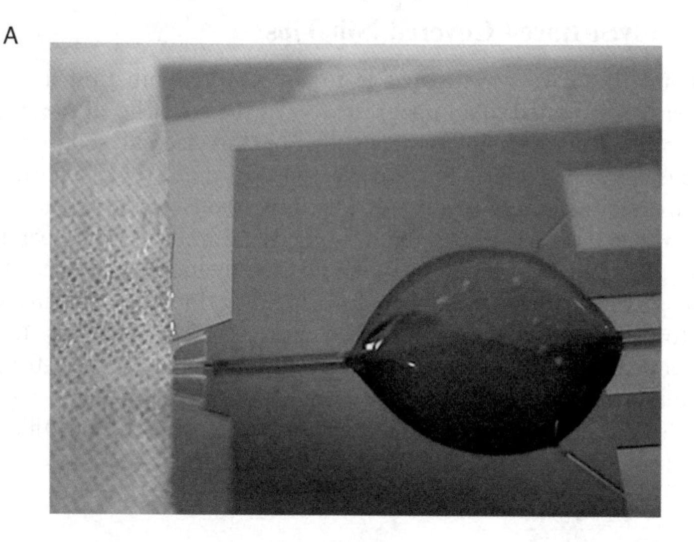

B

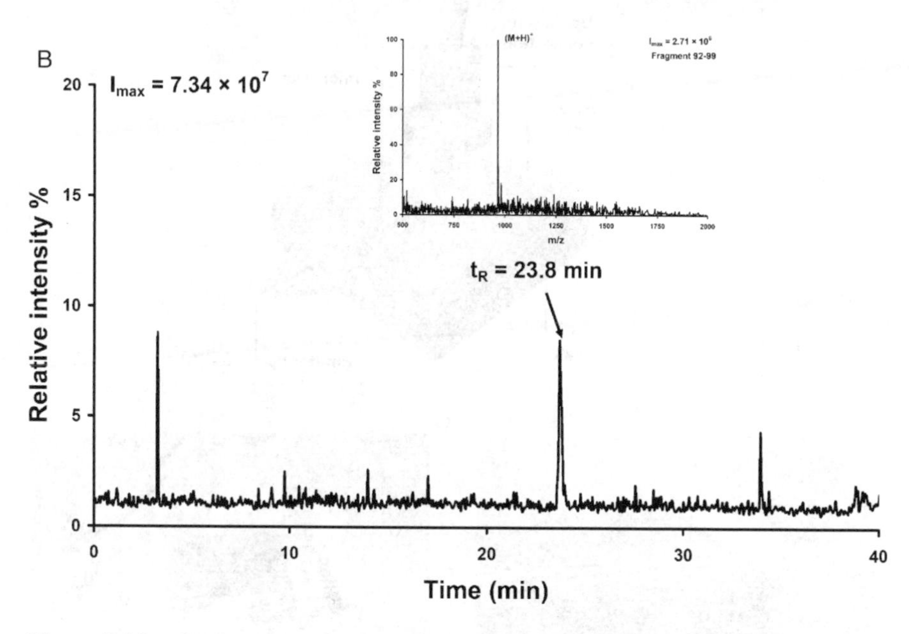

Figure 5.18 On-line analysis. (A) Photograph of a nib-shaped nanoESI interface used for on-line tests. The nib structure is fabricated using the negative photoresist SU-8 and its capillary slot is coupled to standard capillary tubing (10 μm i.d.) for on-line MS analysis. (B) Base peak trace obtained for the separation of 800 fmol of a cytochrome *c* digest using a nanoLC setup coupled to a nib-shaped nanoESI interface. The inset shows the mass spectrum corresponding to fragment 92–99 of cytochrome *c*, which is eluted at 23.8 min (flow rate, 100 nL min^{-1}; HV of 1.5 kV). (Adapted with permission from Ref. 26).

5.5.2 PolySi-Based Covered Nib Tips

The second route we investigated is to use covered nib tips; a capillary is subsequently connected to an inlet reservoir in the cover lid of the system, and thereby on-line analysis is possible.[27,28] For this second solution, we used polySi-based nibs according to the design described above, which can be easily and robustly bonded to a Pyrex lid. The dimensions of the Pyrex cover were chosen so that the glass covers the nib-based feature up to the opening of the slot (sharp part of the nib). This Pyrex lid also includes a pre-etched via hole acting as a conical-shaped reservoir for easy introduction of capillary tubing. Silicon-to-glass bonding was achieved using anodic bonding techniques. Besides, a dedicated holder was designed and fabricated as a user-friendly tool for handling the nib systems and to ensure tight and low dead volume connection to capillary tubing using appropriate Upchurch scientific ferrules.

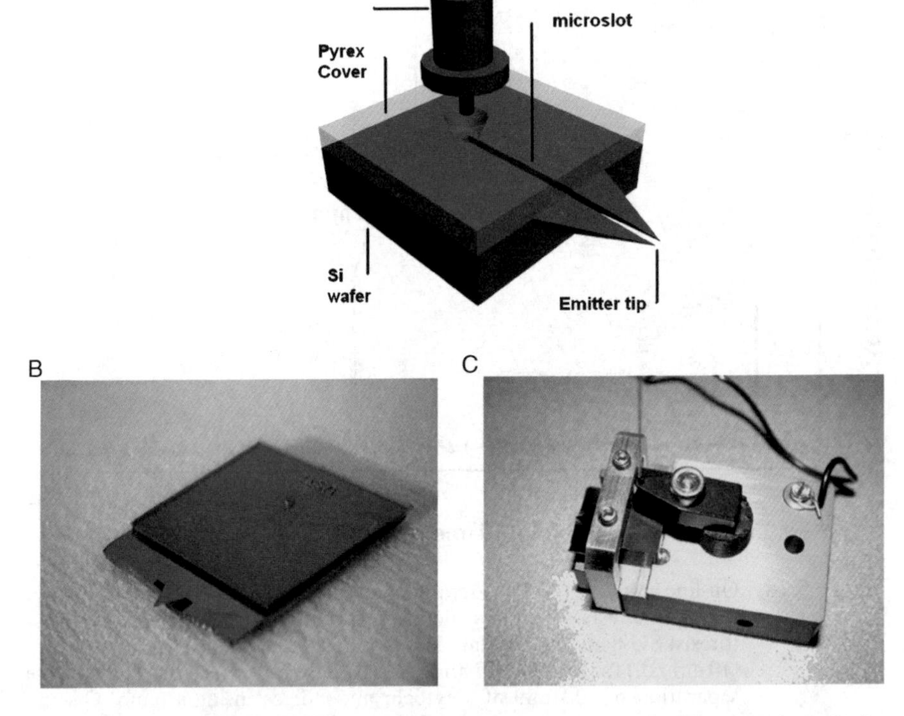

Figure 5.19 PolySi-based nano-tips including Pyrex cover plate and a via hole for easy connection to capillary tubing, and dedicated holder for user-friendly handling. (A) Scheme of the integrated polySi-based nano-tips. (B) Image of an integrated microTAS. (C) Mechanical holder dedicated to the nano-nib microTAS for easy and watertight connection to capillary tubing.

Figure 5.19 illustrates the covered nib tips as well as the dedicated holder for user-friendly manipulation of the nibs.

Unfortunately, we encountered some problems during the fabrication of the integrated nano-nibs; the process was not optimized and the release of the nib tips and their cleavage into independent components was not possible. Indeed, the end of the fabrication proceeds as follows. First, the sacrificial layer is

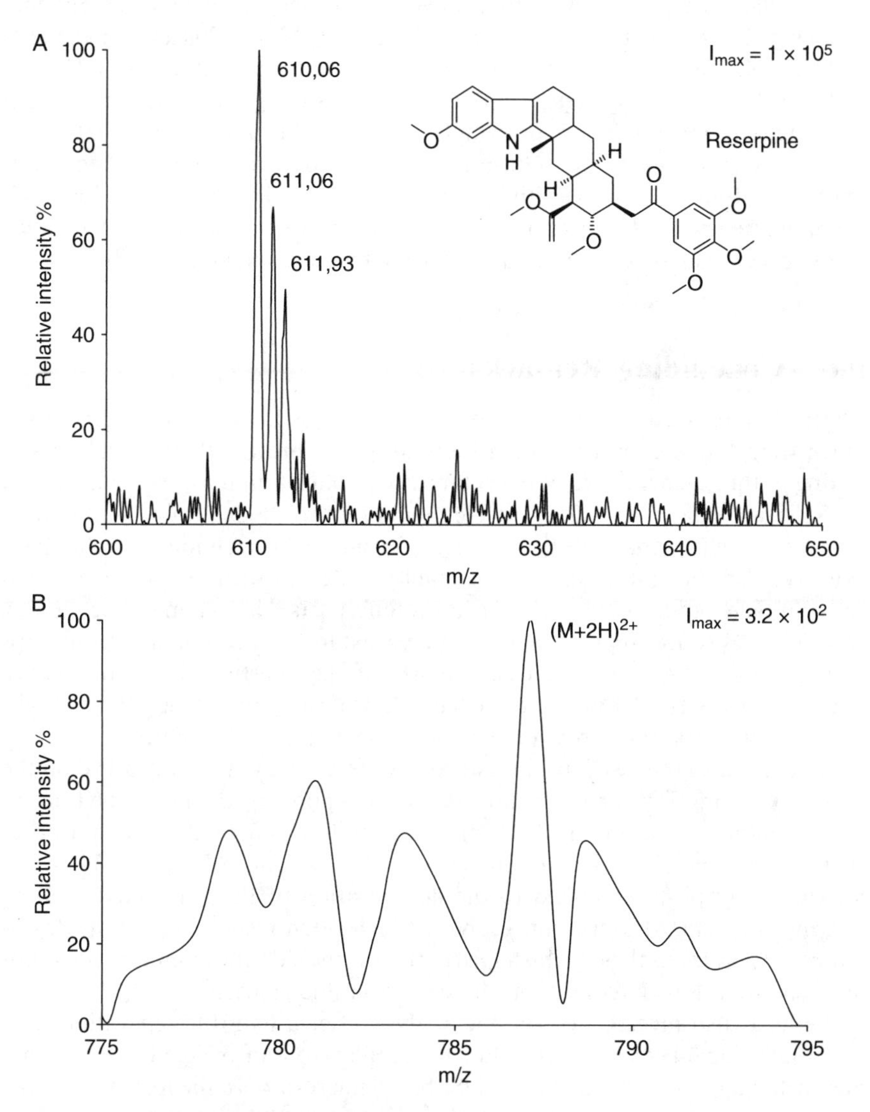

Figure 5.20 Mass spectra obtained using polySi-based nano-nibs and averaged over a period of 1 min. (A) Reserpine sample at 200 μM in 50:50 H_2O–MeOH, 0.1% HCOOH, HV of 1.5 kV; (B) Glu-Fibrinopeptide B sample at 10 pM in 50:50 H_2O–MeOH, 0.1% HCOOH, HV of 1.5 kV.

removed, and then the tips are bonded to a Pyrex wafer using anodic bonding techniques. Finally, the supporting wafer is cleaved so as to free the nib tips and it is subsequently diced to produce independent components. It appeared that the anodic bonding step resulted in adhesion of the polySi cantilever on its silicon support, thus hindering the release of the nib tips.

As a consequence, only simple polySi structures without any cover lid were tested according to the same protocol as before. Two compounds were tested here, a standard peptide, Glu-Fibrinopeptide B, and an alkaloid compound, reserpine. The concentrations were respectively 10 pM and 200 μM for the tests and the solutions were based on methanol and water, as before (MeOH–H_2O 50:50, 0.1% HCOOH). Figure 5.20 shows the resulting mass spectra for both tests, when using only 1 μL of sample. It should be pointed out that a few microlitres was used so the Glu-Fibrinopeptide B was acquired using a few tenths of attomoles. The sensitivity for reserpine which exhibits very low ionization efficiency is also outstanding. We are currently investigated a new fabrication process so as to be able to produce integrated polySi nano-nibs.

5.6 Concluding Remarks

We have developed novel 2D ionization sources whose functioning borrows much from the idea of a common fountain pen. Through the fabrication and testing of three generations of microfabricated ionization nibs, two being based on SU-8 and one on polySi, we have demonstrated the feasibility and performances of such planar nib tips fabricated using microtechnology techniques. We have also shown that their performances are equivalent or even superior to those of their capillary tube-based glass counterparts. Using a microtechnology route enables source reproducibility, in contrast to conventional ionization tips and enables us to easily decrease the size of the aperture so as to enhance analysis sensitivity. For instance, we have shown using three generations of nibs that a smaller aperture provides enhanced ionization performance.

In addition to this, we have investigated different ways to apply a HV on the liquid. Applying HV directly onto the silicon support of the nib turned out to give smoother and more reliable ionization conditions for nanospray purposes than using a metallic wire in contact with the solution. Moreover this route alleviates any problem linked to the deterioration of the conductive coating present on standard ionization sources, which often results in analysis degradation. Therefore, these nib-shaped sources should allow enhancement of analysis conditions in continuous mode for on-line analysis.

We have also investigated on-line analysis after a nanoLC separation step; two routes have been assessed: gluing a capillary on an open SU-8-based nib and including a cover plate on a polySi-based micro-nib. In the former case, the spray was seen to be stable for several nanoLC runs and on-line analysis of the eluent coming out of the nanoLC column was possible.

We plan a future commercialization of such nib-shaped sources as an alternative to capillary-based sources. A patent has already been granted,[29] and a reliable and user-friendly mechanical holder has been developed for dedicated handling of covered polySi-based micro-nibs and their introduction into any mass spectrometer.

References

1. M. Wilm and M. Mann, *Anal. Chem.*, 1996, **68**, 1.
2. G. A. Schultz, T. N. Corso, S. J. Prosser and S. Zhang, *Anal. Chem.*, 2000, **72**, 4058.
3. K. Tang, Y. Lin, D. W. Matson, T. Kim and R. D. Smith, *Anal. Chem.*, 2001, **73**, 1658.
4. J. Sjodahl, J. Melin, P. Griss, A. Emmer, G. Stemme and J. Roeraade, *Rapid Commun. Mass Spectrom.*, 2003, **17**, 337.
5. J. S. Kim and D. R. Knapp, *J. Am. Soc. Mass Spectrom.*, 2001, **12**, 463.
6. C.-H. Yuan and J. Shiea, *Anal. Chem.*, 2001, **73**, 1080.
7. V. Gobry, J. Van Oostrum, M. Martinelli, T. C. Rohner, F. Reymond, J. S. Rossier and H. H. Girault, *Proteomics*, 2002, **2**, 405.
8. J. Kameoka, R. Orth, B. Ilic, D. Czaplewski, T. Wachs and H. G. Craighead, *Anal. Chem.*, 2002, **74**, 5897.
9. S. Le Gac, S. Arscott, C. Cren-Olive and C. Rolando, *J. Mass Spectrom.*, 2003, **38**, 1259.
10. S. Le Gac, C. Cren-Olive, C. Rolando and S. Arscott, *J. Am. Soc. Mass Spectrom.*, 2004, **15**, 409.
11. S. Arscott, S. Le Gac, C. Druon, P. Tabourier and C. Rolando, *J. Micromech. Microeng.*, 2004, **14**, 310.
12. P. Kebarle and H. Yeughaw, in *Electrospray Ionization Mass Spectrometry*, ed. R. B. Cole, Wiley, New York, 1997, pp. 3–63.
13. S. Le Gac, S. Arscott and C. Rolando, *Electrophoresis*, 2003, **24**, 3640.
14. S. Arscott, S. Le Gac, C. Druon, P. Tabourier and C. Rolando, *Sens. Actuators B*, 2004, **98**, 140.
15. S. Le Gac, S. Arscott, C. Rolando and S. Arscott, *J. Am. Soc. Mass Spectrom.*, 2006, **17**, 75.
16. I. V. Chernushevich, U. Bahr and M. Karas, *Rapid Commun. Mass Spectrom.*, 2004, **18**, 2479.
17. M. Brinkmann, R. Blossey, S. Arscott, C. Druon, P. Tabourier, S. Le Gac and C. Rolando, *Appl. Phys. Lett.*, 2004, **85**, 2140.
18. G. J. Van Berkel, G. E. Giles, J. S. Bullock and L. J. Gray, *Anal. Chem.*, 1999, **71**, 5288.
19. G. J. Van Berkel, K. G. Asano and P. D. Schnier, *J. Am. Soc. Mass Spectrom.*, 2001, **12**, 853.
20. V. Kertesz and G. J. Van Berkel, *J. Mass Spectrom.*, 2001, **36**, 204.
21. M. Sanchez and J. J. Sanchez-Aibar, *Analyst*, 1996, **121**, 1581.
22. G. J. Van Berkel, K. G. Asano and M. C. Granger, *Anal. Chem.*, 2004, **76**, 1493.

23. G. J. Van Berkel, V. Kertesz, M. J. Ford and M. C. Granger, *J. Am. Soc. Mass Spectrom.*, 2004, **15**, 1755.
24. R. A. Ochran and L. Konermann, *J. Am. Soc. Mass Spectrom.*, 2004, **15**, 1748.
25. S. Arscott, S. Le Gac and C. Rolando, *Sens. Actuators B*, 2005, **106**, 741.
26. J. Carlier, S. Arscott, V. Thomy, J. C. Camart, C. Cren-Olive and S. Le Gac, *J. Chromatogr. A*, 2005, **1071**, 213.
27. S. Arscott, S. Le Gac and C. Rolando, *Proceedings of the Micro Total Analysis Conference*, 2005, vol. 2, p. 1504.
28. S. Le Gac, S. Arscott and C. Rolando, *Proceedings of the Annual American Society of Mass Spectrometry*, 2006.
29. S. Arscott, S. Le Gac, C. Druon and C. Rolando, Université des Sciences et Technologies De Lille, French patent FR 2862006 and International application WO2005/046881.

CHAPTER 6

Microfabricated Parylene Electrospray Tips Integrated with Cyclo-Olefin Microchips for ESI-MS

YANOU YANG,[a] JACK D. HENION[b] AND
H. G. CRAIGHEAD[c]

[a] Bristol Myers Squibb Research and Development, Pennington, NJ 08534, USA; [b] Advion Biosciences, Ithaca, NY 14850, USA; [c] Applied and Engineering Physics, Cornell University, Ithaca, NY 14853, USA

6.1 Introduction

Microchip electrospray ionization mass spectrometry has become an active research area in recent years, driven by the need for high-throughput analysis in areas such as proteomics, drug screening, pharmaceutical research and environmental analysis. For many modern mass spectrometric-based analyses, the bottleneck limiting sample throughput is the preparation and introduction of the samples into the mass spectrometer. Coupling microfluidic chips to mass spectrometry (MS) can be used to address this bottleneck. It can connect MS analysis to the power, speed and functionality of chips for sample preparation and separation. Microchip electrospray ionization MS could therefore offer rapid analysis, less sample consumption, high throughput and integration into micro total analysis systems (μTAS).

Miniaturization and Mass Spectrometry
Edited by Séverine Le Gac and Albert van den Berg

Published by the Royal Society of Chemistry, www.rsc.org

The coupling of microfluidic chips to electrospray ionization (ESI) works well because the flow rate generated by the microchip is similar to that required for ESI-MS. There are three general types of interface designs for obtaining mass spectral data from continuously flowing samples on a microchip: (1) spray directly from the exposed channel at the edge of the microchip;[1,2] (2) spray from a transfer capillary attached to the microchip;[3,4] (3) spray from microfabricated on-chip electrospray tips made of silicon,[5] polydimethylsiloxane (PDMS),[6] polyimide,[7] SU-8,[8] polycarbonate,[9,10] poly(methyl methacrylate) (PMMA)[10,11] and parylene.[12,13] This chapter focuses on the coupling of cyclo-olefin polymer-based microchips to MS using microfabricated parylene electrospray tips and its applications in bioanalysis.

6.2 Cyclo-Olefin Polymer-Based Microchips

Polymer-based microfabricated devices[14,15] are increasingly utilized for biological and biomedical applications because of low cost, simple manufacturing processes and the availability of a wide range of plastics from which to choose. Resistance of the polymer systems to organic solvents is important, because organic solvents are frequently used in solutions to reduce surface tension during the electrospray process. If the polymer is not stable in the solvent, the device can swell thus distorting the microchannels. In addition, materials dissolved in the ESI solution can cause background chemical noise. Solvent-resistant cyclo-olefin polymer substrates (trade name Zeonor) are used in the manufacture of compact discs and digital videodiscs, and have proven to be suitable for ESI-MS applications.[12–13,16–18] A summary of the physical properties and specifications for this polymer resin may be found on the website of Zeon Chemicals (http://www.zeonchemicals.com).

6.2.1 Fabrication of Polymer Microchips

The microfabrication of cyclo-olefin polymer devices in our laboratory consisted of micromachining of a silicon master, hot embossing and thermal bonding as indicated in Figure 6.1A. The silicon master was fabricated from 3-inch or 4-inch silicon wafers using photolithography and reactive ion etching. First, photoresist was spin coated on the silicon wafer, which was then exposed to ultraviolet (UV) light through a photomask. After development in a chemical developer solution, the silicon wafer with its photoresist was etched by deep reactive ion etching. After the etching process, the photoresist was removed by acetone followed by plasma resist striping. The fabricated silicon master could be used repeatedly for embossing.

In the hot embossing step, Zeonor 1020R plates (15 cm × 10 cm, 2 mm thick) were cleaned with methanol for 2 min in an ultrasonic bath and then cut into chips using a paper cutter. The chip was then pressed against the silicon master with heating (130 °C, 667 N, 10 min, for a 3.3 cm × 3.3 cm chip) on a hydraulic press (model MTP-8; Tetrahedron Associates, San Diego, CA, USA). The cyclo-olefin

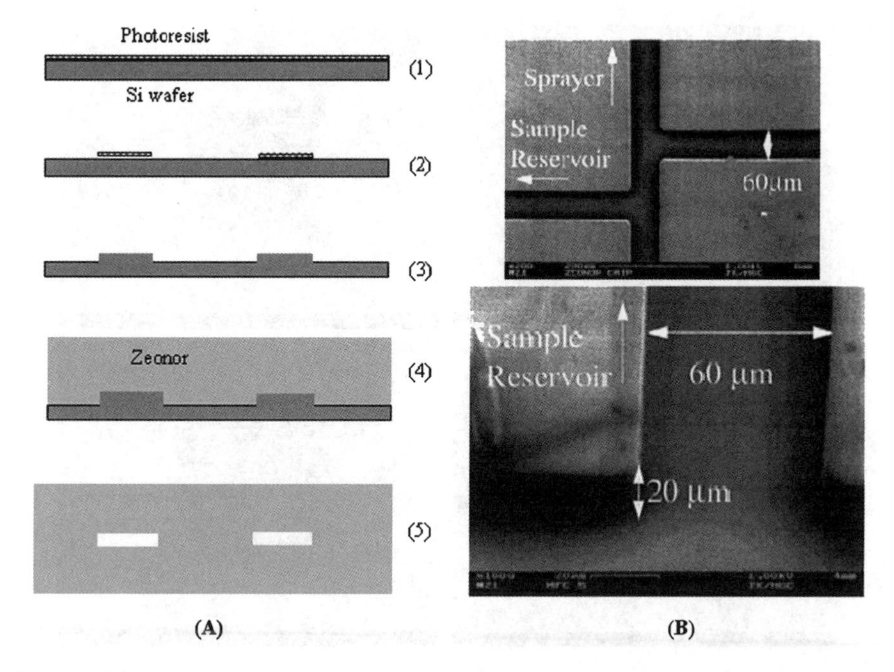

Figure 6.1 (A) Cyclo-olefin polymer device fabrication: (1) spin coating of photoresist; (2) photolithography; (3) deep reaction ion etching and resist striping; (4) hot embossing; (5) thermal bonding. (B) SEM image of hot embossed microchannel on a cyclo-olefin polymer substrate. (From ref. 16.)

chip with microchannels was finally bonded thermally to another polymer chip with heat and pressure (107 °C, 448 N, 10 min, for a 3.3 cm × 3.3 cm chip). The reservoirs for sample introduction were drilled into either the hot embossed chip or the cover chip. The temperature, force and time used in the hot embossing and thermal bonding steps should be experimentally determined for a specific chip design since the optimal conditions are dependent on the final chip size and the microchannel dimensions. Figure 6.1B shows scanning electron microscopy (SEM) images of the hot embossed cyclo-olefin microchannels. One design consideration is the aspect ratio limitation. To prevent the cover plate from collapsing and blocking the channel during thermal bonding, the width of the microchannel should be less than about three times the microchannel depth.

We found that an annealing step could help the cyclo-olefin microchip to withstand a higher flow pressure.[19] The comparison between treated and untreated devices is shown in Figure 6.2. Figures 6.2A and B show SEM images of the microchannel cross-section without annealing treatment at different magnifications. As can be seen in Figure 6.2A, the channel is deformed during the thermal bonding step and there is a separation line between the top and bottom plates (Figure 6.2B). After treating the device for 10 min at a temperature (115 °C) above the glass transition temperature of 105 °C with no pressure, the

Figure 6.2 SEM images of the microchannel cross-section with and without annealing treatment step. (A) Without treatment, ×100 magnification: 1, top cyclo-olefin substrate; 2, bottom cyclo-olefin substrate; 3, separation line. (B) Without treatment, ×500 magnification. (C) With treatment, ×100 magnification. (D) With treatment, ×500 magnification. (From ref. 19.)

top and bottom plate formed one continuous piece around the microchannel (Figure 6.2C). The edge of the microchannel was also smoother (Figure 6.2D).

Other approaches have been used for creating channels in cyclo-olefin polymers. Capillary tubes could also be used as masters for embossing channels in polymer. Tan *et al.* used eight conventional 360 μm fused silica capillaries epoxy-glued on a glass microscope slide as a master.[17] This is suitable for channels with relatively large dimensions and can be used for long channels. Mela *et al.* reported the fabrication of both Zeonor 1020R and 1060R microchips using hot embossing and injection molding with an electroplated nickel stamp.[20]

6.2.2 Surface Properties of the Microchip

The cyclopentadiene structure makes the polymer hydrophobic. The manufacturer's specified contact angle for the polyolefin we used was 94°. The observed contact angle for the untreated Zeonor 1020R in our experiment was $91 \pm 2°$.[21] We found that surface oxidation of cyclo-olefin surface using an oxygen plasma increased the surface hydrophilicity.[22] However, it is clear that commercially obtained source materials may vary and the particular treatment can change the surface charge condition. Various zeta potentials have been found in measurements of electro-osmotic flow.[20,22] We have also found that the oxidized cyclo-olefin surface showed signs of aging over a few days. We found UV-initiated grafting of polyacrylamide could be used to coat the walls of the cyclo-olefin microchannels to increase the hydrophilicity while also maintaining electro-osmotic mobility. This polyacrylamide-coated cyclo-olefin device has been successfully used to separate proteins by isoelectric focusing.[21]

6.3 Integration of Parylene Electrospray Tips with Cyclo-Olefin Polymer-Based Microchips for ESI-MS

The initial approach[16,23] for coupling cyclo-olefin polymer-based microchips to MS utilized a machined micro ion sprayer. However, the use of makeup flow at a rate 2–6 μL min^{-1} for this micro ion sprayer diluted the submicrolitre per minute analyte flow. The liquid junction also introduced a liquid dead volume, which can decrease separation efficiency. A microchip–MS interface which features zero dead volume, no clogging and easy mass production is highly desired for coupling cyclo-olefin polymer-based microchips to MS. Toward this goal, we have fabricated a triangular polymer electrospray tip integrated to a cyclo-olefin polymer microchip for MS coupling. The microfabricated tip localizes and stabilizes the Taylor cone of the ESI tip.

6.3.1 Electrospray Device Fabrication

Parylene C was chosen to fabricate the triangular electrosprayer tip because it was easy to deposit and could be patterned using photolithography and plasma etching. In addition, its resistance to most organic solvent makes it suitable for electrospray applications. Other materials of similar structure should have comparable properties.

The fabrication process for parylene electrospray tips is shown in Figure 6.3A. A film of parylene C with desired thickness was deposited onto a clean silicon wafer using a PDS 2010 deposition machine (Specialty Coating Systems Inc.). The deposited parylene film was patterned by optical lithography followed by oxygen plasma reactive ion etching. The polymeric films patterned on the silicon wafer could then be easily peeled from the silicon surface in an isopropyl alcohol solution. This parylene film containing the parylene electrospray tip was then integrated into the polymer microchip.

The integration of a single parylene electrospray tip with the cyclo-olefin polymer chip is illustrated in Figure 6.3B. First, a microfluidic channel (20 μm wide, 10 μm deep) was formed in one cyclo-olefin polymer substrate by the

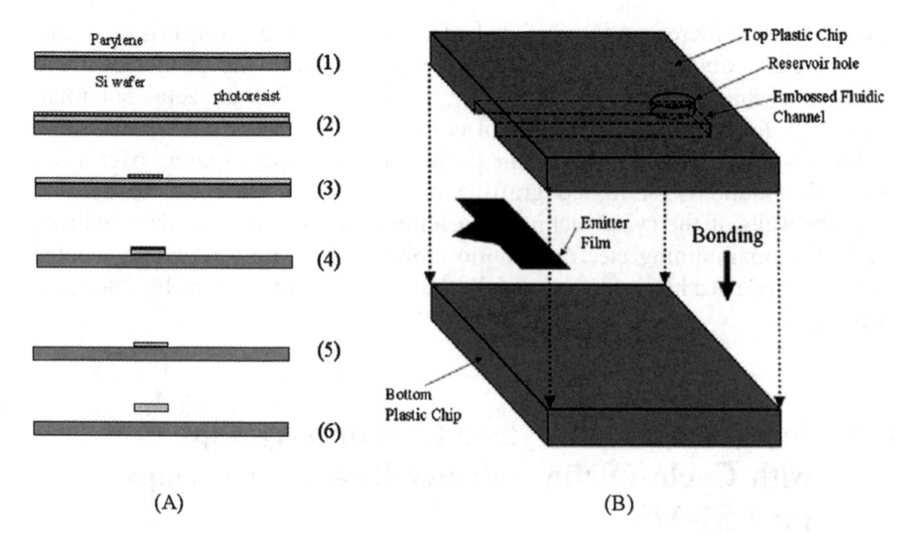

Figure 6.3 (A) Fabrication procedures for parylene electrospray emitter: (1) parylene deposition; (2) spin coating of photoresist; (3) photolithography; (4) reactive ion etching; (5) resist stripping; (6) peel off parylene tip from silicon wafer. (B) Fabrication process of a single triangular tip electrosprayer integrated in a microfluidic channel (from ref. 12).

previously described hot embossing method. The edge of the chip with the microchannel was cut so that the microchannel exit was exposed and the surface of the edge was smooth. A hole was drilled through the channel from the microchannel side with a hand drill (2 mm diameter) for incorporation of sample reservoirs. Then the parylene film was peeled from the silicon surface and aligned to the end of the microchannel while viewing through a microscope. The purpose of this alignment is to ensure the microchannel is at the center of the parylene tip and the width of the tip at the base of the triangle is approximately five times the microchannel width at the channel exit (see Figure 6.4) for optimal performance. The apex angle of the triangular tip was 90°. Another plain cyclo-olefin substrate was placed on top of the chip containing microchannels. Polyimide tape was used to temporally tape two cyclo-olefin substrates together to prevent the relative shifting between the two cyclo-olefin substrates during the thermal bonding step. Finally the three parts were bonded together using the previously described thermal bonding method.

The fabrication procedure for the parylene electrospray tip is relatively simple, but manual alignment of the parylene tip to the end of the microchannel is tedious and should be replaced by a structure with mechanical alignment features. Because the thermal parameters of parylene are different from those of the cyclo-olefin polymer substrate, good control of the thermal bonding conditions is required for proper sealing between two materials without collapsing of the microchannels.

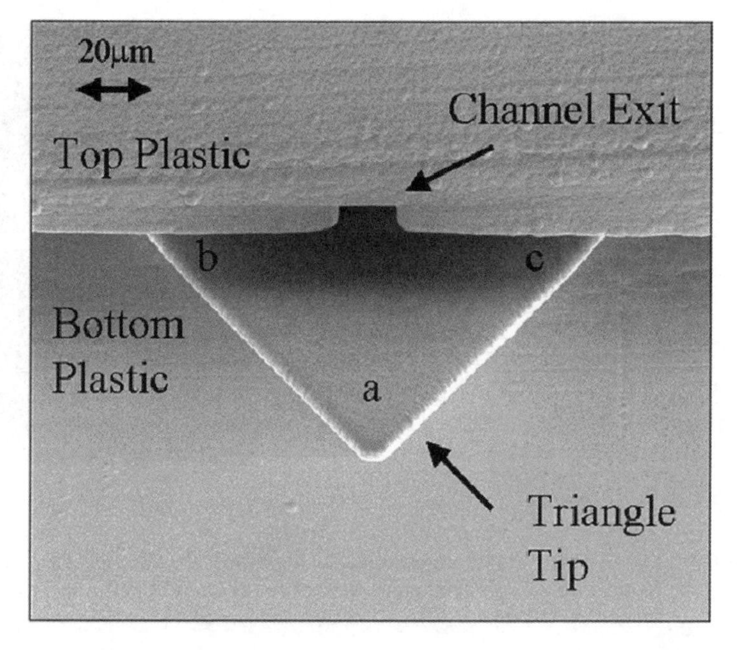

Figure 6.4 SEM image of the triangular tip taken at a 45° tilt angle. The scale bar is 20 μm. Angles: a = 90°, b = 45°, c = 45°. (From ref. 12.)

6.3.2 Taylor Cone Formation

Formation of a Taylor cone is the requirement for stable electrospray. Figure 6.5A shows the experimental configuration for characterization of the Taylor cone. The cyclo-olefin microchip with a parylene electrospray tip at the end of the microchannel was placed under a microscope objective with an aluminium counter-electrode placed 1.0 cm away from the polymeric tip. The aluminium counter-electrode was connected to a picoammeter for measuring the total ion current. A syringe pump was connected to the reservoir, creating a 300 nL min^{-1} flow of buffer solution (50% methanol, 50% deionized water with 0.1% formic acid). The electric potential was applied by inserting a gold wire to the sample reservoir. When the solution from the channel exit wets the two-dimensional parylene triangle tip, it forms a triangularly shaped droplet. The liquid expansion in lateral directions beyond the triangular tip can be controlled by balancing the flow rate and the electrospray potential. A Taylor cone was established at the apex of the triangle tip by applying a 2.0–2.8 kV potential, and electrospray ionization activity was confirmed by monitoring the total ion current. A stable total ion current of 30–40 nA was measured with a picoammeter. An optical image of the triangle prior to activation is shown in Figure 6.5B(a). The image of the Taylor cone and the liquid jet established on the triangular tip after pumping and the application of high voltage is shown in Figure 6.5B(b).

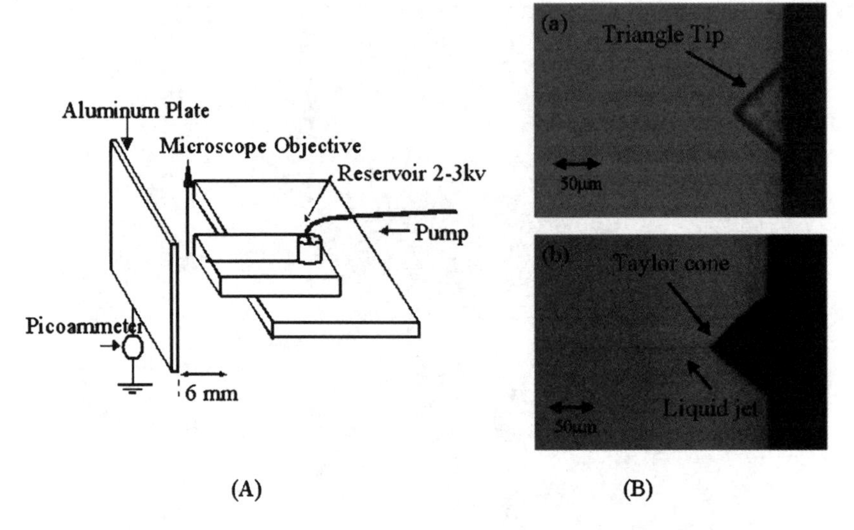

Figure 6.5 (A) Experimental configuration for observing Taylor cone formation. (B) Optical image of a triangular tip before (a) and after (b) the Taylor cone formation (from ref. 12).

Our laboratory[24] fabricated triangular electrospray tips with different apex angles (40°, 60°, 90° and 120°) and found that electrospray could be established for tips with all the angles, but only when the apex angle of the tip was equal to 90° would the electrospray solution be confined to the tip without spreading laterally. In addition, the voltage necessary to initiate the electrospray is the lowest when the apex angle of the triangle tip is equal to 90 °. The relationship between the flow rate and the Taylor cone size was characterized for the triangle tip with apex angle of 90°.[24] It was found that a higher flow rate generated a larger Taylor cone, and consequently a higher potential was required to establish the Taylor cone.

We also studied[12] the relationship between the thickness of the tip (apex angle = 90°) and the electrospray performance and found that the parylene thickness does not affect the spray performance. Spray performance from three tips was compared with different thicknesses (3, 5 and 7 μm) whose sizes were the same as the triangular tip in Figure 6.4. The monitored total ion currents for these three tips were 35.5 ± 1.2 nA for the 3 μm thick tip, 35.3 ± 1.5 nA for the 5 μm thick tip and 35.8 ± 1.4 nA for the 7 μm tip. Liquid droplets were confined to the tips in all cases. This indicates clearly that only the top surface of the triangular tip is involved in the ESI activity.

6.3.3 Interfacing the Electrospray Device to a Time-of-Flight Mass Spectrometer

Figure 6.6 shows the experimental configuration for interfacing a single electrospray device to a Mariner time-of-flight (TOF) mass spectrometer (PerSeptive

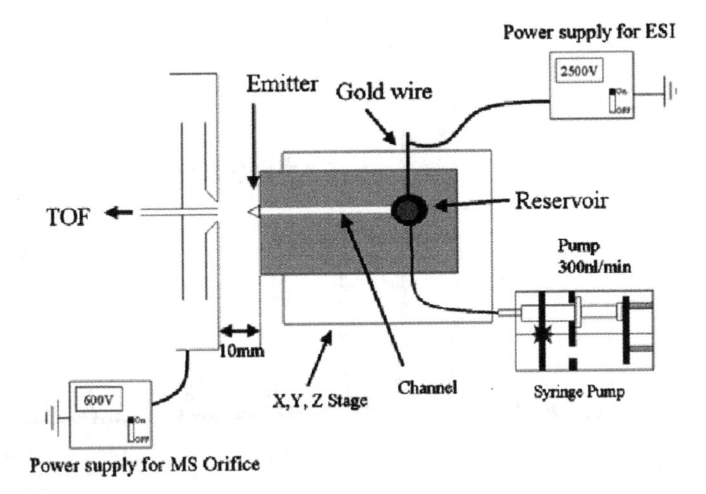

Figure 6.6 Experimental configuration for interfacing a cyclo-olefin polymer device to a TOF mass spectrometer. (From ref. 12.)

Biosystems, Inc., Framingham, MA, USA). The microchip was mounted on an *X*, *Y*, *Z* stage (Newport, Irvine, CA, USA). The position between the emitter and the orifice of the mass spectrometer can be adjusted to maximize the total ion current. The ESI triangular tip was placed about 8 to 12 mm away from the spectrometer orifice. Pipette tips were cut and glued to the reservoir hole. The desired liquid flow rate was generated using a Harvard Apparatus syringe pump (Natick, MA, USA). The pump was connected to the microchip through fused silica capillary tubing via the pipette tip. The flow rate mostly used was 300 nL min^{-1}. A gold wire glued to the reservoir hole under the pipette tip was used to apply the high voltage. A voltage of 2.5–3.0 kV was usually applied between the gold wire and the MS orifice. During mass spectral acquisition, a 600 V potential was applied to the MS orifice, and the heated interface of the mass spectrometer was maintained at 80 °C.

The spray stability performance from coupling a single electrospray device to the TOF mass spectrometer is shown in Figure 6.7. A solution of 5 μM chicken cytochrome *c* in 50% methanol, 50% water and 0.1% formic acid directly infused to the microchip was used to obtain the data. Figure 6.7A shows the mass spectrum for a scan range of the mass-to-charge ratio from *m*/*z* 600 to *m*/*z* 1300. The scan rate across the 700 Da mass range was 1 s. Figure 6.7B shows the total ion current for a 10 min continuous-infusion MS run with an acquisition speed of 1 spectrum per second. As can be seen, the spray current was stable, with a relative standard deviation (RSD) of 1.3% in the ion current.

6.3.4 Electrospray Tip Array

The ability of the parylene electrospray tip to confine the liquid enables the production of high-density ESI tip arrays without cross-channel contamination.

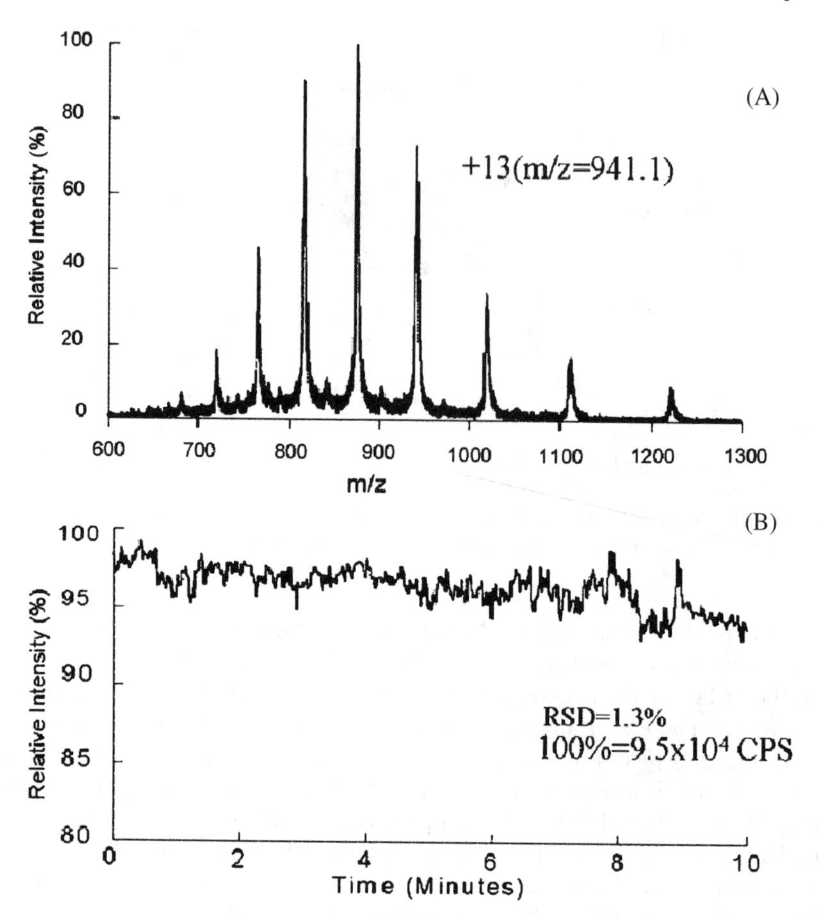

Figure 6.7 Electrospray performance for coupling cyclo-olefin microchip to TOF mass spectrometer. The data are from 5 μM chicken cytochrome *c* dissolved in 50% methanol, 50% water with 0.1% formic acid. The electrospray voltage used was 2.5 kV and the flow rate was 300 nL min^{-1}. (A) Mass spectrum of chicken cytochrome *c*. (B) Total ion current of chicken cytochrome *c* for 10 min. (From ref. 12.)

A four-channel tip array device edge is shown in Figure 6.8 where electrospray no. 2 was actively spraying solution while the other three channels were not in use. The distance between each triangular tip is 80 μm. The same experimental setup and conditions were used to observe the formation of the Taylor cone. Stable electrospray was achieved for all four channels with the application of 2.4–2.6 kV to the reservoirs using a 300 nL min^{-1} flow rate. A stable total ion current of 28–35 nA was measured independently for each triangular tip. There was no indication of liquid spreading at the edge of the device(s).

When interfacing this array device to the TOF mass spectrometer, all four channels were filled with the electrospray solution while only one electrospray

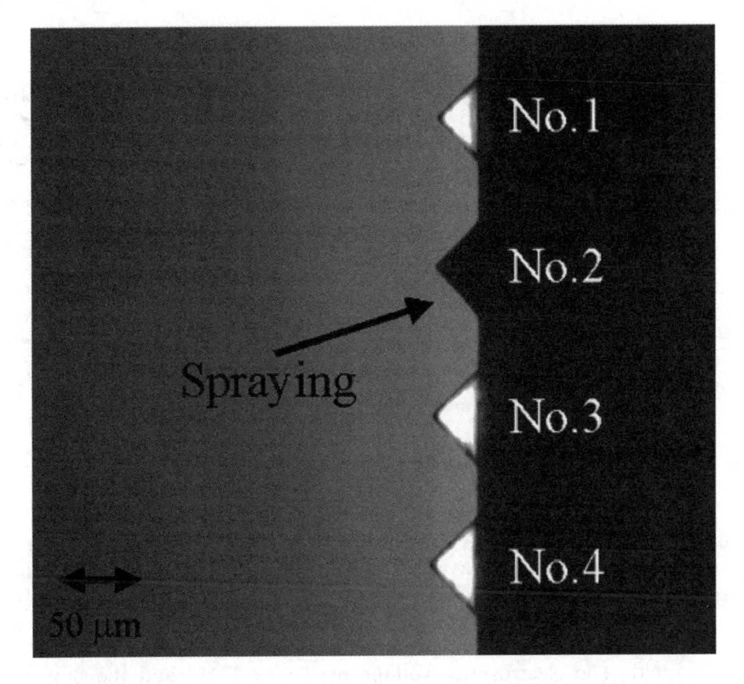

Figure 6.8 ESI array device with four triangular tips. No. 2 tip is spraying. (From ref. 12.)

tip was sprayed at one time. Solutions of two different compounds (1 μM desipramine and 1 μM imipramine dissolved in 50% water and 50% methanol with 0.1% formic acid) were sprayed from four tips to examine the cross-channel contamination. The MS acquisition speed was 1 spectrum per second, scanned from m/z 200 to m/z 400. The electrospray conditions for all four tips were at 2.8–3.2 kV with a flow rate of 200 nL min^{-1} for 20 s. The no. 1 electrospray tip and the no. 3 electrospray tip were used to spray imipramine while the no. 2 electrospray tip and the no. 4 electrospray tips were used to spray desipramine. The mass spectra from the four electrospray tips are shown in Figure 6.9.

6.4 Applications of Two-tip Electrospray Device in Quantitative MS Determination of Methylphenidate Concentration in Urine

6.4.1 Two-tip Electrospray Device

We further investigated the utilization and performance of parylene electrospray tips integrated with cyclo-olefin microchips for quantitative mass spectrometric determination of the small drug molecule methylphenidate (Ritalin).

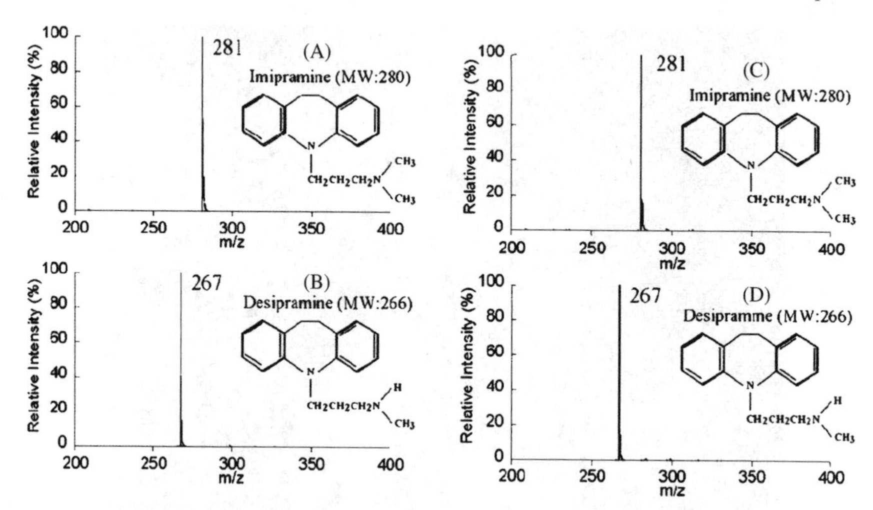

Figure 6.9 Mass spectrum of (A) imipramine sprayed from the no. 1 triangular tip; (B) desipramine sprayed from the no. 2 triangular tip; (C) imipramine sprayed from the no. 3 triangular tip; (D) desipramine sprayed from the no. 4 triangular tip. The data were from either 1 μM desipramine or 1 μM imipramine dissolved in 50% water and 50% methanol with 0.1% formic acid. The electrospray voltage used was 33 kV and the flow rate was 200 nL min^{-1}. (From ref. 12.)

A slightly modified version of the previously reported microchip was used in this study. One modification was to include a microfabricated on-chip gold electrode for applying the electrospray voltage. The drawback associated with inserting a metal wire into the sample reservoir is the possibility of flow perturbation. The fabrication procedure for the gold electrode is shown in Figure 6.10. A gold film with a thickness of 80 nm was deposited onto the cyclo-olefin substrate under 2×10^{-6} torr in a thermal evaporator. A 500 μm gold electrode was then patterned onto the cyclo-olefin substrate using photolithography and lift-off techniques. Another modification was that commercially available Nanoports (Upchurch Scientific, Oak Harbor, WA, USA) were attached in the sample reservoirs instead of pipette tips.

The microchip design is shown in Figure 6.11A. It contained two independent microfluidic channels and corresponding parylene electrospray tips. The two channels on the microchip are closer at the tip end with a 1 mm separation, while they are separated by 2 cm at the inlet end. The purpose of this design was to make alignment of parylene tips to the channel exit easier while the Nanoport connectors can still be accommodated at the input. The total length of the channel is 2.5 cm and the depth 20 to 25 μm. The width of the channel at the sprayer end is 20 μm while it is 60 μm at the reservoir end. The purpose for different microchannel widths was due to the need for different requirements from each end. Manual drilling was used to make the reservoir holes after

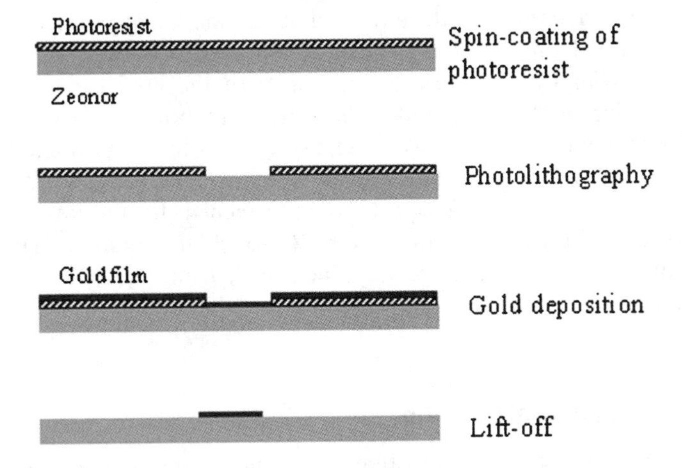

Figure 6.10 Fabrication procedures for gold electrode.

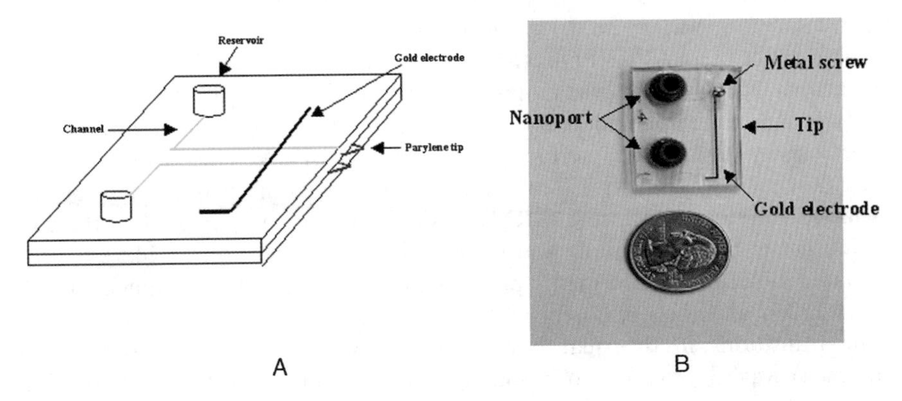

Figure 6.11 Two-tip electrospray device: (A) device design; (B) picture of the device.

embossing, and a wider channel was necessary at the inlet end. However, a smaller channel is desired at the tip end to provide a stable spray. The two parallel parylene tips with apex angle of 90° were fabricated together onto a single film with a distance between them of 1 mm.

The fabrication procedure for the two-tip device is basically the same as the single electrospray device. First, the two independent microchannels were hot embossed onto one cyclo-olefin polymer chip, and the two sample reservoir holes were drilled onto this cyclo-olefin chip at the end of the microchannels. Then, a gold electrode with a width of 500 µm was fabricated onto another cyclo-olefin chip. A hole for gold electrode access was drilled through the other chip surface which contained the microfluidic channels. The parylene films with a thickness of 3–4 μm containing the electrospray tips were fabricated onto a silicon wafer. Finally, the cyclo-olefin chip with the gold electrode and the cyclo-olefin chip with microfluidic channels were bound together with the parylene film containing the two electrospray emitters sandwiched between them. A copper screw was glued into the hole to make contact with the electrode with silver paint. Nanoports were attached to the reservoir holes with TorrSeal adhesive and cured at 60 °C for 12 h. Figure 6.11B shows a photograph of the fabricated device. The finished microchip dimensions are 2 cm × 3 cm × 4 mm.

6.4.2 Experimental Section

Methylphenidate (Ritalin) was fortified in human urine samples and extracted by liquid–liquid extraction. A trideuterated analogue of methylphenidate (methylphenidate-d_3) was used as an internal standard for the analysis. The detailed sample preparation procedure is described elsewhere.[13] The final samples containing both methylphenidate and methylphenidate-d_3 were in 75% methanol, 25% water and 0.1% formic acid.

The orifice of an API-III plus triple quadrupole mass spectrometer (PE Sciex, Concord, ON, Canada) was maintained at a potential of 53 V for methylphenidate and methylphenidate-d_3 to provide optimal ion sampling from the electrospray source. The interface plate voltage was 450 V. The nitrogen curtain gas flow rate was set at 120 mL min^{-1}. The dwell time was 200 ms for the selected reaction monitoring (SRM) mode. Nitrogen gas was used as the collision gas in the SRM mode and the collision energy was set to 24.5 eV. Protonated molecules for the methylphenidate (m/z 234.2) and methylphenidate-d_3 (m/z 237.2) and their major product ions (m/z 84.1) were monitored as separate transitions in the quantitative studies. Since a common product ion (m/z 84) was chosen for both analyte and internal standard, a dummy ion (m/z 250 > m/z 84.1) was included in the SRM sequence of transitions to preclude cross-talk in the collision cell.

The experimental configuration for interfacing the two-tip electrospray device to the API-III plus triple quadrupole mass spectrometer is shown in Figure 6.12. The components used in the setup were similar to those for interfacing the device to a TOF mass spectrometer. The microchip was positioned ~8 mm away from the orifice. The prepared samples with different analyte concentrations were pumped into the microchip at a flow rate of 0.1–0.2 μL min^{-1}, and a voltage of ~2 kV was applied between the MS orifice and gold microelectrode through the metal screw. Only one sample was analyzed at a time. The microchannel with the integrated electrospray tip was washed with clean spray solvent between different samples. SRM data on each sample were collected into different data files. The total selected ion current for each transition was averaged over a period of ~15–30 s and the ratio of analyte to internal standard was calculated.

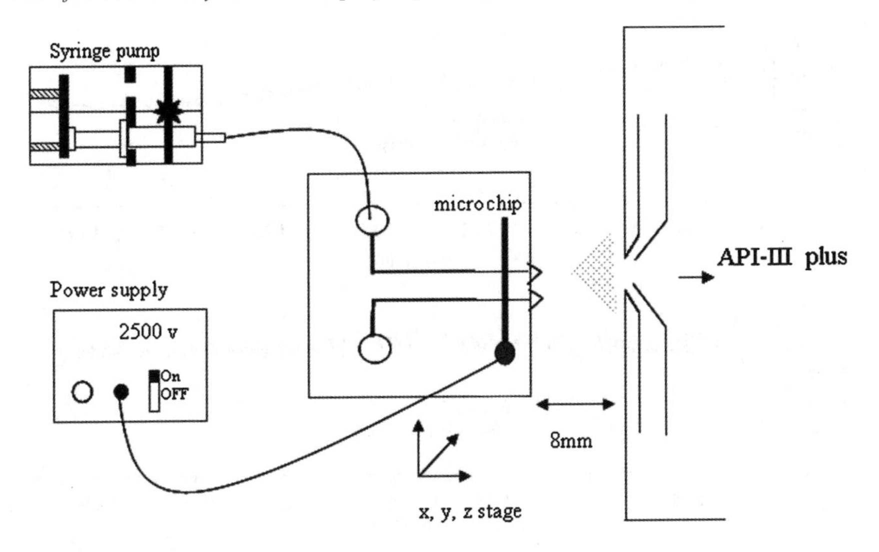

Figure 6.12 Experimental configuration for interfacing a two-tip electrospray device to an API-III plus triple quadrupole mass spectrometer (not drawn to scale).

6.4.3 Electrospray Stability

The electrospray stability of this device was studied with two different samples: 1 μg mL^{-1} imipramine-d_3 directly dissolved in 75% methanol, 25% water and 0.1% formic acid and extracted urine sample containing only methylphenidate-d_3, which corresponds to the blank calibration standards. Figure 6.13A shows the total ion current of full mass range scan (m/z 250 to 350) of the imipramine-d_3 sample over 15 min with a scan speed of 1.16 s per scan while Figure 6.13B shows the extracted ion current of the base peak (m/z 284.2) from Figure 6.13A. The RSD for the total ion current over the 15 min periods was 10.7% and the protonated molecule ion was even more stable with an RSD of 5.7%. Figure 6.13C shows the total selected ion current for the SRM data (m/z 234.2 > m/z 84.1, m/z 234.2 > m/z 84.1) from 150 ng mL^{-1} methylphenidate-d_3 sample. The SRM mode was also very stable with an RSD of 3.05% over a period of 5 min.

6.4.4 Inter-Chip Study

We performed an inter-chip study by comparing the SRM MS data of several devices with the same samples. The same extracted urine sample containing methylphenidate and methylphenidate-d_3 was analyzed by four different electrospray tips from three different microchips. Three measurements were made on each electrospray tip. Between each measurement, the chip was washed with methanol for 5 min at a flow rate of 5 μL min^{-1}. The results are summarized in

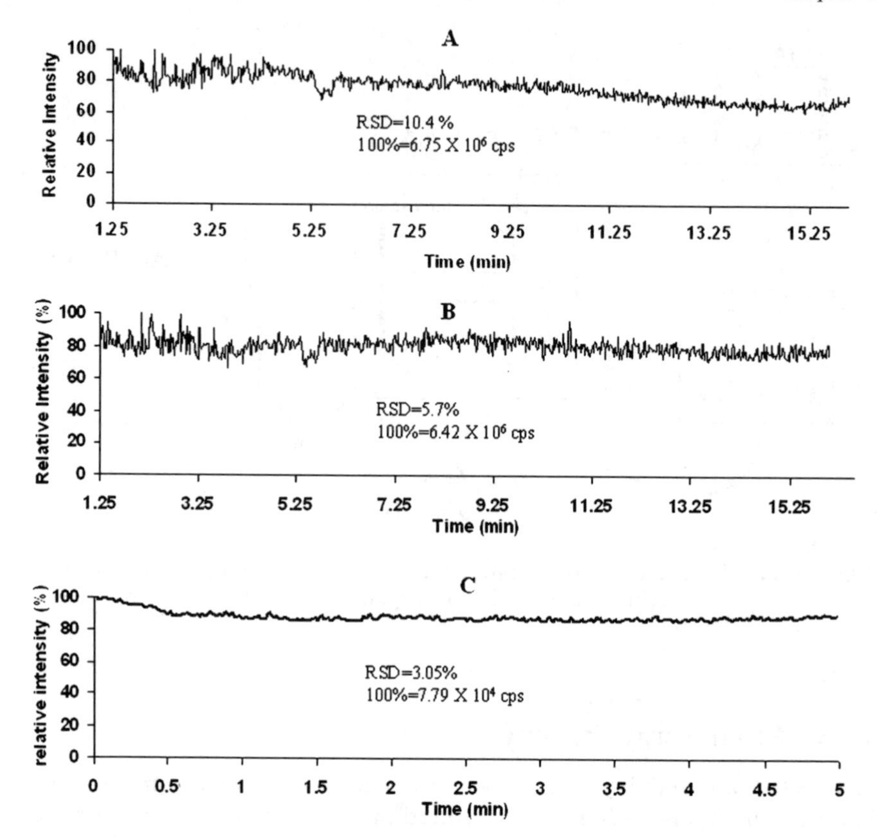

Figure 6.13 Electrospray stability of the device. (A) Total ion current of m/z from 250 to 350 with solution of $1\,\mu g\,mL^{-1}$ imipramine-d_3 in 75% methanol, 25%water and 0.1% formic acid. (B) Extracted ion current of imipramine-d_3 (m/z 284.2) from (A). (C) Selected reaction monitoring of extracted urine sample containing $150\,ng\,mL^{-1}$ methylphenidate-d_3 (from ref. 13).

Table 6.1. As can be seen, the variation between different electrospray tips was very small with an RSD = 1.4%. Different measurements for the same electrospray tip gave the same intensity ratio of the averaged SRM ion current of analyte and internal standards over a period of 15 to 30 s (RSD $\leq$ 0.3%). These results indicate that the quantitative data from one sample are not dependent on the microchip if an internal standard is used. This suggests the possibility for future development of disposable devices for quantitative bioanalysis.

6.4.5 Quantitation Results

A calibration curve was prepared by analyzing the prepared calibration standards using a two-tip electrospray device. The calibration standards were prepared by spiking methylphenidate at known concentrations to a negative

Table 6.1 Inter-sprayer study for methylphenidate fortified in urine sample (from Yang *et al.*[13]).

	Tip 1-1	*Tip 2-1*	*Tip 2-2*	*Tip 3-1*
I/I_0	0.470; 0.470; 0.470	0.468; 0.466; 0.463	0.460; 0.460; 0.460	0.457; 0.455; 0.454
I/I_0 average, $\bar{X}$	0.470	0.466	0.460	0.455
RSD (%)	0	0.3	0	0.3
Statistics of four electrosprayer tips		$\bar{X}=0.463$; RSD = 1.4%		

control human urine sample containing internal standards methylphenidate-d_3 followed by liquid–liquid extraction. Two duplicates of each calibration standard were prepared at seven different concentrations (0, 0.4, 1.6, 40, 200, 400 and 800 ng mL^{-1}). Two duplicates of three quality control (QC) samples (10, 300, 500 ng mL^{-1}) were also prepared from a different stock solution for determining the precision of the method. The prepared calibration standards and QC samples were infused to the microfluidic channel and electrosprayed sequentially into the mass spectrometer through the integrated parylene tips. The device was washed using a syringe pump after each sample analysis to mitigate cross-contamination. The intensity ratio of the averaged SRM ion current of analyte and internal standards over a period of 15 to 30 s for each calibration standard and QC was used to obtain the calibration curve which is shown in Figure 6.14. The resulting calibration curve is linear over the tested range of 0.4–800 ng mL^{-1} with a correlation coefficient $R^2=0.999$. The linear equation for methylphenidate is $y=0.0069x+0.0439$. The dynamic range of the quantitative analysis was 2000. The quantitative data from the two duplicates of the calibration standards with the lowest concentration (0.4 ng mL^{-1}) show good reproducibility, which suggested a subnanogram/millilitre lower limit of quantitation (LLOQ). If one considers the age of the API-III plus, a lower LLOQ may be expected with a more recent mass spectrometer presenting a higher sensitivity. A total of five electrospray tips were used to prepare the calibration curve. The fact that different electrospray tips can be used for preparing one calibration curve with good linearity confirmed the conclusion from the inter-chip study, which is that the MS data appear to be device-independent.

The precision and accuracy were determined for the quantitative analysis with another set of three QC samples with five to six duplicate analyses for each. The QC samples were prepared and analyzed in the same manner as the calibration standards. The concentrations of the QC samples were calculated using the standard curve described above. The accuracy and precision of this analysis method for methylphenidate are summarized in Table 6.2. The precision ranged from 19.1% to 3.2% and the accuracy ranged from 96.3% to 101.6% for three different concentrations, 20, 200 and 667 ng mL^{-1}.

The system carryover was tested by analyzing a blank sample with the same microchannel and tip before and after analyzing the highest concentration of

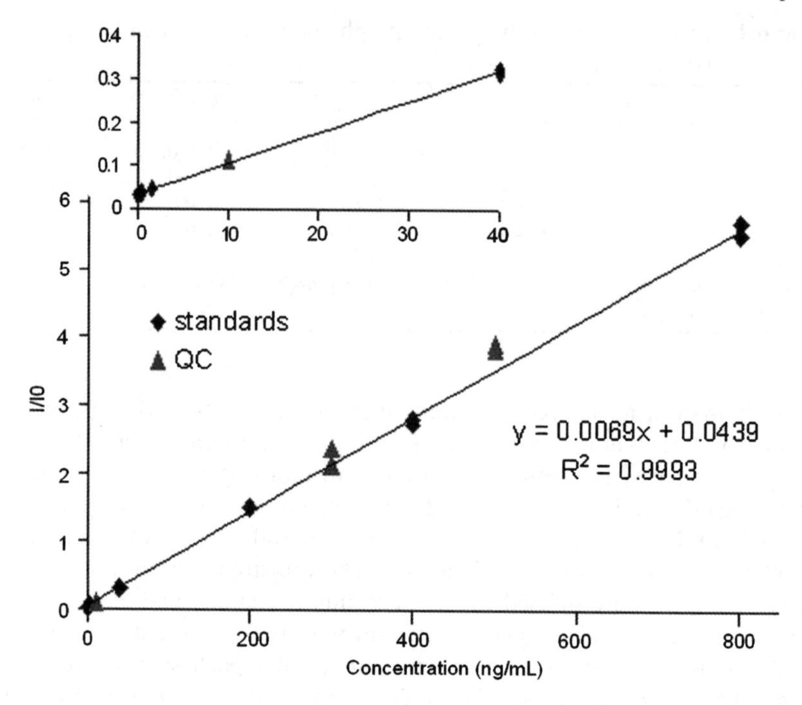

Figure 6.14 Calibration curve of methylphenidate obtained by analyzing calibration standards using a two-tip electrospray device by direct infusion.

Table 6.2 Precision and accuracy summary table for quantitative determination of methylphenidate in urine sample (from Yang *et al.*[13]).

QC conc. ($ng\,mL^{-1}$)	*Calc. conc. ($ng\,mL^{-1}$)*	*Statistics*
20	25.4	$\bar{X} = 19.3$
	18.3	RSD = 19.1%
	17.2	Accuracy = 96.3%
	15.9	
	19.5	
200	206.8	$\bar{X} = 202.3$
	208.5	RSD = 3.2%
	207.6	Accuracy = 101.2%
	201.7	
	193.7	
	195.6	
667	665.9	$\bar{X} = 677.5$
	699.3	RSD = 3.5%
	699.8	Accuracy = 101.6%
	685.6	
	651.6	
	663.0	

calibration standard. The microchannel with integrated electrospray tip was washed with methanol for 5 min at a flow rate of 5 μL min^{-1}. The initial SRM ion current ratio of methylphenidate and methylphenidate-d_3 for the blank sample was 0.0312. After running the highest level of calibration standard, the ratio was 0.0314. These results suggest negligible carryover for the described experiments.

6.5 Concluding Remarks

Polymeric microfluidic systems coupled to a microfabricated planar polymer tip can be used as a stable ion source for ESI-MS. A parylene tip at the end of the microchannel delivers fluid which easily produces a stable Taylor cone at the tip via an applied voltage. The described device appears to facilitate the formation of a stable spray current for the electrospray process and hence offers an attractive alternative to previously reported electrospray emitters. When this interface was employed for the quantification of methylphenidate in urine extracts via direct infusion MS analysis, this system demonstrated stable electrospray performance, good reproducibility, a wide linear dynamic range, a relatively low limit of quantification, good precision and accuracy, and negligible system carryover. We believe polymeric devices such as described in this report merit further investigation for chip-based sample analysis employing electrospray MS in the future.

References

1. Q. F. Xue, F. Foret, Y. M. Dunayevskiy, P. M. Zavracky, N. E. McGruer and B. L. Karger, *Anal. Chem.*, 1997, **69**, 426.
2. R. S. Ramsey and J. M. Ramsey, *Anal. Chem.*, 1997, **69**, 1174.
3. D. Figeys, Y. B. Ning and R. Aebersold, *Anal. Chem.*, 1997, **69**, 3153.
4. J. J. Li, P. Thibault, N. H. Bings, C. D. Skinner, C. Wang, C. Colyer and J. Harrison, *Anal. Chem.*, 1999, **71**, 3036.
5. G. A. Schultz, T. N. Corso, S. J. Prosser and S. Zhang, *Anal. Chem.*, 2000, **72**, 4058.
6. J.-S. Kim and D. R. Knapp, *J. Am. Soc. Mass. Spectrom.*, 2001, **12**, 463.
7. V. Gobry, J. van Oostrum, M. Martinelli, T. C. Rohner, F. Reymond, J. S. Rossier and H. H. Girault, *Proteomics*, 2002, **2**, 405.
8. S. Le Gac, S. Arscott and C. Rolando, *Electrophoresis*, 2003, **24**, 3640.
9. J. Wen, Y. H. Lin, F. Xiang, D. W. Matson, H. R. Udseth and R. D. Smith, *Electrophoresis*, 2000, **21**, 191.
10. M. Svedberg, A. Pettersson, S. Nilsson, J. Bergquist, L. Nyholm, F. Nikolajeff and K. Markides, *Anal. Chem.*, 2003, **75**, 3934.
11. C. H. Yuan and J. Shiea, *Anal. Chem.*, 2001, **73**, 1080.
12. J. Kameoka, R. Orth, B. Ilic, D. Czaplewski, T. Wachs and H. G. Craighead, *Anal. Chem.*, 2002, **74**, 5897.

13. Y. N. Yang, J. Kameoka, T. Wachs, J. D. Henion and H. G. Craighead, *Anal. Chem.*, 2004, **76**, 2568.
14. H. Becker and C. Gartner, *Electrophoresis*, 2000, **21**, 12.
15. H. Becker and L. E. Locascio, *Talanta*, 2002, **56**, 267.
16. J. Kameoka, H. G. Craighead, H. W. Zhang and J. D. Henion, *Anal. Chem.*, 2001, **73**, 1935.
17. A. M. Tan, S. Benetton and J. D. Henion, *Anal. Chem.*, 2003, **75**, 5504.
18. S. Benetton, J. Kameoka, A. M. Tan, T. Wachs, H. G. Craighead and J. D. Henion, *Anal. Chem.*, 2003, **75**, 6430.
19. Y. Yang, C. Li, J. Kameoka, K. H. Lee and H. G. Craighead, *Lab Chip*, 2005, **5**, 869.
20. P. Mela, A. van den Berg, Y. Fintschenko, E. B. Cummings, B. A. Simmons and B. J. Kirby, *Electrophoresis*, 2005, **26**, 1792.
21. C. Li, Y. Yang, H. G. Craighead and K. H. Lee, *Electrophoresis*, 2005, **26**, 1800.
22. J. Gaudioso and H. G. Craighead, *J. Chromatogr. A*, 2002, **971**, 249.
23. T. Wachs and J. D. Henion, *Anal. Chem.*, 2000, **73**, 632.
24. J. Kameoka, Polymeric microfluidic and electrospray devices, Thesis, Cornell University, 2002.

Proteomics Applications of Microfluidics to ESI-MS Coupling

CHAPTER 7

Microfluidic Bioanalytical Platforms with Mass Spectrometry Detection for Biomarker Discovery and Screening

IULIA M. LAZAR

Virginia Bioinformatics Institute and Department of Biological Sciences, Virginia Polytechnic Institute and State University, Washington St. VBI/Bio II, Room 283, Blacksburg, VA 24061, USA

7.1 Introduction

7.1.1 The Proteomics Era and Disease Biomarkers

A fundamental understanding of the basic processes that govern the very essence of life is not possible without a complete characterization of the proteomic complement of a specific genome. To provide, however, a comprehensive description of all protein components in a cell, present during the various stages of a cell lifecycle, is not an easy task. As a consequence, the past few years have witnessed an explosion of the proteomics research arena. The various facets of the problem can be approached using different strategies. Expression proteomics is focused on qualitative and quantitative mapping of cellular protein components to provide information related to the steady-state protein complement of a cell or tissue. Differential expression analysis is

Miniaturization and Mass Spectrometry
Edited by Séverine Le Gac and Albert van den Berg

Published by the Royal Society of Chemistry, www.rsc.org

concerned with monitoring the changes in the temporal distribution of expressed proteins in response to perturbation factors such as pathogen infection, onset of a disease, *etc*. Cell-map proteomics concentrates on determining the spatial distribution and the subcellular location of specific proteins and protein–protein complexes, to elucidate how a cell organizes its protein content. The study of posttranslational protein modifications is focused on determining the number, site and entity of such modifications. There are over 200 documented posttranslational modifications, with phosphorylation and glycosylation being some of the most relevant to essential cellular regulatory processes.

There are a number of challenges associated with the analysis of a proteomic cellular extract. The sample is most often available in a limited amount, *i.e.* a few thousands of cells to start with, or the final subfractions submitted to mass spectrometry (MS) investigations are as small as 1–10 μL at the nM/pM concentration level. The sample is extremely complex, *i.e.* thousands of protein components are present in the mix simultaneously. Prokaryotic proteomes express up to 3000 genes, while eukaryotic proteomes can express as many as 20 000–100 000 genes. Additional difficulties arise from the fact that the total number of genes expressed in a cell at a given time is not well defined (estimates are around 5000–10 000 for mammalian cells), and the dynamic range of the proteins present is very large ($1{:}10^{10}$). Moreover, the total number of expressed proteins/gene is increased by mRNA alternative splicing and posttranslational modifications, especially for higher eukaryotes.

Mass spectrometry has become the most widespread and powerful technique utilized in proteomic investigations, for a number of reasons:[1,2] (i) it generates results with a level of confidence that is comparable to conventional amino acid sequencing of electrophoretically separated proteins; (ii) it is more sensitive, faster and simpler than traditional sequencing protocols; (iii) it is more tolerant towards low molecular contaminants; and (iv) it enables the identification of proteins from complex mixtures. Overall, MS provides the sensitivity, specificity and resolving power necessary to detect low femtomole or even attomole quantities of proteins. The ability to characterize the expressed proteome is challenged though by lengthy and labor-intensive protocols which conclude with the potential generation of hundreds of sample subfractions which must be sequentially analyzed by MS. However, while one MS instrument will always perform spectral acquisition in a sequential format, the parallelization of the sample preparation steps would greatly reduce the overall analysis time. Simple, compact and low-cost devices with parallel processing capabilities would certainly benefit high-throughput investigations.

The recent improvements in MS detection sensitivity and speed of analysis have launched this technique into an effective strategy for the detection of novel disease biomarkers and protein co-expression patterns.[3] Biomarker components relevant to disease diagnosis, prognosis and staging can be found in tissues, blood, cerebrospinal fluid, saliva or urine. Nevertheless, the expression level of these biomarkers can be extremely low, in the pictogram to nanogram per millilitre range. Common biomarker discovery and screening strategies involve the use of immunohistochemical staining, DNA microarray technology, protein chips and multidimensional separations followed by MS detection. These techniques, however, are accompanied by numerous limitations that include: insufficient sensitivity and specificity of assays that rely on the detection of a single biomarker component, insufficient quantitative correlation between gene and protein expression levels, complexity and lengthiness of analytical approaches that attempt comprehensive qualitative/quantitative proteomic profiling, and lack of sufficient sensitivity, specificity and reproducibility of protein chips.[4,5] The development of combined microfluidic-MS strategies that will enable fast, sensitive and reliable MS-MS detection of protein co-expression patterns could potentially surpass these difficulties and greatly advance our capacity for early intervention in disease detection.

7.1.2 Microfluidic Devices and Mass Spectrometry Detection: Brief Overview

We are witnessing a continuous and natural evolution of biological systems, and in a similar manner, our tools, methods and approaches evolve as well. It took almost a century for liquid chromatography and mass spectrometry to reach the performance level that we are enjoying today. In comparison, microfluidics is a relatively new area of research that evolved into analytical instrumentation only during the past 15–20 years. Miniaturization is clearly the very next step that will follow in instrument development and will have to be assimilated in our techniques and practices.[6–8] Why will this have to happen? Because only miniaturization will enable the reliable handling of trace sample amounts and the implementation of innovative operational principles that are not feasible in the macroscale setting.

Microfluidic devices function according to well-established principles, and are characterized by several features that distinguish them from large-scale instrumentation: analytical process integration, high-speed analysis of small sample quantities, multiplexing, automation, and high-throughput processing. Here is what Leroy Hood, director of systems biology in Seattle, had to say in "A personal view of molecular technology and how it has changed biology":[9] "The keys for the future in high-throughput platforms are miniaturization, parallelization, automation, and integration. These requirements are pushing us towards microfluidics and microelectronics and, ultimately towards nanotechnology. The critical objectives are higher throughput, higher quality data, and lower cost per unit. Ultimately, we must move toward single-cell and even single-molecule analysis."

Microfluidic devices comprise a variety of functional elements that perform operations such as, pumping, valving, dispensing, sample clean up, mixing, separation, chemical alterations and detection.[7] These devices typically handle 1–5 μL of sample, and enable the analysis of volumes as low as 1 pL–1 nL. Sample injection, separation, labeling and detection are often performed in a time domain of a few minutes or seconds.[10] A unique advantage of miniaturization is that a variety of innovative configurations become possible only if they are developed in a microfabricated format. Multiplexing is of course an added benefit. For example, the centrifugal disk (CD) performs the analysis of 96 samples in parallel (loading, digestion, preconcentration and sample collection) by spinning a 6-inch diameter microfabricated disk.[11,12] High-speed capillary electrophoresis (CE) can be performed on a microchip device comprising a separation channel of only 200 μm in length and within an 0.8 ms timeframe.[10] High-throughput genotyping can be performed within 325 s by microfabricated devices that comprise 384 CE parallel processing lanes.[13] Moreover, the micro-domain environment supports the emergence of unique physical properties which can be exploited to enhance the functionality of microfluidic systems. As the size of a device decreases, the surface-to-volume ratio increases. As a result, surface-driven phenomena become particularly relevant in the microscale world.

Microfluidic devices have been interfaced to both electrospray ionization (ESI)-MS and matrix-assisted laser desorption ionization (MALDI)-MS.[14–31] Electrospray from microfabricated devices can be generated directly from the chip surface,[14,15] from inserted ESI capillary emitters[16–18] or from on-chip integrated microfabricated emitters.[19–23] Liquid sheath, liquid junction and nano-/microESI interfaces have been developed.[17–27] Microchip interfacing to MALDI-MS was investigated as well, but much less work has been performed in this area. Since MS-MS capabilities were available until recently only with ESI mass spectrometers, the development of ESI sources from the chip has prevailed. The technologies that were developed for microchip-MALDI-MS include microfabricated piezo-actuated flow-through dispensers,[28,29] centrifugal CDs[11,12] and direct MALDI from open CE[30] or pseudo-closed isoelectric focusing channels.[31] The applicability of these microfluidic devices with MS detection for the analysis of proteomic samples was evaluated, demonstrating similar performance to conventional instrumentation and capability to detect peptide components at the low-femtomole/high-attomole level.

The total amount of proteomic sample is usually limited, and biomarker components are often present at the ultra-trace level. In addition to sensitive detection, it is therefore critical to select analysis protocols that maintain sample losses at a minimal level. Microfabricated devices can substantially minimize problems related to sample loss and contamination. The miniaturized and integrated layout reduces the overall surface areas that come in contact with a sample, thus preventing sample losses due to adsorption on the instrumentation surface. At the same time, the multiplexed format enables not only high-throughput analysis, but also parallel processing of a set of samples without contamination from the previous ones. The advanced performance and versatility of microfluidic devices will enable the fabrication of disposable devices with broad applicability in the biological and biomedical arena, and promising capabilities for high-throughput biomarker discovery and screening. In return, the ability to provide a rapid and complete characterization of all cellular components will significantly advance our understanding of the fundamental processes that regulate cell function and lifecycle, and of the molecular networks and biological pathways that distinguish normal and diseased states.

7.2 Microfluidic Devices for Biomarker Discovery and Screening

7.2.1 Microdevice Fabrication, Sample Preparation and Data Acquisition

Microchip devices are commonly fabricated in glass, quartz, polymeric or silicon substrates.[6,8] For the present project, glass chips were utilized. Glass was chosen for several reasons: it has well-known surface chemistries, it is optically transparent and fluidic manipulations can be visualized and optimized with a fluorescent microscope, it is a very good electrical insulator and it has surface properties that enable easy generation of electroosmotic flows (EOFs). Polymeric chips can be cost-effectively fabricated; however, their surface properties are not optimal for the analysis of peptides and proteins which are easily adsorbed on such hydrophobic surfaces. Trace level components are thus completely lost. For our study, glass substrates were purchased from Nanofilm (Shelton, CA, USA) and the microchannel layout of the photomask was prepared with AutoCAD software. The fabrication of the microchips was performed according to standard procedures:[10] the substrate was exposed to ultraviolet light (360 nm) through a photomask for microfluidic pattern imprinting, the photoresist was developed and removed from the exposed area(s), the underlying chrome layer was etched to expose the bare glass, and the glass was etched to the desired depth using buffer oxide etch solution. The substrates were cleaned and bonded to a cover plate by gradual heating to 550 °C. Electrical contact to the microchips was provided through platinum electrodes and computer-controlled power supplies.

MCF7 breast cancer cells (American Type Culture Collection, Manassas, VA, USA) were cultured in an incubator (37 °C, 5% CO_2) and harvested upon confluence. Cells were solubilized in lysis buffer, centrifuged, and the soluble protein extract was processed according to a shotgun protocol.[2] The protein mix was digested with trypsin, fractionated using strong cation exchange chromatography (SCX), and selected fractions were analyzed by microfluidic liquid chromatography (LC)-MS. Mass spectra were acquired with an LTQ mass spectrometer (Thermo Electron, San Jose, CA, USA) and interpreted with the BioWorks/Sequest algorithm. Experimental parameters for data-dependent acquisition are described in detail in previous work.[32] Peptide fragmentation mass spectra were selected by using the X_{corr} *vs.* charge state filtering parameter with values set at $X_{corr} = 1.9$ for $z = 1$, $X_{corr} = 2.2$ for $z = 2$ and $X_{corr} = 3.8$ for $z \geq 3$.[2] Protein matches were performed against a human protein database that was downloaded from the NCBI website.

7.2.2 Microfluidic Functional Elements

In the context of proteomic applications, microfluidic devices must comprise all functional elements necessary to perform essential processing steps prior to MS detection, *i.e.* sample clean-up, preconcentration, affinity selection, digestion and separation. As a result, micropumps, valves, mixers, microreactors, separation elements, *etc.*, must be incorporated within a flexible, comprehensive and effective analysis platform. In addition, to enable MS detection from these chips, microchip-MS interfaces that facilitate efficient sample ionization, such as ESI and MALDI, must be integrated as well. Stand-alone configurations will be particularly desirable in the context of high-throughput applications.

7.2.2.1 Liquid Chromatography System[i]

Pressure-driven separations such as LC, that have demonstrated performance for the separation of peptide mixtures and that enable the loading of large sample amounts, have gained widespread popularity in proteomic applications. In addition, LC eluent compositions are fully compatible with electrospray MS ion sources, making this technique a clearly superior alternative to electrically driven separations which necessitate high-concentration buffer systems that are incompatible with ESI-MS. The proteomic/biomarker chips that are described in this work comprise fully integrated LC systems.[33]

Typical microchip LC devices (Figure 7.1) incorporate two multichannel EOF pumps, a multichannel EOF valve, a separation channel, an on-column preconcentrator and an ESI interface. The separation channel (**5**) is ~2 cm long with a depth of ~50 μm. A slurry of reversed phase packing material, Zorbax SB-C18/ $d_p = 5$ μm (Agilent Technologies), can be loaded manually in the channel from the LC waste reservoir (**11**) with the aid of a 250 μL syringe. The packing material is retained in the separation channel or a preconcentrator with the aid of short (~100 μm in length) multichannel structures, with similar dimensions to the pump (~1–2 μm deep). The two EOF pumps (**1A** and **1B**) each consist of

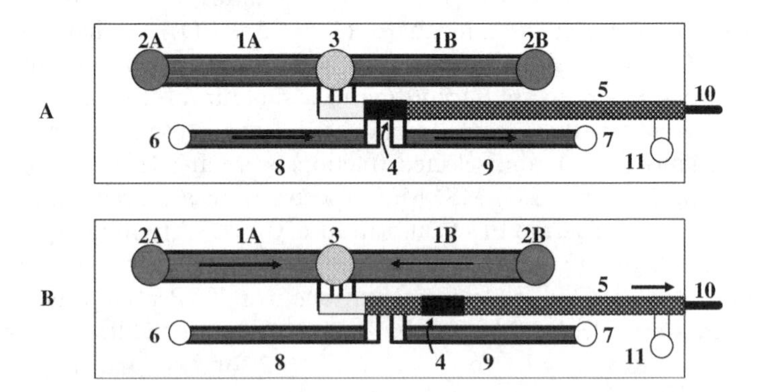

Figure 7.1 Schematic of a microfluidic LC system. (A) Sample loading; (B) sample analysis. 1A and 1B, pumping channels; 2A and 2B, eluent inlet reservoirs; 3, eluent outlet reservoir; 4, double-T injector that contains the sample plug; 5, separation channel; 6, sample reservoir; 7, sample waste reservoir; 8, sample inlet channels; 9, sample outlet channels; 10, ESI capillary emitter; 11, LC waste reservoir. Note: arrows indicate the main flow pattern through the system. (Reprinted with permission from ref. 33).

[i] Adapted with permission from *Anal. Chem.*, 2006, **78**(15), 5513–5524.

200–300 nanochannels (2 cm long, ~1–2 μm deep), and have two inlet reservoirs (**2A** and **2B**) and one common outlet reservoir (**3**). The voltage for EOF generation in the pumps is applied to reservoirs (**2**) and (**3**). The voltage applied to reservoir (**3**) represents also the voltage for electrospray generation. EOF leakage in the outlet reservoir (**3**) is prevented by a porous glass disk (5 mm diameter, 0.8–1 mm width, 40–50 Å pore size; Chand Associates, Worcester, MA, USA) secured to the bottom of the reservoir.[34] The role of this porous disk is to enable the exchange of ions and suppress the bulk eluent flow. Sample loading can be accomplished through a double-T injector (**4**) with the aid of a multichannel EOF valving structure consisting of ~100–200 nanochannels on each arm (2 cm long, ~1–2 μm deep). A fused silica capillary (10 mm long, 20 μm i.d. × 90 μm o.d.) from Polymicro Technologies (Phoenix, AZ, USA), which is inserted into the terminus of the LC channel, provides for stable ESI generation (**10**). Scanning electron microscopy (SEM) images of the LC microchannel packed with 5 μm particles and of the pumping/valving channels are shown in Figures 7.2A and B.[35]

An elegant alternative to particle packing involves the use of low pressure drop, monolithic stationary phases. The fabrication of these polymeric monoliths through a photoinitiated polymerization process facilitates microchannel patterning, and the fabrication of precisely positioned microchip components with distinct chemical functionality. An SEM image of a methacrylate-based monolithic separation medium that was prepared by photopolymerization is shown in Figure 7.2C.[36] These low hydraulic resistance monolithic packing materials have performed adequately on electrochromatography chips that were used for the separations of peptides. It is expected that their future implementation on LC chips will significantly lower the demand on the pumping power of the EOF pumps.

7.2.2.2 EOF Pumping/Valving System[ii]

Fluidic propulsion on the LC chips can be accomplished using an EOF pumping/valving approach.[33,34] Electrically driven flows are routinely utilized

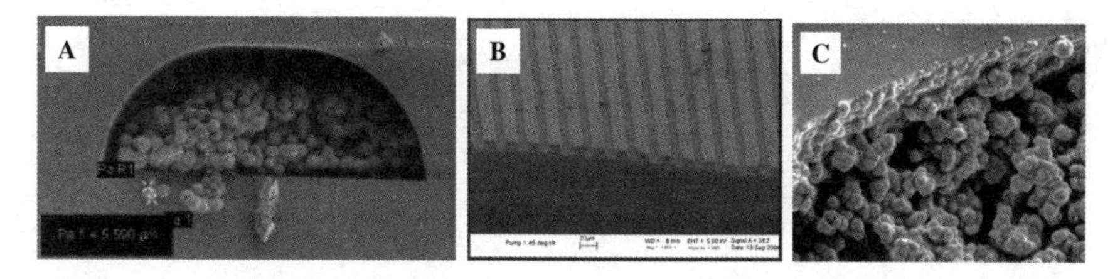

Figure 7.2 SEM images of microfluidic chips. (A) Cross-section through a microfluidic LC channel packed with 5 μm particles. (Reprinted with permission from ref. 35). (B) Pumping/valving channels (~2 μm deep, placed 25 μm apart). (Reprinted with permission from ref. 35). (C) Cross-section through a microfluidic LC channel filled with methacrylate-based monolithic packing material. (Reprinted with permission from ref. 36).

[ii] Adapted with permission from *Anal. Chem.*, 2006, **78**(15), 5513–5524.

for manipulating fluids on a chip, but this fluid transport mechanism is effective only in capillaries with dimensions in the micrometer domain. The principle of operation of the multichannel EOF pump/valve is described in the following. Flow in the EOF pump is generated in hundreds of fine microchannels through an electroosmotic mechanism and then directed to the microfluidic network through a pressure-driven mechanism. Flow in the EOF valve is also generated through an electroosmotic mechanism, but only during sample loading. Key to ensuring the functionality of these chips is the choice of the pumping/valving channel dimensions, which must render these channels with large hydraulic resistance such that pressure-driven flow leakage in the reverse direction is impeded. A schematic diagram shown in Figure 7.3 describes this pumping/valving concept.[33] Consider a channel (**C**) with diameter d_2 that is connected to a few reservoirs (**R**) through a series of shallow microchannels with diameter d_1 (Figure 7.3A). If a potential differential (ΔV) is applied between a reservoir (**R**) and the channel (**C**), EOF will be generated through the microchannels; if the hydraulic resistance of these microchannels is sufficiently high, eluent will be pumped from the reservoir into the channel even if the channel is pressurized, *e.g.* at 10 bar. The large hydraulic resistance of the pumping microchannels will impede flow leakage back into the reservoir. Typical configurations in our designs include microchannels that are ~1–2 μm deep and 5–20 mm long, and which are capable of delivering flow rates in the 10–400 nL min^{-1} range. A valving structure comprised of similar microchannels as the ones used for

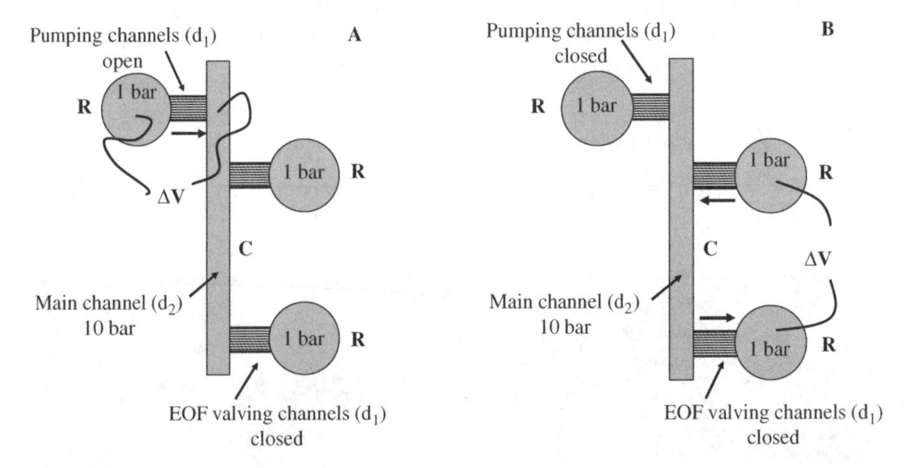

Figure 7.3 Schematic illustrating the pumping and valving capabilities of multiple open channel configurations. A large number (100–1000) of microchannels, with very small diameter (d_1) and large hydraulic resistance, are connecting a series of reservoirs to a large diameter (d_2) channel on the chip. The pressure in the reservoirs is 1 bar and in the main channel is 5–10 bar. As a result of the large hydraulic resistance of the pumping/valving channels, material transport can only occur through an electrically driven mechanism, but not through a pressure-driven mechanism. (Reprinted with permission from ref. 33).

pumping can be used for injecting and processing the sample in a pressurized environment. As the multiple open channel configuration has a much larger hydraulic resistance than any of the other functional elements on the chip, it can basically act as a valve that is open to material transport through an electrically driven mechanism, but is closed to material transport through a pressure-driven mechanism. The same multichannel structure can be used as an EOF pump for eluents, and as an EOF valve for sample introduction into a pressurized microfluidic system. If the depth of the microchannels is small enough, the hydraulic resistance is so large that one set of microchannels can be used for pumping, and several other sets for valving (Figure 7.3B). Sample will be introduced and removed from the main channel on the chip only when a potential differential is applied between adequate sample reservoirs, at appropriate moments during the analysis. Thus, a proper arrangement of pumping and valving channels will enable the implementation of a fully integrated LC system on a miniature device.

The sequence of operations necessary to operate the microfluidic LC chip is provided in the following.[33] Initially (Figure 7.1A), the microfluidic chip is filled with a low organic content eluent. The sample inlet reservoir (**6**) is filled with the sample. When a potential differential is applied between the sample inlet (**6**) and outlet/waste (**7**) reservoirs, the sample will be loaded through the EOF valve inlet microchannels (**8**), will be focused through the double-T injector (**4**) at the head of the separation channel, and the depleted sample eluent will be discarded through the EOF valve outlet microchannels (**9**). During the sample loading process, there is a very small voltage applied to the EOF pumps to eliminate sample diffusion in the direction of the pumps. Once the sample is loaded on the separation channel (**5**) (sample plug **4**), the voltage on the sample reservoirs is removed. Simultaneously, a potential differential is applied between reservoirs (**2**) and (**3**) in order to activate the pumps. Due to the fact that EOF is generated in the pumping channels, but backflow through all the pumping and valving microchannels is minimal, most of the flow is directed towards the LC channel to carry out the separation (Figure 7.1B). The voltage necessary for ESI generation is established through the voltage applied to the pump exit reservoirs (**3**).

7.2.2.3 LC Injector/Preconcentrator

The sample can be injected on an LC system using various microfluidic configurations (Figure 7.4). The inlet/outlet channels of a simple double-T injector (Figure 7.4A), of an on-column preconcentrator (Figure 7.4B) or of a stand-alone preconcentrator (Figure 7.4C) can be connected to a multichannel EOF valving structure. An on-column preconcentrator can be easily fabricated by simply enlarging the head of the separation channel to facilitate the loading of a larger amount of sample within a short length of packing. Stand-alone preconcentrators will facilitate the loading of a sample on a packing material with composition different from the separation packing. For example, a stand-alone

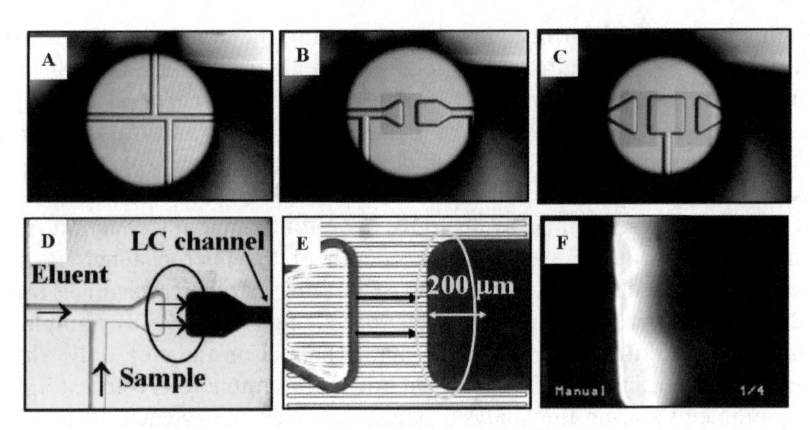

Figure 7.4 Microfluidic LC injector designs. (A) Double-T injector; (B) on-column (head) preconcentrator; (C) stand-alone preconcentrator; (D) microfluidic arrangement for sample loading onto a particle packed on-column preconcentrator; (E) enlarged view of a packed on-column preconcentrator with sample loading area highlighted; (F) Rhodamine fluorescent dye loading at the front of the preconcentrator. (Adapted with permission from Ref. 33).

preconcentrator could be useful for loading hydrophilic phosphorylated peptides on adequate materials such as immobilized metal affinity chromatography particles. Structures such as shown in Figure 7.4A will enable LC systems filled with immobilized monolithic materials, while structures such as shown in Figures 7.4B and C will enable LC systems packed with regular 3–5 μm particles. Sample loading on an LC channel with an on-column preconcentrator is shown in Figures 7.4D, and an enlarged view of this preconcentrator is shown in Figure 7.4E. The sample typically accumulates in the first 100–200 μm segment of the preconcentrator. The visualization of the sample loading process with Rhodamine dye is shown in Figure 7.4F. Sample retention/elution from such a preconcentrator can be easily accomplished with typical solvent systems used in LC applications: low in organic content during sample loading, and high in organic content during sample elution.

7.2.2.4 Mixer

The initial microfluidic LC designs incorporated high-permeability Poros 10 μm packing particles and used isocratic conditions for sample elution. The elution of some hydrophobic peptides from this packing material was, however, rather difficult, even with eluents with high methanol content. These particles were thus abandoned in favor of separation media with more appropriate properties for peptide separations. An example of a peptide mixture separation using Poros 10 μm particles is shown in Figure 7.5. A relatively quick increase in peak width with retention time is observed as a result of the inadequate properties of the packing material and of the lack of a solvent gradient.

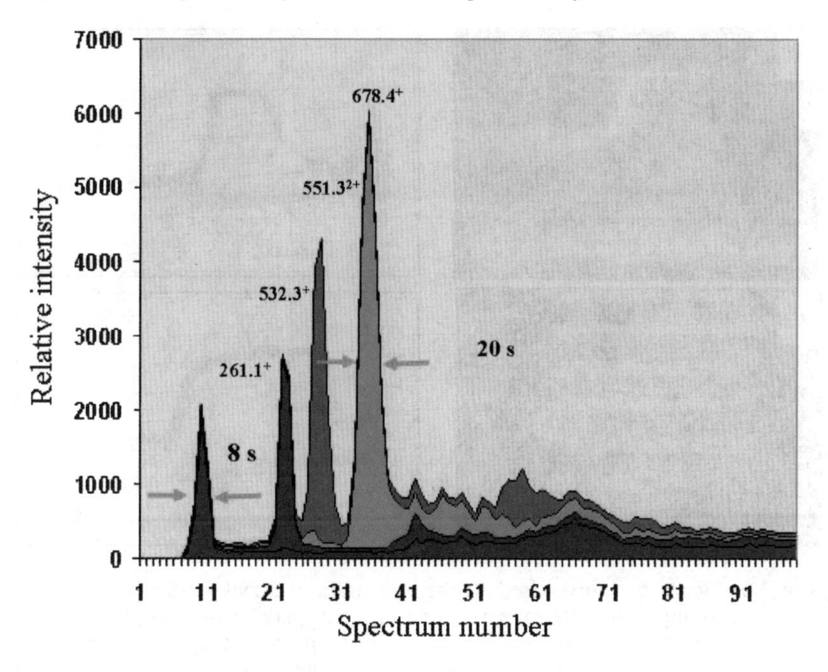

Figure 7.5 Extracted ion chromatogram that illustrates the isocratic separation of a peptide mixture on a Poros particle-filled LC chip: GGK (261.1^+), AAWGK (532.3^+), VDEVGGEALGR (551.32^+), YIPGTK (678.4^+).

Eluent gradient generation within such LC systems can be accomplished, however, with the aid of two independent pumping systems and an efficient mixer element. In micrometer-sized channels the flow profile is laminar. Even small molecules such as solvents, with a diffusion coefficient of approximately $10^{-9}\,m^2\,s^{-1}$, need some time before complete mixing is achieved. A serpentine mixer was developed for these chips. According to the dimensions of the microfluidic channels and the highest nanoLC flow rates ($\sim$150–200 nL min^{-1}), a serpentine channel of 5–8 mm in length will suffice to ensure proper mixing of eluents. The performance of such a mixer is shown in Figure 7.6. One of the eluents contains Rhodamine dye for fluorescent visualization of the mixing effect. Figure 7.6A represents the beginning of the mixer and displays two distinct flows generated by the EOF pumps. The fluorescent intensity profile is non-uniform along the channel cross-section (Figure 7.6B). After mixing (Figure 7.6C), the fluorescent profile is equalized (Figure 7.6D).

7.2.2.5 Mass Spectrometer Interface

The interfacing of the microfluidic chip to MS was accomplished with the aid of a $\sim$10 mm long fused silica capillary that was inserted in the chip and that acted as an ESI emitter. At the edge of the chip, both substrate and cover plates

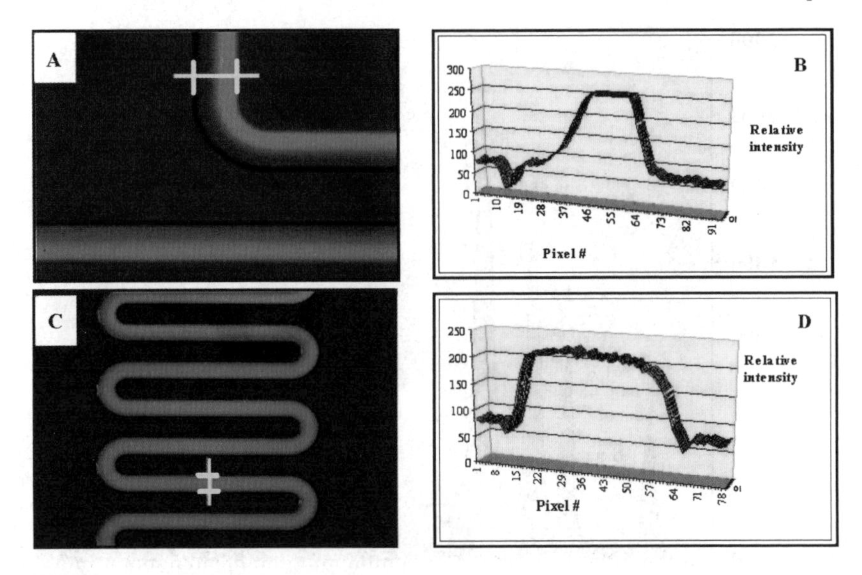

Figure 7.6 Serpentine mixer used for the generation of eluent gradients with LC EOF pumps. (A) Inlet of serpentine mixer displaying two distinct fluid flows; (B) non-uniform fluorescent intensity profile along the channel cross-section; (C) outlet of serpentine mixer displaying the fully mixed flows; (D) equalized fluorescent intensity profile along the channel cross-section.

were etched to a depth of ~50 μm to enable the smooth insertion of the emitter. The junction between the microfluidic channel and the ESI emitter is shown in Figure 7.7, demonstrating the low dead-volume connection that can be accomplished with such an arrangement. The microfabrication of electrospray nozzles that are an integral part of the chip would be clearly beneficial for multiplexed configuration devices. Their fabrication in glass being however problematic, a combination of a glass sample processing chip with a silicon integrated ESI emitter[22] could provide a convenient compromise.

7.2.3 Multiplexed Architectures

Today, routine MS analysis can be performed from sample quantities as low as 10–100 fmol, while state-of-the-art MS has been demonstrated at attomole levels, as well. However, the large number of samples, the time-consuming/labor-intensive experiments, and the very small sample amounts, all point towards an urgent need for developing parallel processing capabilities that would enable fast, sensitive, reliable and high-throughput investigations. Microfabrication enables the implementation of multiplexed microfluidic devices that can easily provide for parallel sample processing and high-throughput analysis. The basic microchip configuration that is shown in Figure 7.1 can be replicated in a parallel format on a single microfabricated structure. The spatial distribution of each functional element can be optimized to allow for maximum

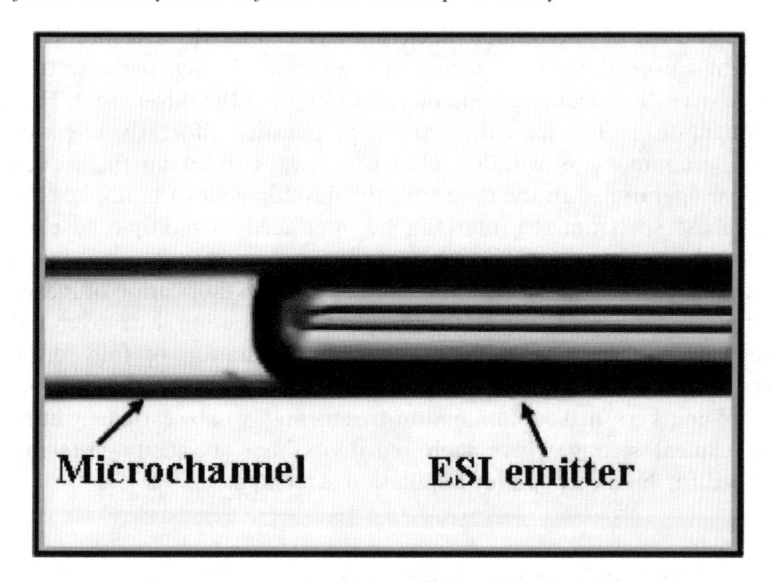

Figure 7.7 Image of a zero dead-volume microfluidic channel–capillary junction for electrospray generation. (Reprinted with permission from ref. 36).

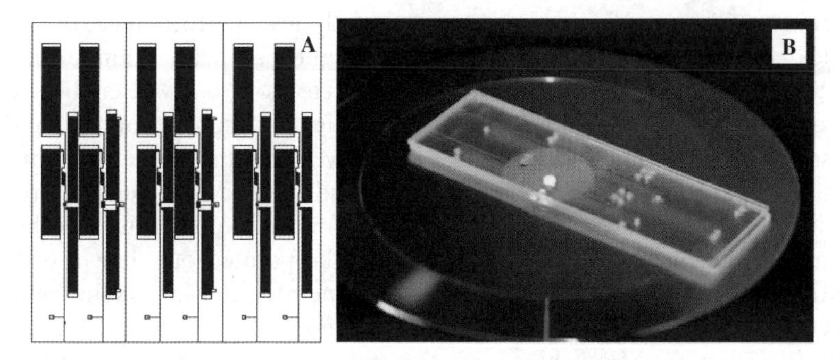

Figure 7.8 High-throughput configuration chips. (A) Drawing of a 3 in × 3 in glass substrate containing six fully integrated LC systems. (Reprinted with permission from ref. 33.) (B) Microfluidic chip (3 in × 1 in) comprising two fully integrated LC systems. (Reprinted with permission from ref. 35).

density structures, and each sample processing line can function independently to avoid cross-contamination or cross-talk between functional elements. The integration of up to 8–10 units on a 100 mm × 100 mm substrate is anticipated with the present microfluidic configurations. A schematic diagram of a multiplexed chip incorporating six microfluidic LC systems on a 75 mm × 75 mm substrate is shown in Figure 7.8A. A glass microfluidic device comprising two fully integrated LC systems is shown in Figure 7.8B. Each processing line comprises two multichannel pumps, a multichannel valve, an LC separation channel and a nanospray emitter.

In the context of microfluidic devices interfaced to MS detection it is often argued that even though the device enables multiplexing, the detection continues to occur in a sequential manner, thus limiting the throughput. However, as throughput involves the entire sample preparation effort, the integration of processing components within a chip will always speed up the sequence of analytical operations. In the long run, the development of multiplexed miniaturized mass spectrometers that can be interfaced to multiplexed chips will obviously result in a very effective analysis strategy. It is expected that parallelization and large-scale integration will result in the fabrication of inexpensive microfluidic platforms that prevent sample contamination, carryover and false positive identifications. Ultimately, the implementation of fully disposable devices is anticipated. Fast sample processing for the detection of multiple cancer biomarkers, in a contamination-free format, will be extremely important in the clinical setting where high sensitivity/high specificity tests must be developed for high-throughput population screening.

7.2.4 Application of Microfluidic Devices for Biomarker Detection

The overall value of the microfluidic-MS strategy for biomarker discovery and screening will arise from the superior analytical performance of microfabricated devices that enable rapid, sensitive and parallel handling of minute amounts of samples, and the power of MS detection that will enable unambiguous detection of trace level components. High-throughput sample handling is essential in proteomic applications. Prior to MS detection there are numerous sample preparation steps that typically conclude with an LC separation. One starting protein mixture can generate tens or hundreds of subfractions that are submitted to sequential LC analysis. It must be pointed out that in order to identify low copy number proteins and to generate relevant results with MS detection, a significant amount of starting material must be considered for analysis (3–5 mg). Microchips are not amenable for handling such large amounts, and the initial sample processing steps will always be performed with bench-top instrumentation. The processes that are amenable for integration on the chip are the ones that handle microlitre and microgram levels of sample. These volumes and amounts are typically produced after a series of preliminary operations have already generated several sample subfractions (*e.g.* preparative LC, isoelectric focusing, SCX). However, since a large number of subfractions are produced from the initial starting material, the utility of microfabricated structures will become especially important in these very final stages of the bioanalysis process.

Microfluidic devices that will enable global proteomic profiling will be particularly useful for biomarker discovery applications where the researcher is interested in comprehensively mapping all peptide/protein components in given biological fluids or tissues. Differential protein expression analysis will enable the discovery of specific protein markers or protein co-expression patterns that differentiate healthy *vs.* diseased states. In the context of biomarker screening applications, however, global proteomic profiling will not be necessary. Key to achieving success will be the capability to analyze relevant sample subfractions that are enriched in a sufficient number of targeted biomarker components.

To establish a successful approach for the qualitative assessment of the proteomic content of a cellular extract, we have performed numerous optimization studies with standard protein mixture digests and MCF7 breast cancer cellular extracts. Using conventional instrumentation, an analytical protocol that provided sensitivities in the high attomole/low femtomole range from pico-/nanomolar level solutions, and that resulted in the identification of ~1900 proteins with $p < 0.001$ in the MCF7 extract, has been implemented.[32] The p-value represents the probability of a random protein match. The experimental false positive rate for such data was <2.4%, when the MS dataset was searched against a reversed human protein database.[32] The MCF7 analysis protocol involved an MS experiment lasting 40 h and the analysis of ~42 μg of protein extract distributed across 16 SCX sample subfractions. From the total number of identified proteins, ~200 were known to be involved in cancer relevant processes, and over 25 proteins were previously described in the literature as potential cancer biomarkers, as they were found to be differentially expressed between normal and cancerous cell states.

To test the performance and applicability of the LC chip for similar biomarker applications, selected SCX sample subfractions were analyzed with the microfluidic-MS platform. A representative data-dependent LC-MS-MS chromatogram of one of the SCX fractions is shown in Figure 7.9. This SCX fraction was fairly rich in biomarker components, and represented an ideal candidate for the evaluation of the microfluidic chips. Peptide peak widths at half height were ~15–30 s, resulting in separation efficiencies of 45 000–180 000 per channel and peak capacities of ~80–100. The micropump that operated the LC system comprised a total of 400 pumping channels (20 mm long and ~1.5 μm deep) and delivered flow rates of approximately 60 nL min^{-1}. The valving system contained 100 microchannels with similar dimensions to the pump. The LC separation was performed using isocratic conditions and an eluent composed of NH_4HCO_3 (15 mM) in H_2O–CH_3OH (40:60) at pH ≈ 8. A high pH eluent was used to ensure high EOF in the pumping and valving system. While this eluent represents an unusual choice for LC separations, it was demonstrated earlier to enable efficient ESI in positive ion mode, using "wrong way round" electrospray conditions.[37,38] The use of 5 μm Zorbax C18 particles as a separation medium, instead of the 10 μm Poros particles, resulted in much better peak shapes and a relatively uniform distribution of peptides along the separation process.

An essential finding that emerged from proteomic experiments was that the number of identified proteins was strongly dependent on the amount of sample consumed for analysis. The LC chip enabled the injection of ~1 μL of sample and resulted in the identification of 77 proteins, of which 39 proteins had $p < 0.001$. A conventional LC system that used conditions similar to the chip (2 cm separation column, ~1 μL sample injection) provided comparable

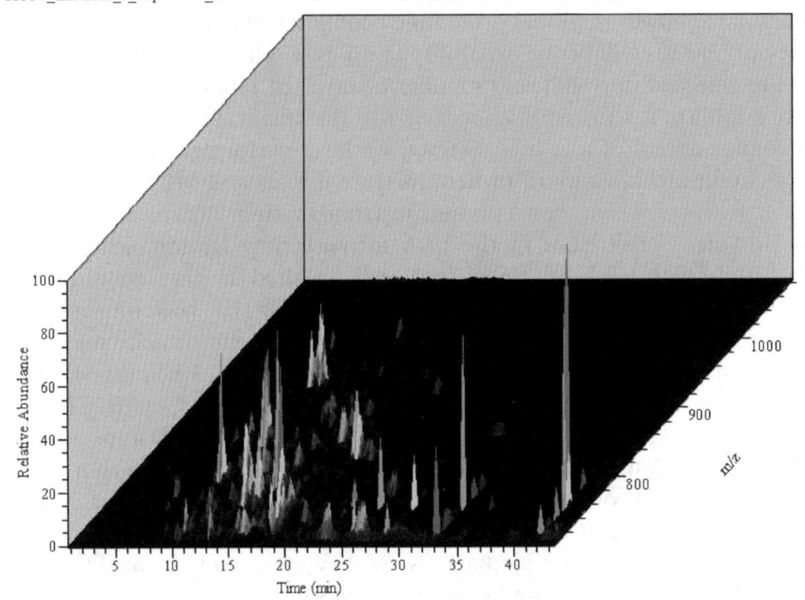

Figure 7.9 Data-dependent microfluidic LC-MS-MS analysis of a protein fraction prepared from the MCF7 breast cancer cell line (SCX fraction eluted with ~50–70 mM NaCl). Conditions: 2 cm long × 50 μm deep channel, Zorbax SB-C18/$d_p = 5$ μm packing material, 2 × 200 pumping channels (~1.5 μm deep), 2 × 100 valving channels, LC eluent NH_4HCO_3 (15 mM) in H_2O–CH_3OH (40:60) at pH ≈ 8, flow rate ~60–70 nL min^{-1}. (Reprinted with permission from ref. 33).

results, enabling the identification of 91 proteins, of which 48 proteins had $p < 0.001$. However, the analysis of 16 μL of sample with conventional instrumentation resulted in the identification of about 10 times more proteins.[32] In addition, most proteins identified from the chip were matched by only one peptide, while conventional LC analysis of large sample volumes resulted in many proteins to be matched by as many as 10–15 unique peptides. A detailed study that compared the microfluidic and bench-top LC systems revealed that the most critical factor that affected the number of identified proteins was the amount of sample consumed for analysis, and not other factors such as column length or eluent composition. It is thus critical to develop capabilities for the analysis of sample amounts with the microfluidic chips. Alternatively, the samples could be large enriched in biomarker components, and a variety of immuno-enrichment techniques are available for this purpose.

Even though the microfluidic-MS analysis was performed with low injection volumes, it enabled the identification of five putative biomarkers:[39,40] proliferating cell nuclear antigen (PCNA), cathepsin D and keratins K8, K18 and K19 (Table 7.1). All corresponding peptides that identified these proteins had $p < 0.001$, confirming the reliability of the match. PCNA is a protein involved

CHAPTER 8

Modular Microfluidics Devices Combining Multidimensional Separations: Applications to Targeted Proteomics Analyses of Complex Cellular Extracts

MIHAELA GHITUN,[a, b] ERIC BONNEIL,[b] CHRISTELLE POMIÈS,[b] MARIA MARCANTONIO,[b] HONGFENG YIN,[c] KEVIN KILLEEN[c] AND PIERRE THIBAULT[a, b]

[a] Department of Chemistry, Université de Montréal, PO Box 6128, Station Centre-ville, Montréal, Canada, H3C 3J7; [b] Institute for Research in Immunology and Cancer, Université de Montréal, PO Box 6128, Station Centre-ville, Montréal, Canada, H3C 3J7; [c] Agilent Technologies, 5301 Stevens Creek Blvd, Santa Clara, CA 95051, USA

8.1 Introduction

The pursuit of sensitive and reproducible protein expression and identification from minute amounts of cell extracts presents significant analytical challenges and new opportunities for the development of efficient separation techniques. The compelling advantages of microfluidics in terms of speed, reduced sample handling and reagent consumption together with the capabilities of conducting different sample processing steps on a device of small footprint offer a promising separation platform for proteomics analyses.[1] These remarkable features

Miniaturization and Mass Spectrometry
Edited by Séverine Le Gac and Albert van den Berg

Published by the Royal Society of Chemistry, www.rsc.org

combined with those available from sensitive mass spectrometry (MS) techniques have led to the development of innovative approaches for trace-level protein analyses; an excellent review describing the technologies and applications of microfluidic-MS was published recently.[2]

Early reports of microfluidic devices interfaced to nanoelectrospray MS were presented almost 10 years ago and described compact separation modules based on zone electrophoresis separation.[3–7] Since then, other separation formats have also been reported to increase the overall peak capacity including electrochromatography,[8,9] micellar electrokinetic chromatography (MEKC)[10,11] and hyphenated approaches combining MEKC[12] or solid-phase extraction[13] to zone electrophoresis separation. Enhancement of sample loading using sample stacking or adsorption preconcentration was later described to alleviate the limitations in concentration detection limits typically observed in electrophoretic-based separation methods and provided sub-nanomolar detection limits.[5,14,15] A microfluidic device using on-line micropipette tip sample desalting for subsequent serial introduction to the mass spectrometer was reported and led to numerous applications including quantitative analyses of drugs in plasma[16,17] and screening of noncovalent protein–protein and protein–ligand interactions.[18]

Although the on-line coupling of microfluidics to MS was largely documented for electrospray ionization, microanalytical platforms using matrix-assisted laser desorption ionization (MALDI) also provided meaningful advantages including the capability of conducting single or parallel sample separation for high-throughput analysis.[19–21] More recently, the combination of electrospray target preparation with off-line MALDI-MS was described and improved the homogeneity of sample morphology for reproducible proteomics analyses compared to mechanical sample deposing technique.[22]

Microfluidic devices combining nanoscale liquid chromatography (LC) separation formats were first introduced using wall derivatization[23] or by ultraviolet-initiated polymerization of monolithic materials into a microchannel.[24–27] These monolith-based microfluidic devices provided high mechanical strength and low flow resistance with reduced risk of clogging but suffered from lower sample loading capacity compared to traditional bead stationary phase. Subsequent embodiment incorporated bead adsorbant into chip channels fabricated out of glass,[28] silicon[29] or polymeric material.[30,31] Recent advances in the field of microfluidic-MS have enabled the incorporation of all necessary fluidic components including electrochemical pumps on a single chip.[32,33] In addition to the embedded nanoelectrospray emitter, these chips included multi-electrolysis-based electrochemical pumps, sample loading and gradient delivery together with a low-volume static mixer, a C_{18} column and integrated frits for bead capture. Examples of applications were demonstrated for the separation and identification of a few hundred femtomoles of protein digests[33] with chromatographic performances comparable to those of commercial nanoflow LC system.

Although the field of microfluidic-MS has bloomed over the past decade, it has not yet received its revolutionary impact. At the root of its overall acceptance lies the requirement of versatile modular components and subsystems, and their integration into complete functional and reliable devices that can be used by non-experts. It is from that perspective that we embarked on the development of an integrated microfluidic device tailored to address the analytical requirements sought for reproducible and sensitive proteomics analyses. Our quest for a functional and integrated microfluidic device for proteomics analyses brought us to examine compact nanoscale LC-MS chip platform (nanoLC-chip-MS) that integrated all the necessary elements for sample preconcentration, separation and ionization into a single chip.[34,35] These integrated microfluidic systems enabled sensitive and reproducible expression profiling of proteins at levels of 2–5 fmol in plasma samples[35] and provide a versatile sample introduction format for phosphoproteome analyses.[36]

In this chapter, we examine the analytical capabilities of the nanoLC-chip-MS system for complex proteome analyses of differentiated cell model systems. Enhancement of chromatographic capacity and sample loading was obtained by combining on-line two-dimensional chromatography separation using strong cation exchange (SCX) and reverse phase C_{18} media. Application of this multidimensional nanoLC-chip-MS system is demonstrated using both ion trap and time-of-flight (TOF) mass spectrometers for expression profiling and protein identification of differentiated U937 monocyte cells following the administration of phorbol ester. Early signaling events resulting from this chemical stimulation were examined following the identification of phosphopeptides from the corresponding protein digests using TiO_2 enrichment.

8.2 Materials and Methods

8.2.1 Materials

HPLC grade water and acetonitrile were purchased from Fisher Scientific (Whitby, ON, Canada). Formic acid (FA), ammonium acetate and ammonium bicarbonate were obtained from EM Science (Mississauga, ON, Canada). Monocytic U937 cells were purchased from ATCC (Manassas, VA, USA). Dimethyl sulfoxide (DMSO), Phorbol 12-myristate-13-acetate (PMA) and acetic acid were obtained from Sigma-Aldrich (Oakville, ON, Canada). Reduced and iodoacetamide alkylated protein digests (bovine serum albumin, rabbit aldolase, yeast alcohol dehydrogenase, bovine catalase, bovine glyceraldehyde 3P dehydrogenase, *E. coli* glycerokinase, human lactotransferrin, bovine lactoperoxidase) were purchased from Michrom Bioresources (Auburn, CA, USA).

8.2.2 Cell Cultures

U937 cells (human monocyte like, hystiocytic lymphoma cells) were grown in RPMI-1640 (Hyclone, Logan, UT, USA) and, supplemented with 10% fetal bovine serum, 1% Pen-Strep at 37 °C in a 5% CO_2 atmosphere. Cells were plated in 150 mm Petri dishes at a density of 0.5 million cells per mL in a total of 250 mL culture medium. Concentrated PMA in DMSO was added to U937 cell cultures to a final concentration of 150 nM and cells were incubated with PMA for 1 h.

8.2.3 Protein Extraction and Digestion

Cells were resuspended in 10 mM Tris, 2 mM $MgCl_2$, 3 mM $CaCl_2$, 300 mM sucrose. Cell lysis was performed using a metal douncer homogenizer. After centrifugation at 10 000*g* for 10 min, the proteins in the supernatant were precipitated with acetone. Proteins were resuspended in 100 mM ammonium bicarbonate reduced in 10 mM D,L-dithioerythiol (Sigma, USA) for 1 h at 56 °C and then alkylated in 55 mM iodoacetamide (Sigma, USA) for 1 h at room temperature. Total protein amount was quantitated by BCA protein assay (Pierce, Rockford, IL, USA). Proteins were digested in 1 M urea, 50 mM ammonium bicarbonate with Lys-C (Wako Chemicals, Richmond, VA, USA) for 4 h at 37 °C and then with Promega trypsin (Fisher Scientific, Whitby, ON, Canada) overnight at 37 °C. Samples were evaporated to dryness. For phosphopeptide enrichment of U937 cell extract, proteins were digested overnight at 37 °C with trypsin in 0.1% SDS, 50 mM ammonium bicarbonate. The digest mixture was acidified with trifluoroacetic acid (TFA; Sigma, USA) and then evaporated to dryness.

8.2.4 Phosphopeptide Isolation

Phosphopeptide enrichment used a protocol similar to that described previously.[37] Sample loading, washing and elution were performed by applying gas pressure from a nitrogen tank to the microcolumn. Each microcolumn was used once to avoid contamination. In-house TiO_2 (GL Sciences, Japan) microcolumns were equilibrated with 10 μL 0.1% TFA in water. The digest mixture was redissolved in 350 mg mL^{-1} 2,5-dihydroxybenzoic acid (DHB; Aldrich, USA), 3% TFA, 70% acetonitrile. An amount of 250 μg (50 μL) of sample was loaded onto the 1.25 mg TiO_2 microcolumn. The column was washed first with 10 μL of 350 mg mL^{-1} DHB, 3% TFA, 70% acetonitrile and then twice with 30 μL of 3% TFA, 70% acetonitrile. The bound phosphopeptides were eluted with 30 μL of 1% ammonium hydroxide in water (Sigma-Aldrich, USA). The eluate was acidified with 2 μL of TFA and 15 μL of the enriched sample was injected on the nanoLC-chip-MS system.

8.2.5 NanoLC-Chip and Two-Dimensional LC Separations

The microfluidic devices all comprised laser-ablated channels (enrichment/trap volume of 40 nL; LC separation channel of $43 \times 0.075 \times 0.050$ mm) packed with Zorbax C_{18} separation media and connected to a nanoelectrospray tip (Figure 8.1). All analyses were performed using a nanoLC1100 system coupled to a TOF or an ion trap XCT mass spectrometer (Agilent Technologies, Waldbronn, Germany). The SCX column (0.5 mm i.d. × 8 cm) was packed with PolyLC stationary phase (5 μm diameter, 300 Å pore size). The SCX column was connected directly to the switching valve and was on-line with the chip C_{18} precolumn during sample loading, and toggled off-line during reversed phase peptide separation on the analytical column. Peptides were sequentially eluted from the SCX column onto the C_{18} precolumn with 8 salt fractions of ammonium acetate (0, 50, 75, 100, 150, 300, 1000, 2000 mM, pH = 3.5). Each 10 μL fraction was loaded at 4 μL min^{-1} for 15 min. Peptides were eluted from the reverse phase column into the mass spectrometer using a gradient from 8% to 60% B over 61 min. The nanoelectrospray voltage was set to 2100 V. Tandem mass spectra were acquired with the Agilent MSD trap XCT mass spectrometer using helium as a collision gas. Multiply charged ions with an intensity above 40 000 counts were selected for MS-MS sequencing. The MS-MS fragmentation amplifier voltage was set to 1.3 V.

8.2.6 Peptide Detection and Clustering

Raw data files (.wiff) generated from the TOF acquisition software were read and processed using in-house peptide detection software to identify all ions according to their corresponding m/z values, retention time, peak widths, intensity and charge state. Intensity values above a user-defined intensity threshold above the background noise (typically 300 counts) were considered. A list of unique peptide ions (peptide cluster) found across relevant replicates of nanoLC-chip-MS analyses was obtained by clustering ions based on their respective charge, m/z and time within user-specified tolerances (typically ±0.05 m/z and ±0.5 min) following alignment of retention times using dynamic time warping. Segmentation analyses were also performed across sample sets from different cell growth conditions (*e.g.* control and challenge) by comparing the intensities of peptide clusters to identify those showing reproducible and statistically meaningful changes in abundance. For two-dimensional nanoLC-chip-MS separation, the intensities of unique and aligned peptide clusters present in contiguous fractions (typically ±1 fraction from most abundant) were summed to obtain the total contribution of each peptide cluster. Segmentation analyses across replicates and conditions were performed as described above.

8.2.7 Database Searching with Mass Spectrometry Data

Database searches were performed against a non-redundant NCBI database using Mascot (Matrix Science, London, UK) selecting human and/or rodent species. Parent ion and fragment ion mass tolerances were both set at ±0.6 Da.

8.3 Results and Discussion

The present microfluidic device differs slightly from similar systems previously reported by our group.[34–36] The chip device integrates a 40 nL enrichment column, a 4.3 cm analytical column (75 × 50 μm cross-section channel) and a 10 μm i.d. nanoelectrospray emitter directly on a polymer surface which is itself encased into an aluminium holder for convenient manipulation (Figure 8.1a). An on-line filter (0.5 μm pore size) is positioned between the valve stator and the nanopump to prevent occlusion of the chip device. The chip is mounted in a

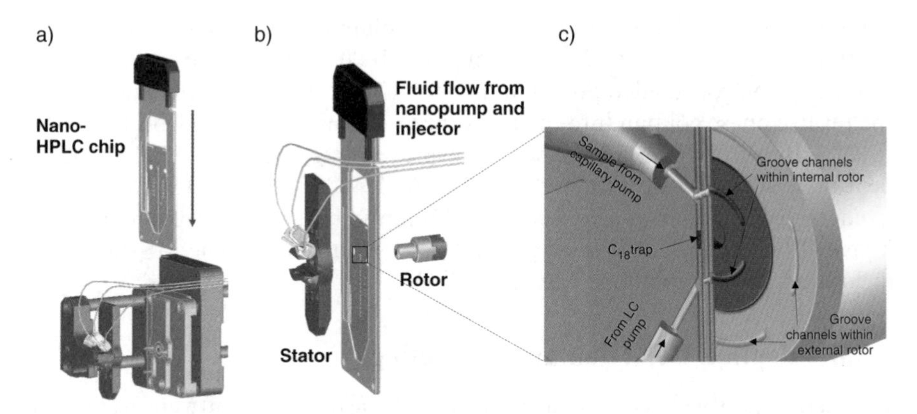

Figure 8.1 Schematic representation of the microfluidic chip cube interface comprising automated chip loading and sealing mechanisms. (a) Detail of the valve assembly below a polymer chip in its carrier (a metal frame). Alignment pins enable proper insertion and chip localization. The arrow indicates the direction of chip insertion. (b) Exploded view showing the chip positioned between the rotor and stator assembly. (c) Cut-out view of the chip device and fluidic connection on the switching valve.

clamping mechanism enabling automatic alignment and proper sealing of the chip within the manifold. The chip is secured in position between a stator–rotor with a dual concentric multiport rotor that provides 360° rotation for both rotors (Figure 8.1b and c). The inner and outer rotors can rotate simultaneously in the same direction or in opposite directions in 1° increments. Pressure sensors on the chip manifold also monitor proper sealing of the microfluidic device with the rotor–stator assembly. Section 8.3.1 describes the figures of merit typically achievable for this device for trace-level protein analyses.

8.3.1 Reproducibility of Chromatography and MS Performances of the LC-Chip-MS System

The analytical performances of this device in terms of reproducibility, sensitivity and linear dynamic range were evaluated in the context of proteomics research using an eight-protein digest of 250 fmol each. It is noteworthy that this sample also contained additional unexpected proteins, the most abundant being chaperonine 10, leucine aminopeptidase, bovine quinolinate phosphoribosyltransferase, bovine enoyl-CoA hydratase precursor, bovine argininosuccinate lyase and *E. coli* dehydrin. This sample was selected to mimic the sample complexity typically observed for in-gel digest or salt fractions from two-dimensional LC separation of protein digests. Chips were packed with 5.0 μm Zorbax particle in the precolumn and 5.0 μm Zorbax particle in the analytical columns and replicate injections ($n = 5$) were performed on the nanoLC-chip interfaced to a TOF mass spectrometer.

Reproducible retention times and peak profiles were obtained as indicated from the overlay ion chromatograms of the five replicate analyses (Figure 8.2a). Comprehensive peptide detection was achieved using an in-house software program to identify all eluting components and their corresponding m/z, elution time, charge state and intensity.[34,36] Segmentation analyses were obtained by grouping each peptide ion across replicate runs using boundaries of ± 0.1 m/z and ± 1.0 min for identical charge state ions. No time alignment was performed for these analyses. Consistent with previous investigations, the vast majority of peptide ions were reproducibly detected across all replicates and a total of 2230 peptide clusters were observed in all 5 replicates. A distribution of their intensities compared to the average values is presented in the scatter plot of Figure 8.2b. Ion intensities are closely nested along the median line with larger variations observed for ions of lower intensity. An average relative standard deviation (RSD) value of 10% was observed for ions less than 2000 counts and decreased progressively to 7% for ions above this intensity level. This distribution is best visualized in the inset of Figure 8.2b which shows a frequency plot of the fold-change variation in intensity for all reproducibly detected peptide ions where 95% of peptide clusters showed less than $\pm 20\%$ (± 1.2 in fold change) variation in intensity. Similarly, mass measurements performed on common peptide ions (data not shown) using external calibration were all within ± 30 ppm across all m/z values attesting of the good reproducibility of this system.

Peak profile analyses were also performed to determine the reproducibility of elution times and the variation of peak widths at 10 and 50% height. The distribution of RSD on retention times (Figure 8.2c) is well within 1% with a median value of 0.22% or ± 3 s. Peak width measurements (50% height) obtained for 360 common peptide ions higher than 10 000 counts showed a distribution extending from 0.07 min to 1.14 min with an average value of 0.29 min. The peak capacity calculated from the average peak width and elution window was 152 consistent with previous reports for similar devices.[36] The observed peak width is directly influenced by the sample loading and diffusion of eluting peptides. While the small dimensions of the present configuration minimizes sample diffusion in transfer lines before and after the analytical column, peptide elution volumes are largely dictated by sample mass loading and zone broadening in the trapping and separation columns. For injections below sample overloading, increased injection volume does not contribute significantly to peak broadening since the effective injection volume of the 40 nL trapping column rarely exceeds 15% of the average peak volume (171 nL). In contrast, sample mass loading was found to have more significant effect on peak broadening. An average peak width of 0.44 min with no significant shift in retention time could be maintained for sample loading up to 200 ng. Beyond this point, peak broadening was accompanied by a gradual decrease in retention and displacement of more hydrophilic peptides. Similarly, peptide ion intensity showed a linear response up to sample loading of 200–400 ng (Figure 8.2d). Although larger sample loading can be made on the present chip configuration, quantitative analyses relying on comparative ion

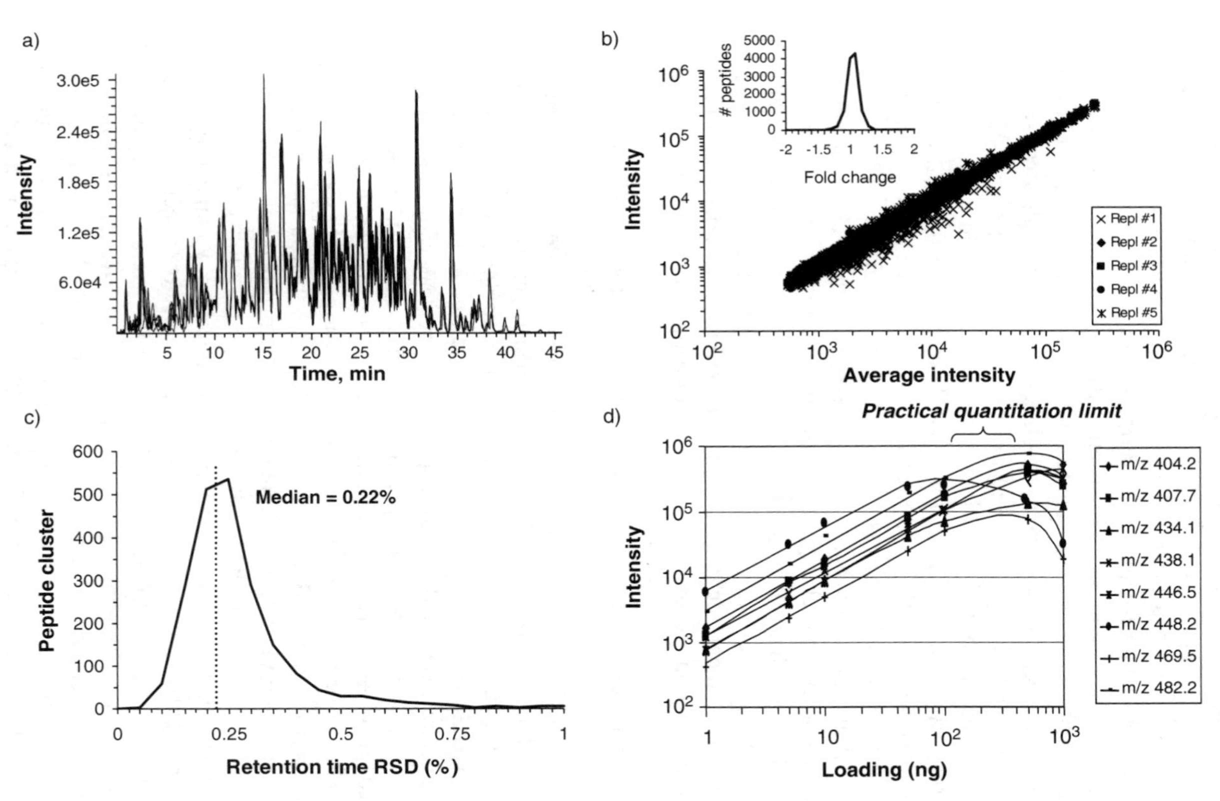

Figure 8.2 NanoLC-chip-MS of replicate injections ($n = 5$) of an eight-protein digest, 250 fmol each. (a) Overlay of 5 total ion chromatograms for the corresponding analyses. (b) Scatter plot of intensity measurements for 5 replicates and 2230 peptide ion clusters. (c) Distribution of RSD values on retention time measurements. (d) Variation of MS response for different tryptic peptides according to sample amount loaded. Conditions: enrichment/trap volume of 40 nL; LC separation channel of 43 × 0.075 × 0.050 mm both packed with Zorbax C_{18} separation media; a 5 μL injection of 80 ng tryptic digest of 8 proteins was performed except for (d) where variable amounts (1–1000 ng) of digest were injected.

abundance measurements are limited to approximately 200 ng, a maximum loading that was maintained in the present investigation. It is noteworthy that an extended dynamic range can be obtained using a trapping column of larger capacity, and 160 nL are now commercially available. In the context of comparative proteome analyses, enhancement of sample loading can be achieved using multidimensional LC separation to resolve sample complexity into simpler fractions of convenient sample size. Alternatively, specific enrichment of target analytes from complex cellular extracts (*e.g.* phosphopeptides, glycopeptides and specific antigens) can be accomplished using affinity purification. Examples of these applications are presented in Sections 8.3.2 and 8.3.3.

8.3.2 Comparative Peptide Detection and Identification using Two-Dimensional Chromatography (SCX/C_{18}) on an LC-Chip-MS System

The reproducible chromatographic and MS performances of the present chip device in terms of retention time, mass and intensity measurements provides reliable operating conditions for comparative proteome analyses. However, the cellular complexity typically observed in total cell lysate combined with the wide dynamic range of protein abundance and corresponding expression changes often dictate the use of separation approaches providing larger peak capacity and sample loading compared to reverse phase LC-MS separation. In this context, we evaluated the analytical capabilities of the on-line two-dimensional separation (SCX/C_{18}) with the nanoLC-chip-MS. Although SCX media could be packed into the microfluidic devices, the relatively large volume requirements for convenient separation of microgram-sized protein digests would exceed the practical bed volume and configuration available of present chip design. Rather, we incorporated an on-line SCX trap (8 cm × 0.5 i.d. mm) in the inlet port of the chip device. This SCX trap provides sample capacity exceeding 2 μg of protein digest.

Complex proteome analysis was evaluated for the total cell lysate of human monocyte U937 cells which provide a convenient cellular model system for differentiation studies.[38] The U937 cells were established from a diffuse histiocytic lymphoma displaying monocyte characteristics which upon chemical stimulation with agents such as PMA can initiate cellular differentiation into macrophage-like cells.

In a preliminary study, we compared the reproducibility and peptide distribution across eight salt fractions (0 to 2 M ammonium acetate, pH = 3.5) for a protein extract of control U937 cells. Figure 8.3a compares the distribution of peptide clusters for individual salt fractions to that observed for the injection of the same sample on a one-dimensional nanoLC-chip-MS system. A total of 2028 unique peptide clusters were obtained for the injection of 100 ng of protein digest for the C_{18}-only separation whereas 13 848 peptide clusters (a 6-fold increase) were detected for 1.5 μg injection on the combined SCX/C_{18} separation. In addition to the increased sample loading capacity of SCX, its complementary separation mode enabled a partition of peptides according to their pI and charge state as illustrated in Figure 8.3a. For example, doubly protonated peptides are predominantly observed in fractions 0 and 50 mM (74 and 78% of total peptides, respectively) and progressively decreased with increasing salt concentration (9% of total peptides for 2 M fraction) whereas the number of pentuply charged peptides is maximized at a salt concentration of 1 M (12% of total peptides).

The reproducibility of the SCX/C_{18} nanoLC-chip-MS system was evaluated for replicate injections of 1.5 µg ($n = 3$) made over a three-day period. Clustering analyses were performed for three replicates to identify unique peptides in all salt fractions. Close examination of these results indicated that totals of 8055, 4246 and 7389 unique peptide clusters were observed across all fractions in 1, 2 and 3 replicates, respectively. Evaluation of unique peptide distribution along contiguous salt fractions revealed a relatively low frequency of spreading. Peptide clusters were observed over a maximum of three salt fractions with an intensity-weighed median value of 1.2 fractions, consistent with previous observations with capillary LC columns.[39] The scatter plot of intensity distribution for unique cluster peptides found in all three replicate injections is shown in Figure 8.3b. As indicated, good reproducibility was obtained across all replicate injections with 95% of all detected clusters showing less than 40% variation in intensity (inset of Figure 8.3b).

The reproducibility of the present two-dimensional nanoLC-chip-MS approach thus provided a meaningful method to detect statistically relevant changes in expression profiles of peptide ions (protein surrogates) across different sample sets. In this context, we compared the changes in the proteome of human monoblastic cell line U937 following PMA treatment. U937 monoblast cells were split in two after 3 days of culture to yield control and PMA-exposed (150 nM PMA, 1 h) cell extracts. Following cytosol protein extraction, reduction/alkylation and BCA quantitation, extracts were digested with Lys C/trypsin and 1.5 µg of the corresponding digests were injected on the two-dimensional nanoLC-chip-MS system. The injections of control and PMA-treated protein digests ($n = 3$) were interleaved such that all eight salt fractions of a given control replicate were analyzed before applying the corresponding PMA replicate sample.

A comparison of all peptide cluster ions detected in at least two replicates for control and PMA-treated samples is shown in Figure 8.4. A total of 16 329 unique peptide clusters were found across all salt fractions and samples, of which 9756 peptide ions were observed in two or more sample replicates. The scatter plot of the corresponding peptide clusters (Figure 8.4a) shows the intensity distribution of cluster ions found across these two protein extracts. The average relative standard deviation for peptide clusters found in all three replicates of either control or PMA-treated samples ($n = 6394$ peptide clusters) varied from 12% at 300 counts and progressively decreased to 6% for ion intensity of 100 000 counts. For convenience, the dotted lines of Figure 8.4a delimit a 3-fold boundary representing a statistically meaningful change in abundance for the corresponding peptide clusters. The expression plot of ion

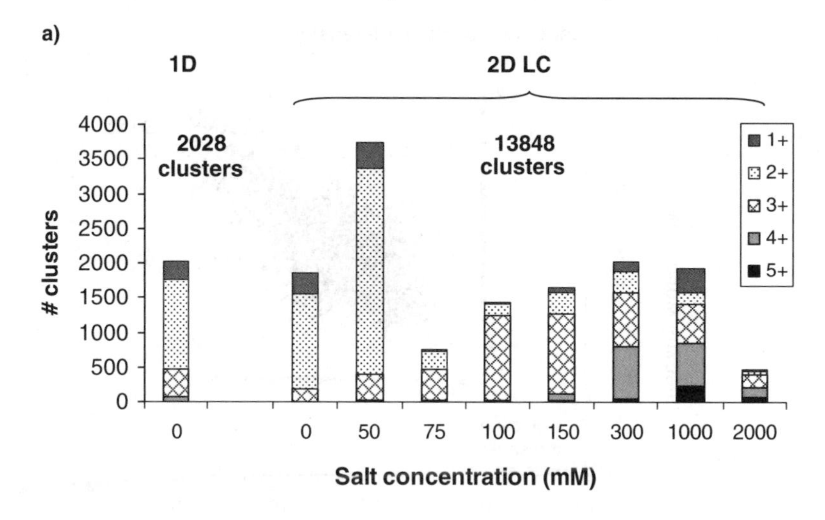

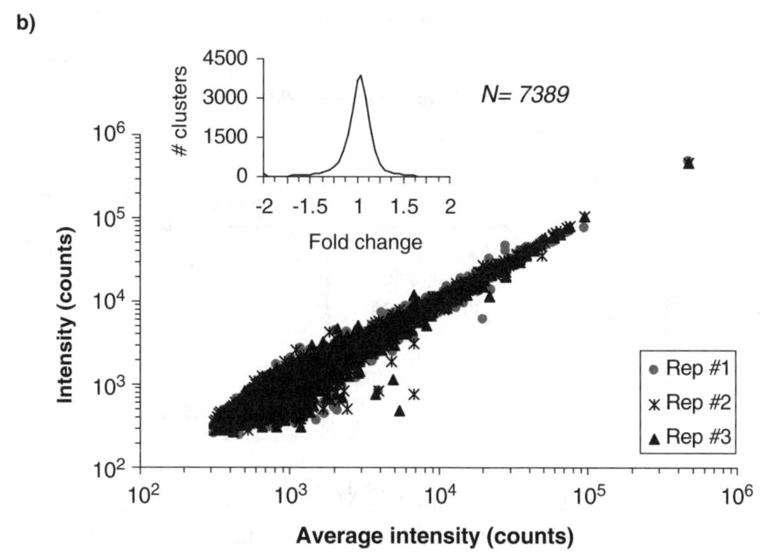

Figure 8.3 Comparison of peptide clusters detected using one- and two-dimensional LC separation using a nanoLC-chip-MS system. (a) Distribution of peptide ions according to charge and salt fractions. (b) Scatter plot of intensity measurements for 3 replicates performed on the combined SCX/C_{18} nanoLC-chip-MS system representing the intensity distribution of 7389 reproducibly detected peptide ion clusters. Conditions: same as for Figure 8.2 except that an SCX column (0.5 mm i.d. × 8 cm) was connected directly to the switching valve and was on-line with the chip C_{18} precolumn during sample loading, and toggled off-line during reversed-phase peptide separation on the analytical column. Peptides were sequentially eluted from the SCX column onto the C_{18} precolumn with 8 salt fractions of ammonium acetate (0, 50, 75, 100, 150, 300, 1000, 2000 mM, pH = 3.5). Injections of 100 ng and 1.5 µg of protein digest from control U937 cells were injected on the one- and two-dimensional LC systems, respectively.

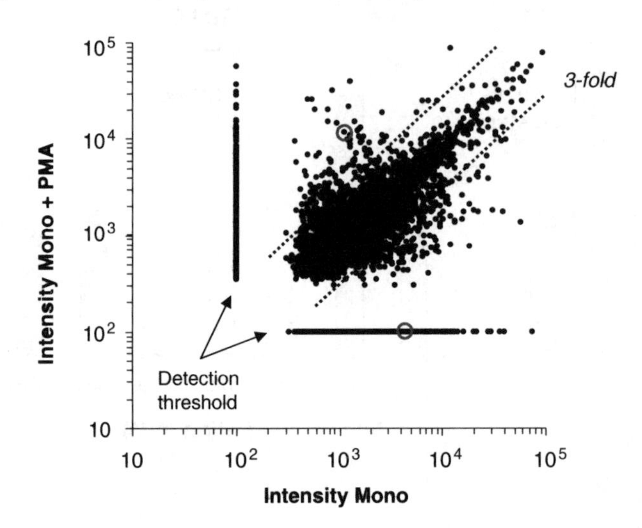

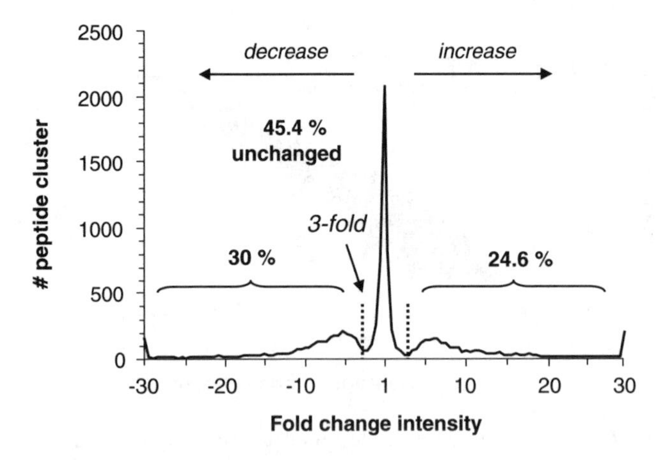

Figure 8.4 Distribution of peptide cluster intensities for control and PMA-treated U937 cells ($n = 3$). (a) Scatter plot of ion intensity for peptide clusters reproducibly detected in at least 2 replicates. The dotted lines show the 3-fold change in abundance. (b) Expression plot showing the number of down-regulated (30.0%), up-regulated (24.6%) and unaffected (45.6%) peptide clusters from a total of 9756 peptide clusters. Conditions: as for Figure 8.3 for SCX/C_{18} separation. Replicate injections of 1.5 μg of protein digest from control and PMA-treated U937 cells were injected on two-dimensional nanoLC-chip-MS systems.

intensity (Figure 8.4b) indicates that 45.4% of observed peptide clusters showed intensity variation within 3-fold across all three replicates suggesting minimal protein expression change upon PMA exposure. In contrast, 30.0 and 24.6% of reproducibly detected peptide clusters showed more than 3-fold decrease and increase in abundance upon PMA treatment, respectively.

In order to determine the identification of peptide clusters, the chip interface was transferred on an Agilent ion trap XCT mass spectrometer and the same samples were re-injected ($n=3$). Results from peptide identification analyses were correlated with the corresponding cluster ions by aligning the retention times and peptide masses found across the different salt fractions using an in-house clustering program. Comprehensive database searches revealed that 196 unique protein entries were identified with average sequence coverage of 22%. Table 8.1 shows examples of proteins with at least two identified peptides for each. For convenience, the sequence coverage, the peptide identity and the average fold change of each protein is also presented in Table 8.1. As indicated, proteins such as HSP84, HSP71, GAPDH and tubulin which were identified with at least five different peptides did not show significant change in expression levels (<3 average fold change) upon PMA treatment consistent with earlier reports.[40–43] In contrast, proteins such as globular-actin and T-complex protein-1 (TCP-1) were down-regulated. TCP-1 proteins are molecular chaperone assisting the folding of newly synthesized cytoskeletal proteins such as α-, β- and γ-tubulins, actin and centractin.[44,45] A concurrent decrease in expression of G-actin and TCP-1 in response to PMA or lipopolysaccharides (LPS) was previously reported,[40,46] and possibly suggests a growth reduction and changes in cell morphology and motility.[47]

Significant up-regulation in expression was observed for a number of proteins including tropomyosin 3, α-enolase and vimentin (Table 8.1, Figure 8.5). Consistent expression measurements were observed from comparative ion profile measurements as indicated for vimentin peptide ions which all showed expression change of at least 6-fold upon PMA treatment. Figure 8.5a shows the reconstructed ion chromatograms for the triply protonated tryptic peptide m/z 568.9 representing an increase of 10-fold in ion intensity upon PMA treatment (circled in Figure 8.4a). The identity of this peptide was confirmed from its MS-MS spectrum which showed abundant sequence-specific ion series corresponding to b- and y-type fragment ions (Figure 8.5b). While α-enolase is an essential glycolytic enzyme involved in cell metabolism, tropomyosin and vimentin are implicated in stabilizing actin filaments and their up-regulation during differentiation and maturation is consistent with the rapid structural changes in size and morphology of monocytes to adopt macrophage-like functional attributes.[42,48]

The comparison of ion profiles for peptides obtained from the same proteins also revealed subtle structural changes taking place during cell differentiation. For example, HSP90-α was identified by two peptides: KHLEINPDH-SIIETLR and ESEDKPEIEDVGSDEEEEKK (Table 8.1). The ion profile ratio of the first peptide was 1.1, suggesting no significant protein expression change, consistent with a previous report.[49] Intensity measurements obtained

Table 8.1 Partial list of identified proteins showing differential expression upon PMA treatment.

Protein	*Mass (kDa)*	*Fold change*[a]	*Sequence coverage*	*Peptides*	*Ref.*
TCP-1	57.8	−12.5	3.6	GIDPFSLDALSK	40
				HTLTQIK	
G-actin	41.6	−5.0	9.4	DLTDYLMK	46
				GYSFTTTAER	
				YPIEHGIITNWDDMEK	
β-Tubulin	49.7	−2.6	17.8	LHFFMPGFAPLTSR	40
				FPGQLNADLR	
				SGPFGQIFRPDNF VFGQSGAGNNWAK	
				IMNTFSVVPSPK	
				ALTVPELTQQMFDAK	
HSP84	83.2	−1.3	10.1	APFDLFENKK	41
				APFDLFENK	
				SIYYITGESK	
				IDIIPNPQER	
				YHTSQSGDEMTSLSEYVSR	
				NPDDITQEEYGEFYK	
HSP71	69.3	−1.1	12.6	NALESYAFNMK	42
				TWNDPSVQQDIK	
				HWPFQVINDGDKPK	
				NQVAMNPTNTVFDAK	
				HWPFMVVNDAGRPK	
				MVNHFIAEFK	
GAPDH	44.5	1.1	20.9	LVINGNPITIFQER	43
				VIISAPSADAPMFVMGVNHEK	
				LISWYDNEFGYSNR	
				VIHDNFGIVEGLMTTVHA ITATQK	
				VGVNGFGR	
HSP90-α	84.5	1.1	5.1	KHLEINPDHSIIETLR	49,51
		−43.7		ESEDKPEIEDVGpSDEEEEKK	
Tropomyosin	28.8	3.8	10.4	IQLVEEELDRAQER	42
				EQAEAEVASLNR	
Vimentin	53.5	6.7	23.6	FANYIDK	48
				KVESLQEEIAFLK	
				VEVERDNLAEDIMR	
				ILLAELEQLK	
				LGDLYEEEMR	
				EEAENTLQSFR	
				ISLPLPNFSSLNLR	
				ETNLDSLPLVDTHSK	
				DGQVINETSQHHDDLE	
α-Enolase	47	9.0	22.1	LAQANGWGVMVSHR	42,46
				IGAEVYHNLK	
				SFIKDYPVVSIEDPFDQDDW GAWQK	
				YDLDFK	
				AAVPSGASTGIYEALELR	
				LAMQEFMILPVGAANFR	

[a]Average fold change determined from all identified peptides.

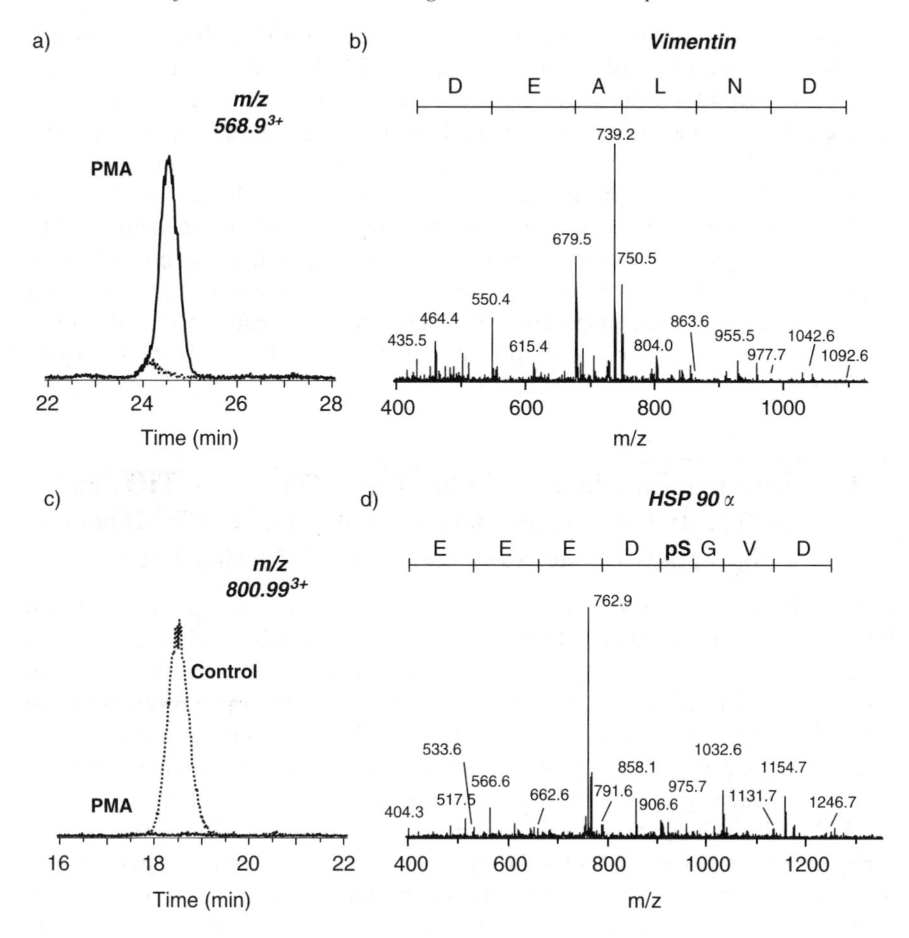

Figure 8.5 Identification of proteins showing differential expression upon PMA treatment. Reconstructed ion chromatograms of (a) m/z 568.9^{3+} and (c) 800.99^{3+} showing differential abundance in control and PMA-exposed cells. Dotted lines correspond to control digests and solid lines to PMA-treated samples. MS-MS spectra of precursor ions m/z 568.9^{3+} and 800.99^{3+} circled on the scatter plot of Figure 8.4a. MS-MS spectra of (b) m/z 568.9^{3+} and (d) 800.99^{3+} confirming identification of vimentin and HSP90-α, respectively. The Ser_{262} phosphorylation site of HSP90-α is shown in (d). Conditions as for Figure 8.4.

on the second peptide showed more than 45-fold decrease in abundance upon PMA treatment (Figure 8.5c). The intensity of this peptide was below the detection of the nanoLC-chip-system set to 100 counts (circled in Figure 8.4a). Closer inspection of the MS-MS spectrum of the corresponding tryptic peptide confirmed its phosphorylation at residue Ser_{262} (Figure 8.5d). Interestingly, HSP90-α can be phosphorylated at Ser_{231} and Ser_{262} by casein kinase II *in vitro*.[50] A recent report indicated that HSP90-α strongly associates with IKappa kinase-γ (IKK-γ) possibly contributing to its stabilization, activation and/or shuttling IKKs to the plasma membrane, and that this interaction is dramatically reduced following exposure to PMA.[51] While the expression levels of HSP90-α appeared constant upon PMA stimulation, the rapid change of Ser_{292} phosphorylation suggests that its interaction with IKK-γ could be mediated by phosphorylation. The comparison of protein expression and phosphoproteome profiles thus provides valuable information to probe transient protein interactions that are modulated through signaling events. Obviously, such comparison relies on the meaningful identification of phosphoproteins and their phosphorylation sites, and section 8.3.3 examines a phosphopeptide enrichment strategy based on TiO_2 beads to identify trace-level phosphoproteins and their corresponding changes in expression following PMA treatment.

8.3.3 Selective Enrichment of Phosphopeptides using TiO_2 and Differential Phosphoproteome Analysis of U937 Monocyte Cells Following the Administration of Phorbol Ester

Comprehensive phosphoproteome analysis represents a significant analytical challenge in view of the low abundance of phosphoproteins within cell extracts and the instrumental requirements to achieve meaningful identification. Indeed, in human cells it is estimated that approximately one-third of all proteins can be phosphorylated at any one time by a subset of 575 different protein kinases.[52,53] Protein phosphorylation primarily takes place on serine and threonine residues and less frequently on tyrosine amino acids (pSer : pThr : pTyr in vertebrate cell is 1800:200:1).[54] Furthermore, the phosphorylation stoichiometry or site occupancy is relatively low and the dynamic range of protein phosphorylation can span over more than three orders of magnitude of abundance often exceeding the detection range of most sensitive analytical instruments. More importantly, phosphorylated residues represent $<1\%$ of the whole protein sequence and the isolation of peptide segments comprising these modifications requires further fractionation and enrichment.

Recent reviews have compared the respective merits and limitations of different phosphopeptide enrichment approaches including immunoaffinity precipitation, immobilized metal affinity chromatography (IMAC) and targeted chemical labeling through β-elimination/Michael's addition.[54–57] For practical reasons, the selection of the most appropriate enrichment method is often guided by the sample availability, the recovery yield, the extent of side reaction products and the inherent complexity arising from the sample preparation. Enrichment approaches involving less sample manipulation steps are generally preferred to maximize recovery, and previous reports from our group and others described the use of IMAC for the isolation of phosphopeptides.[36,58–60] More recently, selective enrichment media such as TiO_2 and ZrO_2 have been introduced and hold significant promises for reliable and reproducible enrichment of phosphopeptides from complex cell extracts.[61,62] As part of this study, we evaluated the analytical potentials of the TiO_2 enrichment approach and compared the phosphopeptides expression profiles of U937 treated or not with PMA using the nanoLC-chip-MS system.

Preliminary experiments on TiO_2 enrichment media were first performed to compare the recovery yields and detection limits of phosphopeptides with increasing amounts of α-casein tryptic digests spiked into complex protein digests. These experiments indicated that unambiguous identification of phosphopeptides present at levels of 25 fmol in 250 μg of protein digest was successfully achieved with good linear dynamic response ($r^2 > 0.98$) over the concentration range 25 fmol–2.0 pmol of α-casein (data not shown). It is noteworthy that phosphopeptides present in the original cell extract remained constant across all spiked samples examined.

Similarly to that described in Section 8.3.2, we evaluated the capability of the nanoLC-chip-MS system to monitor changes in phosphoproteome of U937 cells following chemical stimulation with PMA and enrichment with TiO_2 media. Reduced and alkylated protein extracts from U937 monoblastic cells exposed or not to 150 nM PMA were digested with trypsin and 250 μg of the corresponding digests were subjected to TiO_2 isolation. Enriched phosphopeptide samples were then analyzed in triplicate on the nanoLC-chip-TOF system. On average, a total of approximately 1500 peptide ions were detected in each analysis out of which 1791 peptide clusters were reproducibly observed in at least two replicates. Figure 8.6a shows a representation of the ion intensity distribution for reproducibly detected peptide clusters in both control and PMA-treated cells. For convenience, the dotted lines shown in Figure 8.6a represent a threefold average standard deviation (3×-Stdev) on ion intensity of peptide clusters and delimit the boundary of statistically relevant change in expression across the intensity domain. Accordingly, a total of 1467 peptide clusters (82% of all reproducibly detected clusters) did not show significant change in abundance upon PMA treatment. These ions were unaffected and exhibited less than 3-fold change across 2.5 orders of magnitude in intensity with a normal distribution centered on unity. The intensity distribution of the corresponding peptide clusters is represented in the central portion of the clustergram of Figure 8.6b which regroups replicate clusters according to a logarithmic grayscale with intensity values transitioning from white to black to indicate ions of increasing intensity.

Significant changes in expression were observed in both up-regulation (303 peptide clusters, 17% of total) and down-regulation (21 peptide clusters, 1% of total) for ion clusters distributed outside of the dotted lines boundary. It is noteworthy that this trend is somewhat different from that observed for U937 protein extracts not subjected to TiO_2 enrichment (24.6% up-regulated, 30.0% down-regulated and 45.4% unaffected; Figure 8.4b) suggesting that variations in phosphoproteome are distinct from those of overall protein expression changes. This observation is not entirely unexpected, as previously described in Section 8.3.2 for HSP90-α. Obviously, such changes in differential phosphorylation must consider variations in actual protein abundance in order to determine whether or not the increase in kinase activity for given substrate is independently regulated from their synthesis or translocation. For challenge

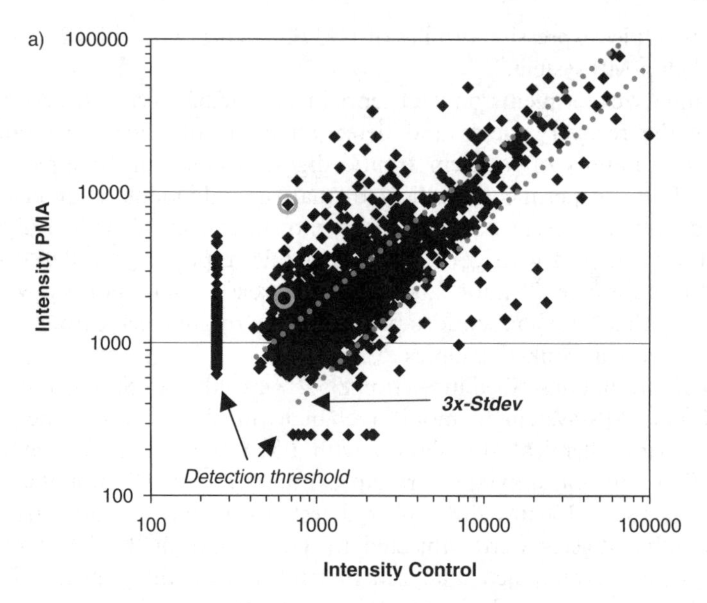

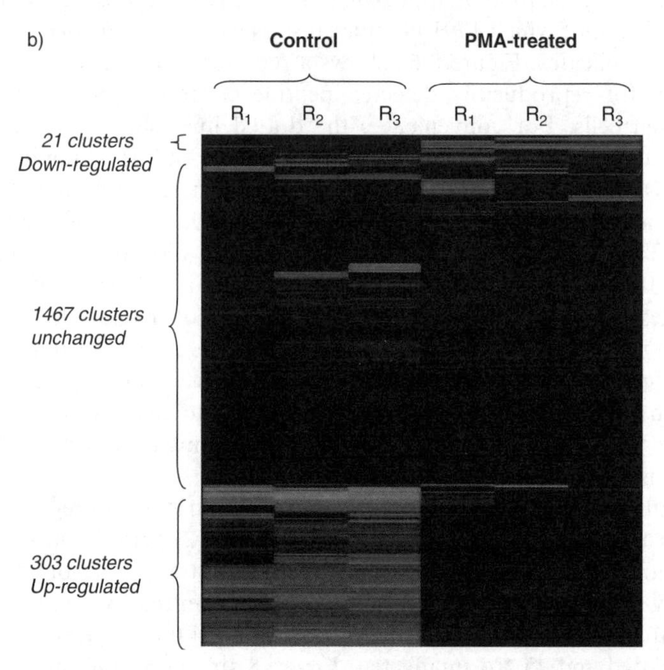

Figure 8.6 Differential phosphoproteome analysis of U937 cells following PMA challenge and TiO_2 isolation. (a) Scatter plot ion intensity for peptide clusters reproducibly detected in at least 2 replicates ($n = 1791$ peptide clusters). The dotted lines represent the delimitation of ion abundance showing ±3-fold change in standard deviation. (b) Clustergram of intensity distribution for 1791 peptide clusters in control and PMA-exposed samples. Conditions: as for Figure 8.2. An aliquot (50%) of TiO_2-enriched phosphopeptides from U937 protein digests (250 μg) corresponding to approximately 150 ng injection was injected on the nanoLC-chip-MS system.

experiments taking place under short time duration (<2 h) the changes in protein abundance due to newly synthesized proteins are minimized.

Following expression analyses, nanoLC-chip-MS-MS experiments were performed using an ion trap mass spectrometer to identify suspected phosphoproteins showing up- and down-regulation upon PMA treatment. Examples of data obtained are illustrated in Figure 8.7 for two peptide clusters circled on the scatter plot of Figure 8.6a. The reconstructed ion chromatogram for m/z 851.8^{2+} (Figure 8.7a) shows a consistent up-regulation by at least 3-fold in the PMA-treated cells. Tandem mass spectrometry analyses of the corresponding doubly charged peptide ion (Figure 8.7b) provided a partial sequence

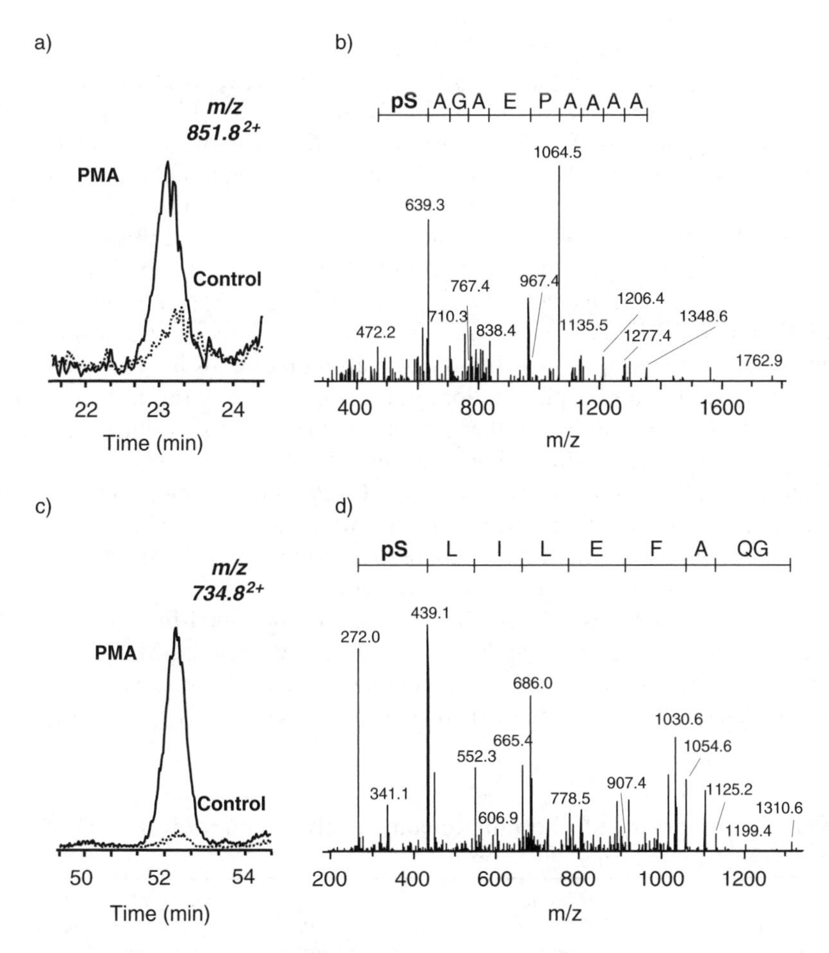

Figure 8.7 Identification of differentially expressed phosphoproteins from U937 cells following PMA administration. Reconstructed ion chromatograms of (a) m/z 851.8^{2+} and (c) m/z 734.8^{2+} circled on Figure 8.6a and showing increased abundance upon PMA treatment. MS-MS spectra of (b) m/z 851.8^{2+} and (d) m/z 734.8^{2+} confirming the protein identification and the site of phosphorylation. Conditions as for Figure 8.6.

assignment for AAAAAPEAGA**pS** which upon data search using Mascot was identified to myristoylated alanine-rich C-kinase substrate (MARKS). MS-MS spectrum of the corresponding ion showed a fragment at m/z 472.2 and a weak loss of 98 Da from m/z 639.3 consistent with a phosphorylated Ser_{100} residue. Phosphorylation of this residue was not reported previously but showed homology to a cyclin-dependent kinase (CDK) motif. Interestingly, MARKS binds to calmodulin, actin and synapsin and is a cellular substrate for protein kinase C (PKC). Also, this protein becomes significantly phosphorylated upon PMA treatment.[63] Sites of PKC phosphorylation on MARKS could not be confirmed by the present MS-MS experiments as the resulting tryptic peptides located in a highly basic region between residues 151 and 164 were too small or hydrophilic to be detected efficiently.

Another example of differential phosphorylation is shown in Figure 8.7c and d for the doubly protonated tryptic peptide m/z 734.8 showing more than 6-fold change in abundance upon PMA treatment. The MS-MS spectrum of this precursor ion (Figure 8.7b) enabled the identification of stathmin based on the partial sequence GQAFELIL**pS**. The location of the modified residue was assigned to Ser_{24}, based on the observation of a fragment at m/z 272.1 and a loss of 98 Da from m/z 439.1. A previous report indicated that this residue can be phosphorylated by CDK1 and MAPK.[64] Stathmin is implicated in the regulation of microtubule stability and dynamics, and PMA induces its phosphorylation and stimulates membrane ruffling and cell motility. Phosphorylation regulates the spatial distribution of stathmin–tubulin interaction within cells and a stathmin phosphorylation gradient is necessary for correct spindle formation in mitotic cells.[65] Other important mediators of cellular transformation and organization including L-plastin, HSP90, c-Myc-responsive protein Rcl, microfibrillar-associated protein 1, ras-GTPase-activating protein binding protein 1, 14-3-3 and Src substrate cortactin were also identified as part of this study and highlight the analytical capabilities of the nanoLC-chip-MS systems for expression and identification measurements.

Table 8.2 summarizes the protein identification obtained for the U937 differentiated cells. Database search performed on the acquired MS-MS spectra revealed a total of more than 106 unique protein clusters, among which approximately 80% corresponded to phosphoproteins. It is noteworthy that

Table 8.2 Summary of phosphoproteome analysis obtained on U937 differentiated cells.

Sample type (TCL)	*Average no. of detected peptide ions*	*No. of protein clusters identified*	*No. of phosphoprotein clusters (peptide sequences)*
Monocyte	1513	64	49 (93: 91pS 10pT 2pY)
Monocyte + PMA	1466	77	67 (120: 99pS 19pT 3pY)

comparable analyses performed by our group on similar protein extracts from U937 using IMAC (Ga^{3+}) isolation provided an enrichment level of approximately 25%,[36] and the use of TiO_2 provided a marked advantage for phosphoproteomics studies. Furthermore, the TiO_2 enrichment media enabled the identification of several phosphorylation sites from the same proteins and almost two phosphosites/unique phosphoprotein clusters were identified on average. For example, phosphoproteome analyses performed on 250 μg of U937 cells treated with PMA enabled the identification of 67 unique phosphoprotein clusters corresponding to 120 different phosphopeptides. These peptides showed the expected distribution of phosphorylated residues favoring Ser and Thr residues with significantly lower proportion of phosphorylated Tyr residues. Our study enabled the identification of 152pS 25pT 4pY sites on phosphopeptides, more than 40% of which were unreported in the literature. Altogether these data provided valuable biological insights to probe the subtle change in cellular phosphoproteome arising from environmental stimulation.

8.4 Concluding Remarks

Microfluidic devices integrating an enrichment trap, a separation column and a nanoelectrospray emitter offer a convenient protein expression and identification platform for proteomics discovery programs. The modularity, ease of use, reproducibility and ruggedness conferred by these devices facilitate the identification of differentially abundant proteins from cell model systems or biomarker studies. The present system provides good reproducibility in terms of retention time, m/z and intensity measurements for both one- and two-dimensional nanoLC separations. Indeed, peak measurements taken on more than 2200 different peptide clusters from replicate injections ($n = 5$) of an eight-protein standard provided RSD values of less than 0.5%, 0.003% and 10% for retention time, m/z and intensity, respectively. A linear dynamic response of ion intensity was maintained over more than three orders of magnitude for sample loading up to 200 ng using a 40 nL enrichment trap and a separation column of 4.3 cm length (75 × 50 μm channel).

Improvement of sample loading up to 2 μg was achieved by combining an SCX trapping column prior to the C_{18} chip device. The application of the combined SCX/C_{18} was evaluated for the analysis of complex protein extracts of U937 human monocytic cells, and good reproducibility was obtained across all replicate injections over a three-day period with 95% of all reproducibly detected clusters ($n = 7389$) showing less than 40% variation in intensity. The remarkable sensitivity and reproducibility of the present system enabled the identification of proteins such as TCP-1, G-actin, tropomyosin and vimentin, all showing relevant changes in abundance upon PMA administration, consistent with the rapid size and morphology changes expected during the transformation of monocytes into macrophage-like cells.

More detailed profiling of signal-induced changes of U937 cells was achieved by comparing the phosphoproteome of protein extracts obtained from control and PMA-exposed cells following TiO_2 enrichment and nanoLC-chip-MS analyses. Expression profiling of enriched phosphopeptide extracts revealed a subset of 18% of the entire population that showed a reproducible increase in abundance upon PMA exposure. Among these were phosphopeptides from MARKS, stathmin, L-plastin, HSP90, c-Myc-responsive protein Rcl, micro-fibrillar-associated protein 1, ras-GTPase-activating protein binding protein 1, 14-3-3 and Src substrate cortactin, all representing important mediators of cellular transformation and organization. These analyses also enabled the identification of 106 protein clusters with approximately two phosphosites/unique phosphoprotein clusters on average. Differential phosphoproteome analyses from 250 µg protein extract from U937 cells identified 152pS 25pT 4pY sites on phosphopeptides of which approximately 40% were unknown. The approach described here represents a meaningful analytical strategy to profile and identify important cellular markers of the U937 phosphoproteome using relatively modest protein amounts. More importantly, this approach is anticipated to be of practical use in cancer research programs to compare the phosphorylation signatures associated to normal and cancerous cells.

The capability to profile complex data sets with sensitivity, reproducibility and comprehensiveness using microfluidic devices coupled to nanoelectrospray MS open new horizons for proteomics research. Indeed, the ability to conduct expression profiling of unlabeled protein digests and to identify important markers of cellular changes is of significant interest to MS-based proteomics platforms involved in biomarker studies, drug discovery and cancer research programs. Although the field of microfluidic-MS is still in its infancy, advances made in the automation, the reproducibility and the versatility of these systems will accelerate their acceptance as meaningful discovery tools that can be used by non-experts.

Acknowledgements

This study was supported by the National Science and Engineering Research Council of Canada. We thank Agilent Technologies for providing early access to the prototype nanoLC-chip-MS system and the MSD ion trap XCT mass spectrometer. We are also grateful to Agilent staff members Georges L. Gauthier for valuable discussion and support, Debbie Ritchey and Reid Brennan, Linda Côté and Sébastien Marchand for technical support. Finally, we acknowledge G. Jaitly (IRIC) and N. Jaitly (PNNL) for invaluable assistance with the peptide detection and clustering algorithms.

References

1. G. M. Whitesides, *Nature*, 2006, **442**, 368.
2. I. M. Lazar, J. Grym and F. Foret, *Mass Spectrom. Rev.*, 2006, **25**, 573.
3. Q. Xue, F. Foret, Y. M. Dunayevskiy, P. M. Zavracky, N. E. McGruer and B. L. Karger, *Anal. Chem.*, 1997, **69**, 426.

4. R. S. Ramsey and J. M. Ramsey, *Anal. Chem.*, 1997, **69**, 1174.
5. D. Figeys and R. Aebersold, *Anal. Chem.*, 1998, **70**, 3721.
6. J. Li, P. Thibault, N. H. Bings, C. D. Skinner, C. Wang, C. Colyer and J. Harrison, *Anal. Chem.*, 1999, **71**, 3036.
7. I. M. Lazar, R. S. Ramsey, S. C. Jacobson, R. S. Foote and J. M. Ramsey, *J. Chromatogr. A*, 2000, **892**, 195.
8. D. Bandilla and C. S. Skinner, *J. Chromatogr. A*, 2004, **1044**, 113.
9. I. M. Lazar, L. J. Li, Y. Yang and B. L. Karger, *Electrophoresis*, 2003, **24**, 3655.
10. J. D. Ramsey and G. E. Collins, *Anal. Chem.*, 2005, **77**, 6664.
11. C. T. Culbertson, S. C. Jacobson and J. M. Ramsey, *Anal. Chem.*, 2000, **72**, 5814.
12. J. D. Ramsey, S. C. Jacobson, C. T. Culbertson and J. M. Ramsey, *Anal. Chem.*, 2003, **75**, 3758.
13. A. P. Dahlin, S. K. Berrgström, P. E. Andrén, K. E. Markides and J. Bergquist, *Anal. Chem.*, 2005, **77**, 5356.
14. J. Li, C. Wang, J. F. Kelly, D. J. Harrison and P. Thibault, *Electrophoresis*, 2000, **21**, 198.
15. J. Li, T. LeRiche, T.-L. Tremblay, C. Wang, E. Bonneil, D. J. Harrison and P. Thibault, *Mol. Cell. Proteomics*, 2002, **1**, 157.
16. J. T. Kapron, E. Pace, C. K. Van Pelt and J. Henion, *Rapid Commun. Mass Spectrom.*, 2003, **17**, 2019.
17. L. A. Leuthold, C. Grivet, M. Allen, M. Baumert and G. Hopgartner, *Rapid Commun. Mass Spectrom.*, 2004, **18**, 1995.
18. C. A. Keetch, H. Hernandez, A. Sterling, M. Baumert, M. H. Allen and C. V. Robinson, *Anal. Chem.*, 2003, **75**, 4937.
19. D. P. Little, T. J. Cornish, M. J. O'Donnell, A. Braun, R. J. Cotter and H. Köster, *Anal. Chem.*, 1997, **69**, 4540.
20. J. Preisler, F. Foret and B. L. Karger, *Anal. Chem.*, 1998, **70**, 5278.
21. J. Preisler, P. Hu, T. Rejtar and B. L. Karger, *Anal. Chem.*, 2000, **72**, 4785.
22. Y.-X. Wang, Y. Zhou, B. M. Balgley, J. W. Cooper, C. S. Lee and D. L. DeVoe, *Electrophoresis*, 2005, **26**, 3631.
23. J. P. Kutter, S. C. Jacobson and J. M. Ramsey, *J. Microcolumn Sep.*, 2000, **12**, 93.
24. C. Yu, M. H. Davey, F. Svec and J. M. J. Fréchet, *Anal. Chem.*, 2001, **73**, 5088.
25. C. Yu, M. Xu, F. Svec and J. M. J. Fréchet, *J. Polym. Sci. Polym. Chem. A*, 2002, **40**, 755.
26. S. Le Gac, J. Carlier, J.-C. Camart, C. Cren-Olivé and C. J. Rolando, *J. Chromatogr. B*, 2004, **808**, 3.
27. Y. Yang, C. Li, J. Kameoka, K. H. Lee and H. G. Craighead, *Lab Chip*, 2005, **5**, 869.
28. I. M. Lazar, P. Trisiripisal and H. A. Sarvaija, *Anal. Chem.*, 2006, **78**, 5513.
29. M. McEnery, A. Tan, J. Alderman, J. Patterson, S. C. O'Mathuna and J. D. Glennon, *Analyst*, 2000, **125**, 25.

30. J. D. Xu, L. Locascio, M. Gaitan and C. S. Lee, *Anal. Chem.*, 2000, **72**, 1930.
31. M. Svedberg, M. Veszelei, J. Axelsson, M. Vangbo and F. Nikolajeff, *Lab Chip*, 2004, **4**, 322.
32. J. Xie, Y. Miao, J. Shih, Q. He, J. Liu, Y. C. Tai and T. D. Lee, *Anal. Chem.*, 2004, **76**, 3756.
33. J. Xie, Y. Miao, J. Shih, Y. C. Tai and T. D. Lee, *Anal. Chem.*, 2005, **77**, 6947.
34. M.-H. Fortier, E. Bonneil, P. Goodley and P. Thibault, *Anal. Chem.*, 2005, **77**, 1631.
35. H. Yin, K. Killeen, R. Brennen, D. Sobek, M. Werlich and T. van de Goor, *Anal. Chem.*, 2005, **77**, 527.
36. M. Ghitun, E. Bonneil, M.-H. Fortier, H. Yin, K. Killeen and P. Thibault, *J. Sep. Sci.*, 2006, **29**, 1539.
37. M. R. Larsen, T. E. Thingholm, O. N. Jensen, P. Roepstorff and T. J. D. Jørgensen, *Mol. Cell. Proteomics*, 2005, **4**, 873.
38. G. Rovera, T. G. O'Brien and L. Diamond, *Science*, 1979, **204**, 868.
39. E. Bonneil, S. Tessier, A. Carrier and P. Thibault, *Electrophoresis*, 2005, **26**, 4575.
40. X. Zhuang, Y. Kuramitsu, M. Fujimoto, E. Hayashi, X. Yuan and K. Nakamura, *Electrophoresis*, 2006, **27**, 1659.
41. M. V. Metodiev, A. Timanova and D. E. Stone, *Proteomics*, 2004, **4**, 1433.
42. S. R. Pereira, V. M. Faça, G. G. Gomes, R. Chammas, A. Maria Fontes, D. T. Covas and L. J. Greene, *Proteomics*, 2005, **5**, 1186.
43. D. Chatterjee, P. Pantazis, G. Li, T. A. Bremner, E. A. Hendrickson and J. H. Wyche, *Oncogene*, 2000, **19**, 4108.
44. P. Liang and T. H. MacRae, *J. Cell Sci.*, 1997, **110**, 1431.
45. J. Grantham, L. W. Ruddock, A. Roobol and M. J. Carden, *Cell Stress Chaperones*, 2002, **7**, 235.
46. K. C. M. Verhoeckx, S. Bijlsma, E. M. de Groene, R. F. Witkamp, J. van der Greef and R. J. T. Rodenburg, *Proteomics*, 2004, **4**, 1014.
47. J. Grantham, K. I. Brackley and K. R. Willison, *Exp. Cell Res.*, 2006, **312**, 2309.
48. P. Benes, V. Macečková, Z. Zdráhal, H. Koneènά, E. Zahradníčková, J. Mužík and J. Šmarda, *Differentiation*, 2006, **74**, 265.
49. J. Galea-Lauri, D. S. Latchman and D. R. Katz, *Exp. Cell Res.*, 1996, **226**, 243.
50. S. P. Lees-Miller and C. W. Anderson, *J. Biol. Chem.*, 1989, **264**, 2431.
51. K. Ah Park, H. Sun Byun, M. Won, K.-J. Yang, S. Shin, L. Piao, J. Man Kim, W.-H. Woon, E. Junn, J. Park, J. H. Seok and G. M. Hur, *Carcinogenesis*, 2006, E-Pub, 14 June.
52. S. Zolnierowicz and M. Bollen, *EMBO J.*, 2000, **19**, 483.
53. J. C. Venter, *et al., Science*, 2001, **291**, 1304.
54. M. Mann, S.-E. Ong, M. Grønborg, H. Steen, O. N. Jensen and A. Pandey, *Trends Biotechnol.*, 2002, **20**, 261.

55. K. M. Loyet, J. T. Stults and D. Arnott, *Mol. Cell. Proteomics*, 2005, **4**, 235.
56. E. Salih, *Mass Spectrom. Rev.*, 2005, **24**, 828.
57. J. Reinders and A. Sickmann, *Proteomics*, 2005, **5**, 4052.
58. L. Andersson and J. Porath, *Anal. Biochem.* 1986, **154**, 250.
59. M. C. Posewitz and P. Tempst, *Anal. Chem.*, 1999, **71**, 2883.
60. S. B. Ficarro, M. L. McCleland, P. T. Stukenberg, D. J. Burke, M. M. Ross, J. Shabanowitz, D. F. Hunt and F. M. White, *Nat. Biotechnol.*, 2002, **20**, 301.
61. M. W. Pinkse, P. M Uitto, M. J. Hilhorst, B. Ooms and A. J. R. Heck, *Anal. Chem.*, 2004, **76**, 3935.
62. H. Kyong Kweon and K. Håkansson, *Anal. Chem.*, 2006, **78**, 1743.
63. A. Tanabe, Y. Kamisuki, H. Hidaka, M. Suzuki, M. Negishi and Y. Takuwa, *Biochem. Biophys. Res. Commun.*, 2006, **345**, 156.
64. U. Marklund, G. Brattsand, V. Shingler and M. Gullberg, *J. Biol. Chem.*, 1993, **268**, 15039.
65. C. I. Rubin and G. F. Atweb, *J. Cell. Biochem.*, 2004, **93**, 242.

On-line Chemical Investigations

CHAPTER 9

Simple Chip-based Interfaces for On-line Nanospray Mass Spectrometry

MONICA BRIVIO,[†] WILLEM VERBOOM AND DAVID N. REINHOUDT

Laboratory of Supramolecular Chemistry and Technology, University of Twente, PO Box 217, 7500 AE Enschede, The Netherlands

9.1 Introduction

Mass spectrometry (MS) is one of the most powerful detection techniques used in liquid-phase analyses,[1] mainly due to the ease of interfacing with separation techniques such as capillary electrophoresis (CE)[2,3] and high-performance liquid chromatography (HPLC).[4] Due to its sensitivity and applicability to a wide variety of chemical and biochemical species, MS is also used for the analysis of (bio)chemical molecules processed in microfluidics devices.[5,6] Electrospray ionization (ESI)[7–10] is often used to transfer samples from microfluidics chips to a mass spectrometer, involving analyte ionization directly from solutions and operating at flow rates typically used in microfluidics devices.[11] Due to its effectiveness, the use of chip-MS coupling has rapidly spread in many research areas with bioanalytical applications,[12] such as the

[†]Present address: MIC-Department of Micro and Nanotechnology, Technical University of Denmark-building 345 east, 2800 Lyngby, Denmark.

Miniaturization and Mass Spectrometry
Edited by Séverine Le Gac and Albert van den Berg

Published by the Royal Society of Chemistry, www.rsc.org

fields of genomics[13] and proteomics.[14] The first example of the coupling of a microfluidics device with ESI-MS was reported by Xue *et al.*[15,16] in 1997. They used glass chips as a platform to carry out enzymatic digestions and to deliver standard peptide and protein samples to an ESI-MS system. In the same year Ramsey and Ramsey[17] extended this method using electroosmotic flow (EOF) to pump the sample into the mass spectrometer. Figeys *et al.*[18] achieved detection limits of the order of fmol μL^{-1} (*i.e.* 10^{-9} M) using a hybrid capillary/micromachined microfluidics device for sequential infusion of peptides into a mass spectrometer, which demonstrates the potential of chip-MS coupling.

Since these first chip-MS-based lab-on-a-chip devices were presented, there has been much research activity on the development of hyphenated interfaces for coupling microfluidic devices with ESI-MS.[6,19] As a result, several coupling systems have been exploited,[20–31] based on two main designs: those that spray liquid samples directly from an exposed channel, as first proposed in 1997 by Xue *et al.*[15] and Ramsey and Ramsey,[17] and those in which a capillary is attached in various ways to the chip, as first presented by Figeys *et al.*[18] In the early days researchers working on chip-MS coupling were mainly focused on design and microfabrication, aiming to develop a spraying chip able to generate a good-quality Taylor cone. More recently, although the optimization of the chip-ESI interface in terms of signal stability is still a key issue,[29–33] more attention is paid to the integration of multiple functions on-chip,[34,35] aiming to demonstrate the high-throughput potential of the chip-MS coupling.[36–40] However, most of the interfaces reported in the literature have been designed for biochemical applications. This implies a significant limitation in terms of applicability to organic chemical systems, due to the use of glue and polymeric construction materials. Therefore the development of an integrated chip-MS interface that allows rapid on-line study of (bio)chemical reactions without solvent limitations is highly attractive.

In this chapter a simple and efficient coupling system for on-line MS from glass chips using a nanoflow electrospray ionization (NESI) interface is described.[41] Two integrating concepts were studied, based on both a monolithic and a modular approach and their ionization capabilities were characterized using reserpine as a reference compound in a 1:1 mixture of acetonitrile and water. The modular integration approach was chosen for the final setup, making use of glass chips comprising mixer(s) and reactor(s). The mixing capacity of the glass microchip was optimized by exploiting its small dimensions and merging the flows in a unique way.

The modular NESI-chip system was used (i) to qualitatively as well as quantitatively study two well-known supramolecular systems, *i.e.* the Zn–porphyrin complexation with nitrogen-containing ligands in acetonitrile and the inclusion of small organic molecules into the cavity of β-cyclodextrin in water,[41] and (ii) to study the kinetics of the reaction of 4-nitro-7-piperazino-2,1,3-benzoxadiazole (NBDPZ) with isocyanates as a model system.[42]

9.2 Microfluidics

9.2.1 Materials and Methods

9.2.1.1 Microreactors

All glass (Schott Borofloat 33) chips were fabricated at the cleanroom facilities of the MESA$^+$ Institute for Nanotechnology at the University of Twente. The process sequence is as follows: (i) channels are isotropically etched in one or in both glass wafers with an HF solution and a Cr–Au mask; (ii) through-holes for the fluidic connections to the channels are powder blasted in the top wafer; (iii) the processed wafer pair is joined together via fusion-bonding. Two main chip designs were realized having channel cross-sections of $100 \times 50\,\mu m^2$ and $50 \times 20\,\mu m^2$, respectively. In order to provide a broader variety of residence times than those available by adjusting the pump flow rates only, various fluidic path shapes were exploited, resulting in variation of the channel lengths from 100 to 200 mm (corresponding to reaction volumes ranging between 0.5 and 1 μL for the 100 μm wide by 50 μm deep channels) and from 13 to 71 mm (corresponding to reaction volumes ranging between 13 and 71 nL for the 50 μm wide by 20 μm deep channels), respectively.

9.2.1.2 Mixing Simulation

Mixing simulation was carried out by finite volume computations with the multi physics software package CFDRC ACE+.[43] The mixing time was estimated assuming diffusion of water molecules in water (injected at a total flow rate of $2\,\mu L\,min^{-1}$) across the microchannel.

9.2.1.3 Mixing Characterization

Studies of mixing dynamics were carried out according to literature procedures.[44] The mixing dynamics in both microchannels was evaluated using an Olympus CK40M inverted microscope with a fluorescence kit.

9.2.1.4 Fluidic Handling

Sample solutions were mobilized in all experiments by means of a dual CMA/102 microdialysis pump on which 100 μL flat tip Hamilton syringes were mounted. Syringes were connected to fused silica capillaries by means of Upchurch™ connectors.[45,46]

9.2.2 Results and Discussion

Various types of chips were designed and produced to be used in combination with the NESI-chip interface. Since the interface was designed for the study of organic reactions, glass was chosen as material for the microreactors. Glass is chemically inert, it has good optical properties and it is durable. Glass microreactors are often reused multiple times and efficiently cleaned with a piranha solution (H_2SO_4 and H_2O_2, 3:1 v/v). Furthermore, the choice of borosilicate glass as material has also the practical advantage of being compatible with the standard glass chip fabrication[47] at the MESA$^+$ Institute.

Solutions can be pumped in and out of the chip by using Upchurch Nanoport™ connectors or by using an EOF when platinum electrodes are sputtered through a shadow mask in the inlet and outlet cavities after bonding.[48]

The microreactors, which are operated in a continuous flow fashion, consist of a long, meander-shaped channel, the length of which can be designed to meet the required residence time to yield a fully developed reaction product. Heating or cooling of the reactor takes place by heaters or Peltier elements and can be controlled by temperature sensors (*i.e.* thermocouples or platinum elements). The chips comprised externally placed heaters and sensors, but could also be applied by sputtering afterwards, similar to the electrodes for EOF.

At Reynolds numbers of the order of 0.1 in microchannels mixing takes place by diffusion only.[49] Of the numerous mixer designs that have been proposed in the literature,[50] the grooved mixer of Stroock *et al.*[51] has a big advantage because of its simplicity and good efficiency. In glass processing directional etching of such structures with reactive ion etching (RIE) requires very long etching times with high ion energies. Since the mask–substrate etching selectivity for such situations is limited, a very thick mask layer of polysilicon is required. Consequently, we decided to implement a different mixer type that meets our technological requirements.

The diffusion time (t_d) as a function of the diffusion coefficient (D) and diffusion length (L) can be roughly estimated using the equation

$$t_d = \frac{L^2}{2D} \tag{9.1}$$

This means that a reduction of the diffusion length will enhance the mixing speed with a quadratic order. Most micromixers use diffusion in the horizontal direction by merging two channels next to each other, parallel to the wafer surface. From a technology point of view, this is a good solution since lithography gives the designer most freedom to design structures in the planar orientation. Since isotropically etched channels are much wider than they are deep, a vertical merging would be more efficient. In the latter case, very high width/depth aspect ratios can be chosen such that the diffusion lengths can be considerably reduced, while maintaining the cross-sectional area, and thus the same residence times.

Both horizontal and vertical diffusion-based mixer designs were tested in the 100 μm wide by 50 μm deep and 50 μm wide by 20 μm deep reaction channels. In the first chip design the two 100 μm wide by 50 μm deep inlet channels merge in-plane, resulting in the 100 μm wide and 50 μm deep reaction microchannel. A diffusion time across the channel width of about 5 s is estimated when diffusion coefficients are assumed of the order of 10^{-9} m^2 s^{-1} (for small molecules).

In the 50 μm wide and 20 μm deep microchannels, the improved mixer is based on vertical diffusion. The two reagent streams merge together over a length of 100 μm (channel depth is 2 × 20 μm), after which one channel stops and consequently the diffusion length is reduced to 20 μm. With this design mixing is theoretically enhanced 6.25 times compared to the horizontal diffusion design with identical dimensions and 25 times compared to the 100 μm channels. Assuming a diffusion coefficient $D = 1 \times 10^{-9}\ m^2\,s^{-1}$, based on Equation (9.1), a diffusion time $t_d = 200$ ms is calculated in the 20 μm deep channels. This value drops to $t_d = 20$ ms with $D = 1 \times 10^{-8}\ m^2\,s^{-1}$ for small ions (*i.e.* H^+) in water.

The performances of the new mixer design were evaluated by computations. In the theoretical simulation complete mixing of water molecules in water injected at a total flow rate of 2 μL min^{-1} is achieved within a few tens of milliseconds. Figure 9.1a and b show the typical concentration profiles inside the mixer. From these results it can be concluded that indeed a good "lamination" takes place which improves the mixing.

A study of the mixing dynamics in both 100 μm wide by 50 μm deep and 50 μm wide by 20 μm deep channels was performed by injecting a solution of fluorescein in water (pH = 8.4) in inlet A and a solution of hydrochloric acid in water (pH = 4) in inlet B of the chip at total flow rates ranging between 0.2 and 2 μL min^{-1}. Monitoring the pH-dependent quenching of the fluorescence of solution A, induced by mixing with solution B, allows one to study the on-chip mixing dynamics.[44] The results of this study in the 50 μm wide by 20 μm deep channels are reported in Figure 9.2. At a total flow rate of 0.2 μL min^{-1} (Figure 9.2a), complete mixing is observed within the first 100 μm of the reaction channel (where the depth is 40 μm), which correspond to a residence time $t_{res} = 60$ ms. As evident from the images of Figure 9.2, the merging zone with the two overlapping channels is rather long, compared to the total length of the fluorescence tail. This

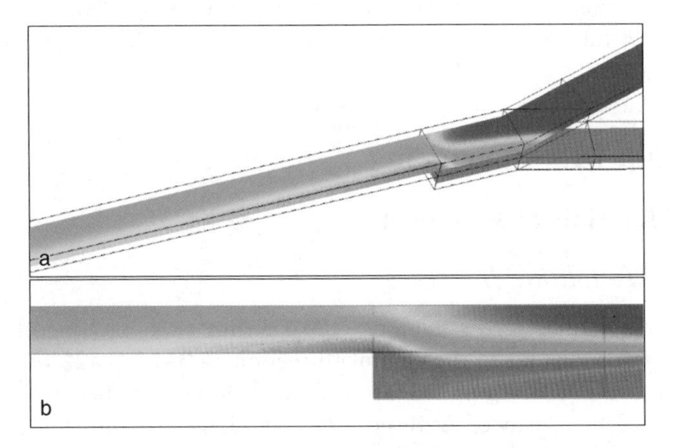

Figure 9.1 (a) Steady-state simulation of the mixing of two laminated flows in the microchannel, top view half-way in the channel. (b) Side view of the mixing profile at the junction of the two channels; because of the channel height reduction diffusion is increased.

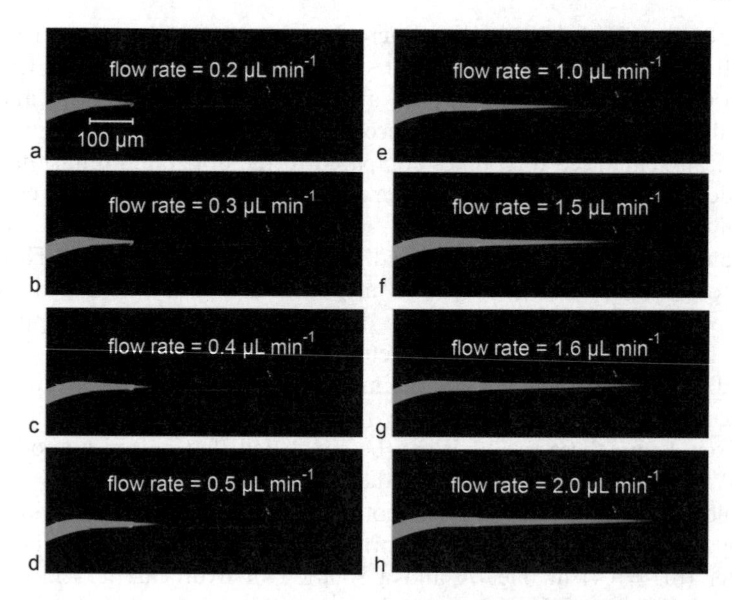

Figure 9.2 Sequence of fluorescence microscope images describing the mixing dynamics in the 50 μm wide by 20 μm deep channels. Quenching of the fluorescence of a solution of fluorescein indicates complete mixing of the reagents injected at total flow rates ranging between 0.2 and 2 μL min^{-1} (from a to h). The double amount of fluorescein in the 40 μm deep channel (first 100 μm) compared to that in the 20 μm deep channel is clearly visible in the images, due to the higher fluorescence intensity.

means that the mixer might be further improved by reducing the merging zone. As expected, based on the larger diffusion distance, a much longer mixing time (of the order of seconds) was observed in the 100 μm wide by 50 μm deep channels. The results of this experiment are in good agreement with the theoretical simulation (see above).

9.3 Chip-MS Interfacing

9.3.1 Materials and Methods

9.3.1.1 Monolithic Interface

The spraying channel (cross-section = 100 × 50 μm^2) was obtained by dicing a glass chip perpendicular to the channel direction. A 3M FC-722/FC-40 fluoro-containing coating was applied around the spraying orifice to change the wettability of the surface. A fused silica capillary for infusing the sample solutions into the channel was fitted into the powder-blasted inlet reservoir using Araldite Rapid epoxy adhesive (Ciba-Geigy). Nebulizing argon gas was connected to the chip *via* a Teflon tube, screwed into a glued Plexiglas piece by means of a Minstack™ (Lee) type of connector.

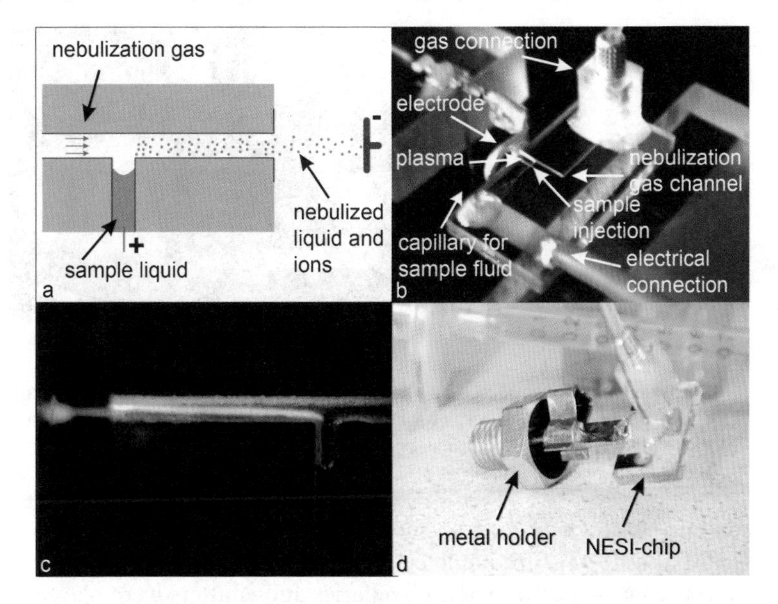

Figure 9.3 (a) Principle of the monolithically integrated NESI chip: the sample solution is injected into a nebulization gas such that a nebula of small droplets is created that ionize much more easily with less chance on large droplet formation. (b) Ionization test setup with the gas nebulization chip. (c) At high field strengths, stable plasmas could be obtained. (d) Chip glued onto the metal holder used to interface the microreactor to the MS instrument.

9.3.1.1.1 Ionization Rate Evaluation Setup. The chip was mounted in the setup depicted in Figure 9.3b. An 8.3×10^{-3} M phosphate buffer of pH = 7.4 was dispensed at flow rates in the range of $\mathrm{nL\,min^{-1}}$ to tens of $\mathrm{\mu L\,min^{-1}}$. Various potentials were applied (up to 1000 V) *via* a platinum electrode sputtered in the inlet cup. A counter electrode was located at very close proximity (~5 mm) to the exposed channel at the chip's edge.

9.3.1.1.2 ESI-MS Process Evaluation Setup. The chip glued onto the metal holder depicted in Figure 9.3d was screwed on the Nanospray™ interface of the mass spectrometer. Cone gas (automatically controlled via the ESI-MS software) was set at a value of $260\,\mathrm{L\,h^{-1}}$. Nebulizing argon gas was dispensed at various pressures up to about 8 bar.

9.3.1.2 Modular Interface

Microreactor chips were mounted on a dedicated holder (Figure 9.4a and b), placed on the Nanospray™ interface of the mass spectrometer using a metal plate. Solutions were introduced on-chip via fused silica capillaries (o.d. = 360 μm) of 100, 40 or 20 μm (i.d.), depending on the microchannel cross-sectional

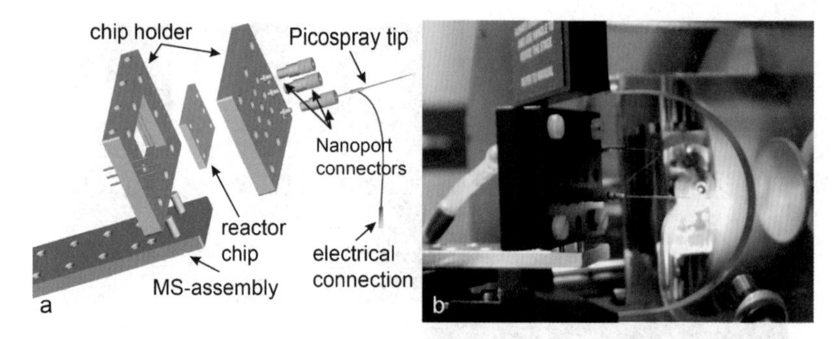

Figure 9.4 (a) Exploded view of the modular NESI-chip approach: the fibers and spray tip are connected to the chip using Nanoport™ connectors and the developed chip holder. (b) Picture of the mounted system in the mass spectrometer (turned for a better view).

dimension and on the flow rates. New Objective Picotip™ emitters (o.d. = 360 μm; i.d. = 20 and 40 μm; tip diameter = 5 and 10 μm) were used to spray the sample into the mass spectrometer. Capillaries and emitters were placed in the inlet and outlet cups, respectively, and kept in place by means of Upchurch Nanoport™ assemblies. The potential, controlled via the ESI-MS software, was applied through an electrical connection attached to the holder (see Figure 9.4).

9.3.1.2.1 ESI-MS Process Evaluation Experiments. ESI-MS experiments were performed using a Micromass (Manchester, UK) LCT electrospray time-of-flight mass spectrometer (ESI-TOF-MS). A 1.6×10^{-4} M solution of reserpine in a mixture of acetonitrile and water (1:1 v/v) was used as a standard. Acetic acid (0.1%) was added to the solution in order to enhance analyte protonation. For both monolithic and modular interfaces, samples were infused at flow rates ranging between a few tens of $\mathrm{nL\,min^{-1}}$ to a few $\mathrm{\mu L\,min^{-1}}$. Spectra were acquired at various capillary, cone and extraction voltages and optimized for each of the interfaces.

9.3.2 Results and Discussion

9.3.2.1 Monolithic NESI-chip MS Interface

In the monolithically integrated NESI-chip (Figure 9.3) the spray is generated directly at the edge of the chip. Although very attractive because of its simplicity, this design is known to suffer from a poor ionization signal stability, which is mainly caused by the formation of large droplets at the outlet.[16,17] This problem becomes more prominent at high flow rates when the sample flow exceeds the maximum ionization rate. The hydrophilicity of the chip surface at the outlet also increases this effect substantially, since the liquid sample will start to spread out over the chip surface such that the Taylor cone will destabilize. Attempts to minimize this effect, as reported in literature, include hydrophobic coating of the chip outlet[17] or pneumatically assisting the spraying process.[52]

In order to avoid this effect in the monolithic NESI-chip, active nebulization by injecting the sample solution in a gas flow (nitrogen) was used, combined with the application of a fluoro-containing coating around the chip outlet to make it hydrophobic. Due to the injection of the solution into the nebulization gas flow, the liquid sample is already nebulized into very small droplets before exiting the chip, such that the critical charge density of the droplets is more easily obtained and consequently the ionization is enhanced and the chance of large droplet formation at the outlet is reduced. A schematic of the principle is shown in Figure 9.3a. An additional advantage of the forced nebulization is the possibility to adjust the nebulization and thereby the ionization grade separately from the liquid flow rate, which might provide more freedom to obtain a perfect ionization over a wider sample flow range.

The nebulization and ionization performances were evaluated in a setup as shown in Figure 9.3b. Phosphate buffer was fed into the chip and nebulized with nitrogen gas. The sample was electrically charged via a sputtered electrode in the connection cavity for feeding the sample flow. At potential differences of about 1000 V DC between this electrode and the counter electrode, placed at 2.5 mm from the chip outlet, stable plasmas could be observed as shown in Figure 9.3c. Interestingly, these plasmas occurred at atmospheric conditions and ignited virtually directly at the injection point. So the plasmas were located 2.5 mm inside the chip (distance between the injection point and outlet) and 2.5 mm outside. Although corona discharge is a well-known unwanted phenomena in MS analysis, the ease with which the plasma could be obtained in a controlled manner indicates that the setup may be employed for optical spectral analysis of liquids.[53]

In order to test whether the monolithic NESI-chip interface would give a stable m/z MS signal, the chip was mounted on a standard Nanospray™ source of the mass spectrometer, using the holder depicted in Figure 9.3d. A standard solution of reserpine was injected into the chip at flow rates varying between a few hundred nL min^{-1} and a few tens of μL min^{-1}. ESI-TOF mass spectra were acquired at a capillary voltage varying between 1500 V and 3500 V. The nebulizing gas (argon) flow turned out to be the most important parameter influencing sample ionization. As a result a stable analyte signal could be obtained at a broad range of sample infusion flow rates by simply tuning the gas pressure.

An example of a total ion current (TIC) spectrum and the corresponding mass spectrum is shown in Figures 9.5a and b, respectively. The graphs were obtained at 10 μL min^{-1} flow rate and 6.5 bar argon nebulization gas pressure, 260 L h^{-1} cone gas flow and 3300 V, 48 V and 15 V for the capillary, cone and extraction voltages, respectively. Over a time frame of 4 min the measured standard deviation of the total ion current was less than 8%, which is comparable with values for chip-MS reported in the literature.[29,30]

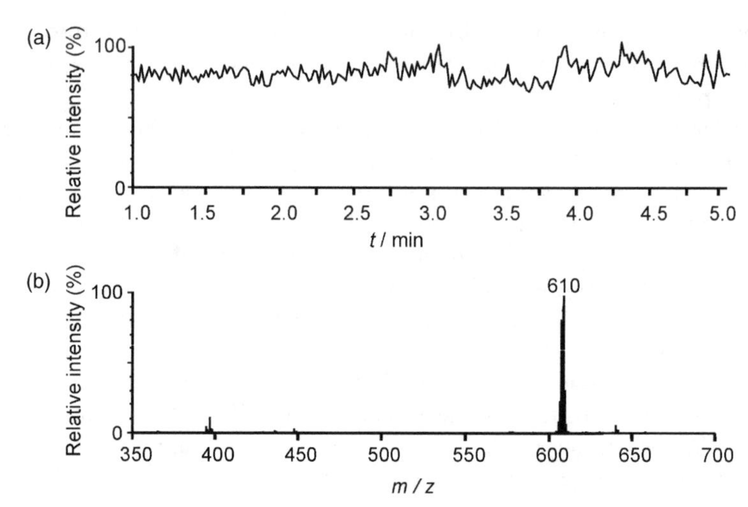

Figure 9.5 (a) Total ion current and (b) mass spectrum of a 1.6×10^{-4} M solution of reserpine in a mixture of acetonitrile and water (1:1 v/v) with 0.1% acetic acid, obtained with the monolithically integrated NESI-chip interface.

Although the results indicate a good ionization stability, the problem of the unwanted formation of relatively large droplets could not be completely solved. This is likely due to the injection geometry: since liquid is injected from one side, a liquid film occurs at one side of the nebulization channel. Further optimization of the injection geometry, such that the liquid is enclosed by the nebulization gas at both sides, might reduce this unwanted effect.

9.3.2.2 Modular NESI-chip MS Interface

Modular chip to MS interfaces include micromachined nozzles[20,29] attached to the chip and conventional needles attached to the microchannel in various ways.[19] These designs generally offer several advantages, such as high spray stability and higher sensitivity and resolution of the MS analysis over monolithically integrated devices. Nevertheless, they often require the use of glue or other materials that, upon prolonged contact with organic solvents, may pollute the mass spectra of the reaction mixture. Furthermore, the modular integration may introduce large dead volumes between the reactor and the needle or nozzle, due to the relatively large volumes of the through-wafer connection holes.

A solution for these drawbacks was found in using commercially available New Objective Picospray™ emitters with distal coating, mounted onto the chip with Upchurch Nanoport™ connectors. To facilitate mounting of the needles and to connect the chip with capillary fibers to the mass spectrometer, a holder was designed as shown in Figure 9.4. The electrical connection to the liquid was realized by using a metal-coated needle, electrically contacted *via* a gold-plated miniature connector. Electrical contact to the fluid *via* sputtered electrodes, similar to the monolithically integrated chip, could be realized as well. The Nanoport™ connectors are designed to seal at the chip surface when using cylindrically drilled holes of 1 mm diameter. However, since the chips used contain conically shaped holes, created by powder blasting, the chemically inert perfluoro elastomer (Perlast™) ferrules will seal inside the holes as well, leaving a dead volume of less than 11 nL, which could be further reduced by using 1.5 mm long ferrules, thinner glass wafers and a further optimized hole shape.

Experiments were carried out by injecting a standard solution of reserpine into the mass spectrometer through the chip-based interface. A stable spray was obtained for a wide range of flow rates between 20 and 100 nL min^{-1} and between 100 and 600 nL min^{-1} when using emitters with tip diameters of 5 and 10 μm, respectively. An example of a TIC spectrum and the corresponding mass spectrum is shown in Figures 9.6a and b, respectively. The small standard deviation of the ionization current of only 1%, calculated over 2 min, indicates a remarkably high ionization stability. The use of specific Picotip™ emitters to generate the spray guarantees a high-quality Taylor cone during the ionization process, which certainly leads to an optimal signal. Moreover, the use of Nanoport™ fittings in combination with the fabricated holder and the choice of low-volume connectors and a highly controllable flow-driven pumping method allow the high quality of the ionization process to be maintained even after coupling of the tips with the microreactor unit. Besides the good performance capability, the modular approach offers several advantages. The sprayer can in

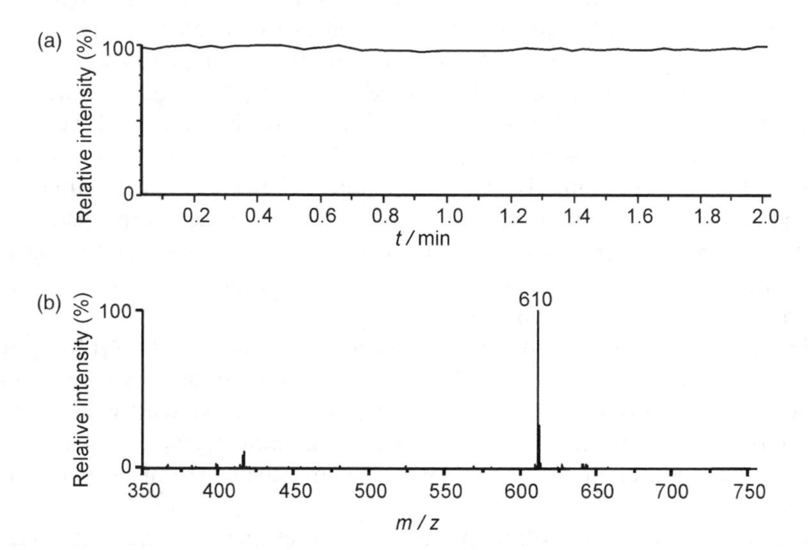

Figure 9.6 (a) Total ion current (TIC) and (b) mass spectrum of a 1.6×10^{-4} M solution of reserpine in a mixture of acetonitrile and water (1:1 v/v) with 0.1% acetic acid, acquired using the modular NESI-chip MS interface.

fact easily be replaced independently from the chip in case of clogging or rupture. Also, the interface can easily be adapted to study a variety of chemical reactions by simply mounting a suitable microreactor in the holder. Furthermore, the modular interface can be connected to analytical instruments other than a mass spectrometer by simply replacing the needle at the outlet with a fused silica capillary.[45]

Due to the better performances and the versatility offered by the design, the modular interface, based on the use of commercially available Picotip™ emitters, was used in the final setup. The modular integration allows the replacement of the spraying tips, independently from the more expensive chips, as often as required to avoid sample cross-contamination. More importantly, the microreactors can be easily replaced, offering the possibility to study a wide variety of chemical reactions as demonstrated, for example, in the qualitative and quantitative determination of the binding strength of supramolecular interactions as well as the study of the kinetics of NBDPZ.

9.4 On-line Monitoring of Supramolecular Interactions by NESI-MS

9.4.1 Introduction

Supramolecular chemistry,[54,55] the chemistry beyond the molecule, has rapidly penetrated into many areas. It involves the study of processes based on noncovalent interactions and mimics the role of nature in controlling biochemical processes by means of weak interactions such as hydrogen bonding, van der Waals forces, hydrophobic interactions and electronic interactions.

Over the past decade, ESI-MS has been successfully used to study and characterize supramolecular complexes,[56–58] after the detection of noncovalent receptor–ligand complexes by ESI-MS was first reported by Ganem *et al.* in 1991.[59] A number of methods have been reported in the recent literature for the quantitative determination of noncovalent binding interactions using soft ionization mass spectrometry.[60,61] However, to the best of our knowledge, supramolecular interactions have not yet been studied in a microreactor interfaced to a mass spectrometer.

The work described in this section will demonstrate that the NESI system provides a powerful tool to qualitatively as well as quantitatively study supramolecular interactions. Two well-known supramolecular systems were studied: Zn–porphyrin complexation with nitrogen-containing ligands in acetonitrile and the inclusion of small organic molecules into the cavity of β-cyclodextrin in water.

The Zn–porphyrin coordination reactions (Scheme 9.1) were studied in 100 μm by 50 μm microchannels (Figure 9.7a), while the β-cyclodextrin (β-CD) encapsulation reactions (Scheme 9.2) were studied using 50 μm by 20 μm reaction channels (Figure 9.7b).

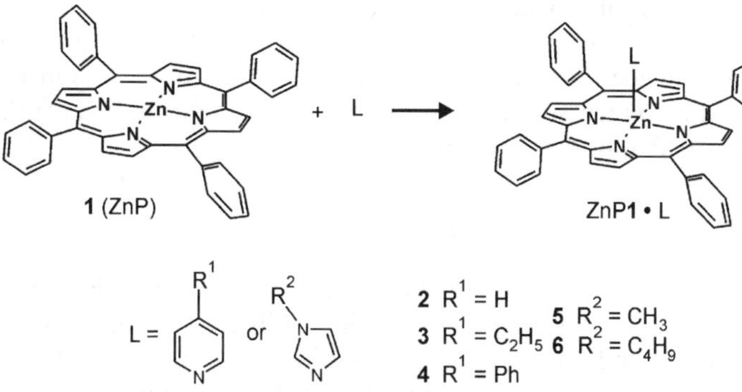

1 (ZnP) + L → ZnP1 • L

L = (4-substituted pyridine, R^1) or (N-substituted imidazole, R^2)

2 R^1 = H
3 R^1 = C_2H_5
4 R^1 = Ph
5 R^2 = CH_3
6 R^2 = C_4H_9

Scheme 9.1 Coordination of nitrogen-containing heterocycle ligands (L).

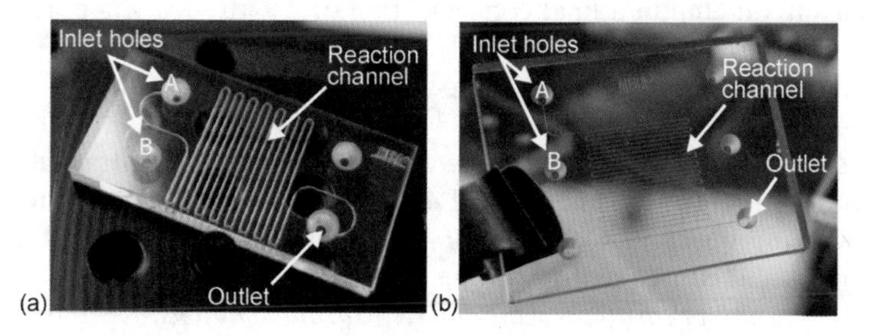

Figure 9.7 Photographs of the glass microreactors used: (a) the larger (100 μm wide by 50 μm deep) channels and (b) the smaller (50 μm wide by 20 μm deep) channels.

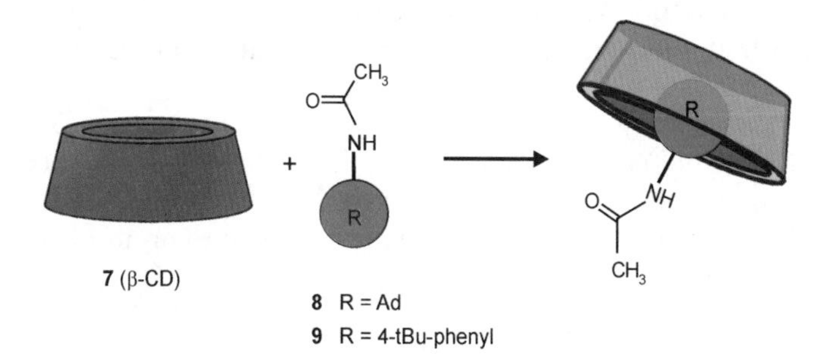

Scheme 9.2 Encapsulation of *N*-(1-adamantyl)acetamide (**8**) and 4-*tert*-butylacetanilide (**9**) in the hydrophobic β-CD (**7**) cavity.

9.4.2 Materials and Methods

9.4.2.1 Chemicals

All reagents were purchased from Sigma-Aldrich (Steinheim, Germany) and used without further purification. All solutions were prepared using HPLC grade solvents (Biosolve Valkenswaard, The Netherlands). Acetonitrile and a mixture of DMSO and water (1:9 v/v) were used as solvents to study the Zn–porphyrin **1** (Scheme 9.1) and β-CD (**7**; Scheme 9.2) complexation reactions, respectively, due to their suitability for both ESI-MS analysis and for the target complexation reaction.

9.4.2.2 On-chip Experiments

On-chip experiments were performed by injecting reagent solutions into the microreactor at flow rates ranging from 100 to 400 nL min^{-1} (100 μm by 50 μm channels), and 20 and 100 nL min^{-1} (50 μm by 20 μm channels), by means of 100 μL syringes mounted on a CMA/102 microdialysis pump.

Qualitative studies of both types of complexation reactions were performed injecting solutions of host and guest molecules in a 1:2 and 1:1 ratio, resulting upon mixing on-chip in a final concentration of 2×10^{-4} M Zn–porphyrin (**1**) and 2×10^{-5} M β-CD (**7**), respectively.

Quantitative binding studies were performed in both cases by on-chip mass spectrometric titration experiments.

(a) Zn–porphyrin titrations by on-chip dilution. On-chip titrations of Zn–porphyrin with both pyridine (**2**) and 4-phenylpyridine (**4**) were performed by simply varying injection speeds of 10^{-3} M solutions of reagents from 100 to 20 nL min^{-1} and consequently varying the mixture composition.

(b) Zn–porphyrin titrations with different titrant solutions. Keeping the total flow rate constant at 400 nL min^{-1} and the ratio between injection speed at the two inlets at 1:1, two titration experiments were performed by injecting stock solutions of Zn–porphyrin at concentrations of 2×10^{-4} M and 4×10^{-4} M, respectively, in inlet A. Solutions of pyridine (**2**) at concentrations ranging from 4×10^{-5} M to 1×10^{-3} M and from 8×10^{-5} M to 2×10^{-3} M were injected in inlet B, in both titrations resulting in on-chip final **1**/**2** ratios of 5:1, 2:1, 1:1, 1:2 and 1:5.

(c) β-CD titrations by on-chip dilution. On-chip titrations of β-CD ($C_{\text{on-chip}} = 5 \times 10^{-6}$ M) were performed by injecting solutions of *N*-(1-adamantyl)acetamide (**8**) at concentrations ranging from 2.5×10^{-6} to 1×10^{-4} M. For all β-CD encapsulation reactions, formic acid (0.1%) was added to promote protonation of the otherwise neutral molecules.

9.4.2.3 ESI-MS Detection

The complexation behavior was studied by spraying the reaction mixtures directly into the mass spectrometer, allowing real-time reaction monitoring. All complexations were followed by simultaneously monitoring the signal intensities of the singly charged free host (H) and host–guest (HG) complex. Spectra were acquired in the positive ion mode at a capillary voltage ranging from 1.2 kV to 2.0 kV, at a sample cone of 15 V to prevent complex dissociation, and an extraction cone voltage of 2 V. The source temperature was kept constant at 100 °C and the cone gas between 80 and 90 L h^{-1}. The fractions of complexes

$f_{(HG)}$ were calculated as the ratio between the intensities (I) of the complex and the total host (free and bound) according to the equation

$$f_{(HG)} = \frac{I(HG)}{I(HG) + I(H_f)} \tag{9.2}$$

Association constants were calculated by fitting the experimental titration curves to a 1:1 binding model based on K_a (Equation (9.3)) and the mass balance (Equations (9.4) and (9.5)), by applying a least squares optimization routine:

$$K_a = \frac{[HG]}{[H][G]} \tag{9.3}$$

$$[H]_t = [HG] + [H]_f \tag{9.4}$$

$$[G]_t = [HG] + [G]_f \tag{9.5}$$

$[H]_f$ and $[G]_f$ are the concentrations of free host and free guest, respectively, and $[H]_t$ and $[G]_t$ are the total host and guest concentrations, respectively.

9.4.2.4 Ultraviolet/Visible Titrations

Ultraviolet/visible (UV/vis) titrations were performed with a Varian Australia Pty Cary 3E UV/vis spectrophotometer. Titrations of Zn–porphyrin **1** were carried out by adding 5 μL aliquots (0.5 equiv.) of 10^{-2} M solutions of pyridine **2** or **4** to 1 mL of a 10^{-4} M Zn–porphyrin **1** solution, up to a maximum of 10 pyridine equivalents. Association constants were calculated by fitting the experimental titration curves to a 1:1 binding model based on K_a (Equation (9.3)), on the mass balances (Equations (9.4) and (9.5)) and the Lambert–Beer law by applying a least squares optimization routine.

9.4.3 Results and Discussion

9.4.3.1 On-chip Complexation Studies

Metalloporphyrins, which include important complexes such as chlorophyll and hemoglobin, represent a vast and unique group of compounds. By varying the metal center, the complexation behavior and characteristics of the metalloporphyrin can be dramatically affected. This great diversity is one of the reasons why metalloporphyrins have found broad application in the production of dyes, semiconductors and catalysts, as well as in numerous studies spanning the chemical and biological fields.[62]

In Zn–porphyrin **1** the metal center has one binding site available for coordination of a guest molecule such as pyridine (**2**), 4-ethylpyridine (**3**), 4-phenylpyridine (**4**), *N*-methylimidazole (**5**) and *N*-butylimidazole (**6**), to yield the corresponding 1:1 complexes. Experiments were carried out by injecting solutions of 2×10^{-4} M Zn–porphyrin **1** ($C_{on-chip} = 10^{-4}$ M) and 10^{-3} M pyridine (**2**) ($C_{on-chip} = 5 \times 10^{-4}$ M) in acetonitrile into the microreactor at decreasing total flow rates (400, 200 and 100 nL min^{-1}) and keeping constant the ratio between the flow rates at the two inlets (1:1). Figure 9.8 shows the spectra of the mixture after 2.5 min, 4.9 min and 9.9 min. The calculated fractions of Zn–porphyrin–pyridine complex (**1·2**) formed on-chip after these residence times are reported in Table 9.1. The fraction of complex formed after 2.5 min is identical to that at the longest residence times, indicating that thermodynamic equilibrium is already reached after 2.5 min (*i.e.* at a flow rate of 400 nL min^{-1}). This result was confirmed by control experiments performed by analyzing aliquots withdrawn from a mixture of solutions of 10^{-4} M Zn–porphyrin **1** and of 5×10^{-4} M pyridine (**2**) at different times.

It is known[63] that pyridine derivatives, such as **2**, **3** and **4**, have a comparable binding affinity for **1**, whereas that of imidazoles **5** and **6** is higher. The behavior

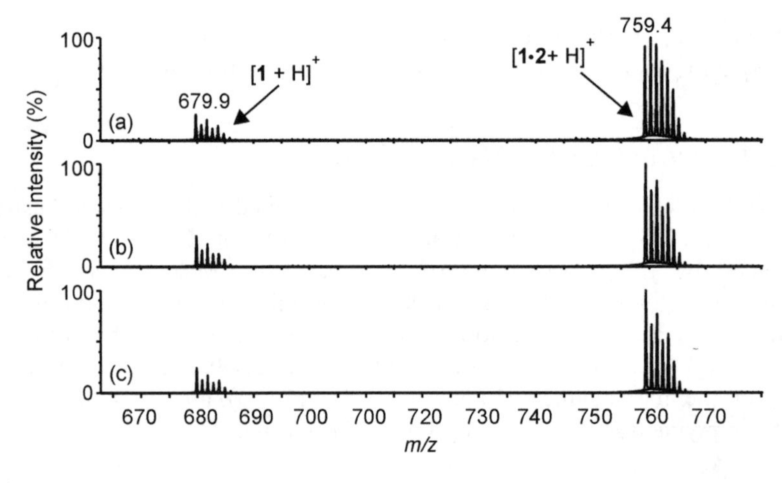

Figure 9.8 ESI-TOF mass spectra showing free Zn–porphyrin **1** (*m*/*z* 679.9) and the **1·2** complex (*m*/*z* 759.4) formed on-chip by mixing **1** and pyridine (**2**) after (a) 2.5 min, (b) 4.9 min and (c) 9.9 min residence time.

Table 9.1 Fractions of Zn–porphyrin–pyridine complex **1·2** $f_{(\mathbf{1}\cdot\mathbf{2})}$ formed on-chip at three residence time values.

Flow rate (nL min^{-1})	*Residence time (min)*	$f_{(\mathbf{1}\cdot\mathbf{2})}$ [a]
400	2.46	0.78 ± 0.01
200	4.93	0.77 ± 0.01
100	9.85	0.79 ± 0.01

[a] $f_{(\mathbf{1}\cdot\mathbf{2})} = I_{\mathbf{1}\cdot\mathbf{2}}/(I_{\mathbf{1}\cdot\mathbf{2}} + I_{\mathbf{1}})$.

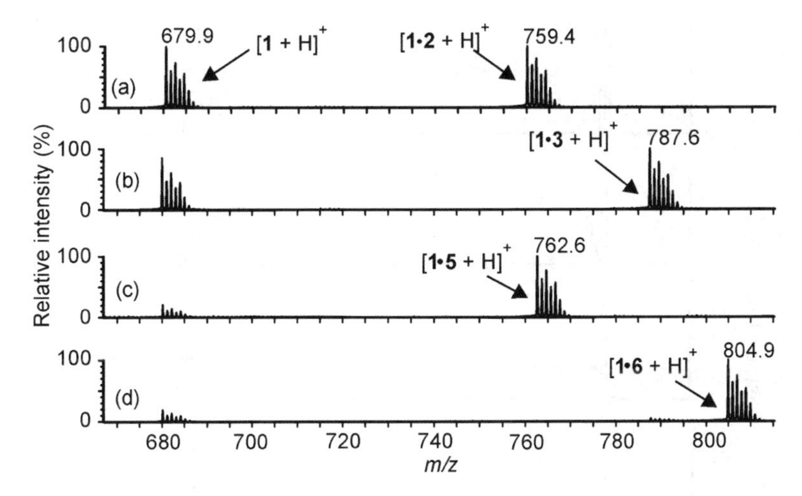

Figure 9.9 ESI-TOF mass spectra showing the free Zn–porphyrin **1** (m/z 679.9) and the complexes formed on-chip when injecting a solution of 2×10^{-1} M of **1** in inlet A and 4×10^{-4} M solutions of (a) pyridine (**2**), (b) 4-ethylpyridine (**3**), (c) *N*-methylimidazole (**5**) and (d) *N*-butylimidazole (**6**) in inlet B.

is evident from the ESI-MS spectra shown in Figure 9.9 where the relative intensities of the signals of the imidazole complexes **1·5** and **1·6** (Figures 9.9c and d) are significantly higher than those of the pyridine complexes **1·2** and **1·3** (Figures 9.9a and b). The four spectra were acquired in a total time of about half an hour, by mixing on-chip a 2×10^{-4} M solution of **1** ($C_{\text{on-chip}} = 10^{-4}$ M) injected in inlet A with 4×10^{-4} M solutions of, respectively, **2**, **3**, **5** and **6** ($C_{\text{on-chip}} = 2 \times 10^{-4}$ M) injected in inlet B consecutively. Before switching between different guest solutions, solvent was flushed into inlet B. During all the experiments, the flow rate at each inlet was kept constant at 200 nL min^{-1}. These experiments clearly demonstrate the ability of the NESI-chip interface to provide a very fast screening tool, giving qualitative information about the relative binding strength of different ligands in a high-throughput fashion.

A quantitative determination of the binding strength of pyridines **2** and **4** with **1** in acetonitrile was achieved by on-chip titration experiments. Figure 9.10 shows a typical ESI-TOF mass spectra sequence, collected during an on-chip titration experiment performed by varying the ratio between the injection speed of **1** and **2** stock solutions. Each ESI-TOF mass spectrum was recorded after an equilibration time of about 10 min between two different speed ratios. The experimental titration curves (data points) and the corresponding fit to a 1:1 model (solid lines) for both pyridines **2** and **4** are reported in Figure 9.11. For **2** and **4**, K_a values of $(4.6 \pm 0.4) \times 10^3$ M^{-1} and $(6.5 \pm 1.2) \times 10^3$ M^{-1}, respectively, were calculated.

For comparison, the same complexations were studied in acetonitrile by UV/vis spectrophotometry. Fitting the data to a 1:1 binding model, K_a values of 1.1×10^3 M^{-1} and 1.5×10^3 M^{-1} were found for pyridines **2** and **4**, respectively.

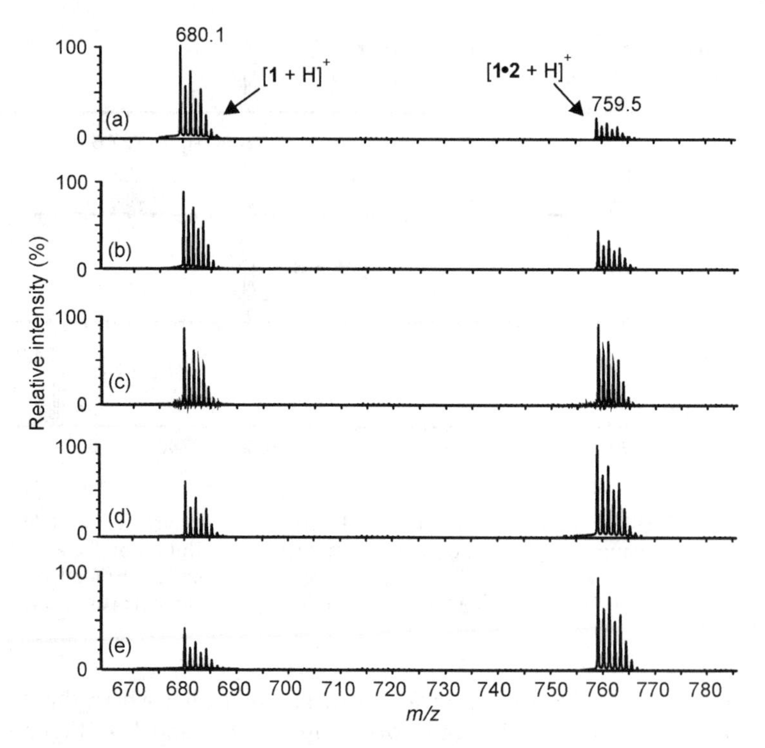

Figure 9.10 On-chip ESI-MS titration of Zn–porphyrin **1** with pyridine (**2**) performed at varying reagent injection speed ratios: (a) 9:1, (b) 8:2, (c) 1:1, (d) 3:7 and (e) 1:9.

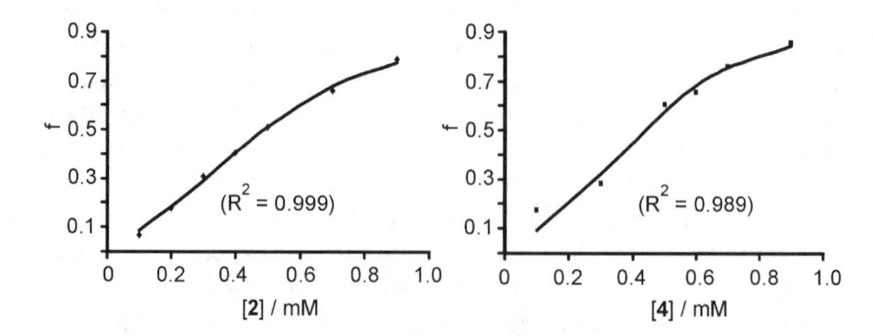

Figure 9.11 Experimental (points) host–guest fractions as a function of the guest concentration calculated from on-chip ESI-TOF-MS titrations of Zn–porphyrin **1** with pyridine (**2**; left) and 4-phenylpyridine (**4**; right), and corresponding fits to a 1:1 binding model (solid lines).

The K_a values of the complexes of pyridines **2** and **4** with Zn–porphyrin **1** determined using the on-chip ESI-MS method are about four times higher than those determined by UV/vis spectrophotometry. This difference indicates that, in the investigated metal–ligand complexation, the ESI-TOF mass spectra do not reflect factually the concentration of the species in solution. Differences in the relative vaporization and ionization efficiencies[60] of the free and bound Zn–porphyrin **1** are very likely to be responsible for the deviation between solution and gas-phase equilibria. In ESI-MS a number of instrumental[64] and chemical[65] parameters are known to influence the ionization efficiency and therefore the signal intensity of a given ion. Accounting for all these parameters, a transfer coefficient t_X can be defined for a given compound X, which corrects the deviations of the ESI-MS signal intensity (I_X) from the real concentration of the species in solution [X] *via* $I_X = t_X \times [X]$.[60]

Most of the quantitative studies reported in the literature assume similar transfer coefficients for free host and host–guest complex, in particular when the guest is much smaller than the host. However, this assumption is not always correct as demonstrated by a few studies reported in a recent review by Daniel *et al.*[60] In particular, the difference in solvation energy between free host and host–guest in metal–ligand type of interactions has a large influence on the ionization efficiency of the species in solution.[65]

A higher ionization efficiency of the complexed Zn–porphyrins would explain the higher K_a values determined by ESI-MS as compared to those determined by UV/vis. However, in both cases the relative binding strength of the different ligands remains the same (ESI-MS: K_a **1·4**/K_a **1·2** = 1.4; UV/vis: K_a **1·4**/K_a **1·2** = 1.4).

Competition experiments were performed, mixing on-chip a 10^{-3} M solution of **1** with a mixture of pyridines **2** and **4** (10^{-3} M total concentration) at 400 nL min^{-1}, 200 nL min^{-1} and 100 nL min^{-1} flow rates. In all cases the ratio between the fractions of the **1·2** and **1·4** complexes is 1.40 ± 0.39, which is consistent with the ratio found in the independent experiments (see above).

The experiments described so far have been carried out by diluting reagent solutions on-chip in a continuous flow fashion. Because of the absence of a rinsing step between different speed ratios, a possible limitation of the on-chip titration method might be the contamination between solutions at different concentrations. To verify the reliability of the method, titration experiments of **1** with **2** were performed keeping the Zn–porphyrin concentration on-chip constant at 1×10^{-4} M and 2×10^{-4} M (respectively) and injecting solutions of **2** at different concentrations in the second inlet. The experimental titration curves (data points) and the corresponding fits (solid lines) are shown in Figure 9.12. The calculated K_a value of $(5.3 \pm 0.8) \times 10^3$ M^{-1} is in good agreement with that determined with the titration performed by varying the reagents flow ratios (see above). The values of the correlation coefficient between experimental and calculated sets of data ($R^2 = 0.991$ and $R^2 = 0.996$) indicate in both cases a good fitting of the experimental data to the theoretical model. These results clearly show that, upon changing the velocity ratios within a single on-chip titration experiment, no significant contamination takes place, probably due to the very efficient mixing in the microchannel.

To study the validity of the method a second type of supramolecular host–guest complex was investigated, namely the encapsulation of small organic molecules into the cavity of β-CD (**7**) in aqueous media (Scheme 9.2), driven by

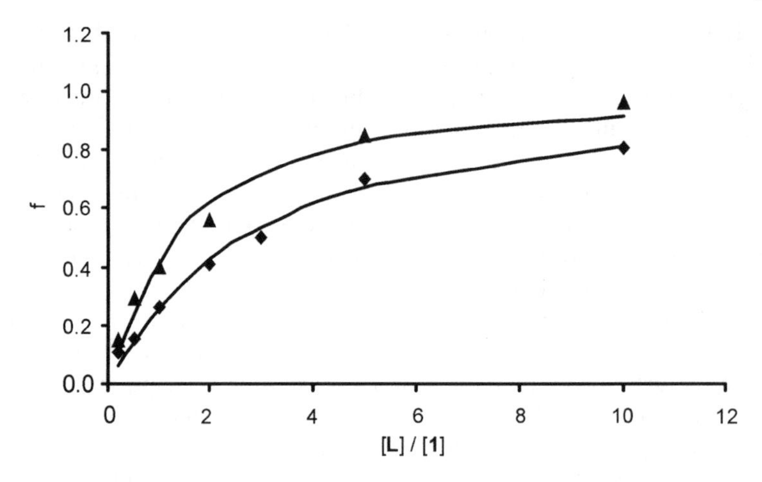

Figure 9.12 Experimental curves (points) determined by on-chip ESI-TOF-MS titrations of 10^{-4} M (▲) and 2×10^{-4} M (◆) Zn–porphyrin **1** with pyridine (**2**), and corresponding fits to a 1:1 binding model (solid lines).

the so-called "hydrophobic effect".[66] Due to their ability to encapsulate guests in their hydrophobic cavity, cyclodextrins[67] are interesting molecules for applications in drug carrier systems,[68] the food industry,[69] cosmetics[69] and various industrial processes.[70]

p-tert-Butylphenyl and adamantyl derivatives[71] are well known guests for β-CDs in aqueous solution. As guests, *N*-(1-adamantyl)acetamide (**8**) and 4-*tert*-butylacetanilide (**9**) were selected in order to minimize the difference between the ionization properties of the free neutral host (β-CD) and the host–guest complex.

First, an aqueous solution of 4×10^{-5} M β-CD (**7**; $C_{\text{on chip}} = 2 \times 10^{-5}$ M) was injected in inlet A of the chip. Equimolar solutions of **8** and, after flushing the channel with solvent, of **9** were injected in inlet B (Figure 9.13) at a flow rate equal to that of the β-CD solution. Table 9.2 summarizes the host–guest complex fractions $f_{(7 \cdot G)}$ determined by ESI-MS and those calculated based on microcalorimetric experiments.[72] It is evident that the ESI-MS data are in good agreement with those obtained with microcalorimetry.

Figure 9.14 shows the experimental titration curve (data points) and the corresponding fit (solid line) to a 1:1 binding model for the complexation of β-CD (**7**) with **8**. The K_a value of $(3.6 \pm 0.3) \times 10^4$ M^{-1} is in good agreement with that determined by microcalorimetry (Table 9.2). The somewhat lower K_a value found by on-chip ESI-MS is probably due to the 10% DMSO present in the reaction mixture to enhance the solubility of **8**.

Compared to the extensive studies on cyclodextrin inclusion complexes in solution,[67] corresponding studies in the gas phase are still very limited.[73,74] During the solution-to-vacuum transfer in the mass spectrometer, the effect of solvent and counter ions no longer occurs. A main consequence is that electrostatic interactions might become dominant, resulting in "unspecific binding" between host and

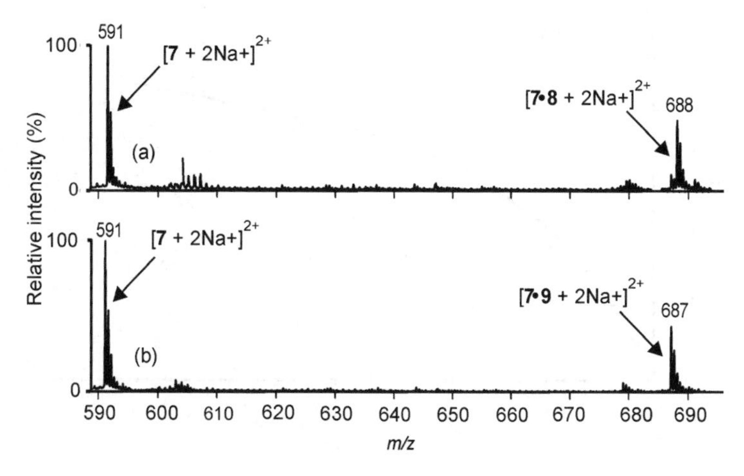

Figure 9.13 ESI-TOF mass spectra of the complexes formed on-chip when injecting equimolar solutions (4×10^{-5} M) of β-CD (**7**) and (a) *N*-(1-adamantyl) acetamide (**8**) and (b) 4-*tert*-butylacetanilide (**9**).

Table 9.2 Molar fractions of the β-cyclodextrin–guest complexes $[f_{(7\cdot G)}]^a$ determined by ESI-MS on-chip experiments and data based on microcalorimetric studies.

On-chip ESI-MS experiments		*Microcalorimetric data*[72]	
Guest	$f_{(7\cdot G)}$	K_a (M^{-1})	$f_{(7\cdot G)}$
7	0.39 ± 0.02	6.8×10^4	0.43
8	0.31 ± 0.02	3.0×10	0.30

$^a f_{(7\cdot G)} = I_{7\cdot G}/(I_{7\cdot G} + I_7)$.

guest, rather than the expected inclusion complex.[75] However, ESI-MS studies demonstrated that β-CD complexes with amino acids in the gas phase are in fact inclusion complexes.[76,77] In general, "a peak at the mass of an alleged complex is not a sufficient criterion to prove its specificity".[60] Compared to β-CD (**7**), the corresponding α-cyclodextrin has one α-D-glucose unit less, resulting in a smaller cavity. In experiments injecting on-chip equimolar (2×10^{-5} M) solutions of α-CD and guests **8** and **9**, no complex could be detected by ESI-MS. This clearly proves the size-selectivity of β-CDs towards the guests **8** and **9**.

9.5 On-line Study of Reaction Kinetics by NESI-MS

9.5.1 Introduction

To study reaction kinetics, real-time analysis of a reaction mixture composition is commonly carried out by quenched-flow[78,79] and stopped-flow[80–83] methods.

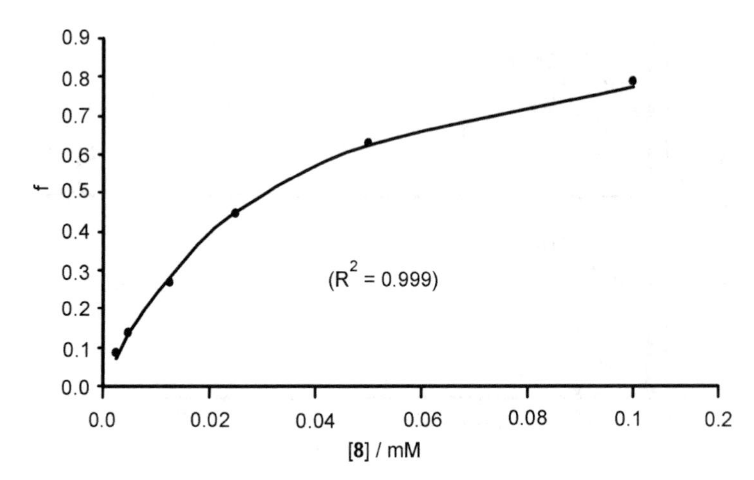

Figure 9.14 Experimental (points) host–guest fractions determined by on-chip MS titrations of 5×10^{-6} M β-CD (**7**) with *N*-(1-adamantyl)acetamide (**8**), and corresponding fit (solid line) to a 1:1 binding model.

Fast kinetic measurements are essential to study processes such as protein folding,[84] enzymatic kinetics[78–83] and chemical reactivity.[85,86] Both quenched- and stopped-flow methods use turbulence to induce fast mixing by means of very powerful tangential mixers. These methods require high flow rates (milliliters per second),[87] and suffer from poor reaction control and mixing efficiency due to the large reaction volumes involved. However, it has recently been demonstrated that a turbulent flow is not a requirement.[88] The first theoretical model to describe mixing under laminar flow conditions was proposed by Taylor[89] in 1953. More recently, in a theoretical study of reaction kinetics under laminar flow, Konermann[90] demonstrated that the distortion of the kinetic data due to the laminar regime is "surprisingly small".

Microreactors offer a good alternative to conventional laboratory-scale equipment for the study of reaction kinetics, because they allow low reagent consumption and fast mixing[50,51] as well as a continuous flow operative mode and real-time analysis.[91,92] Microfluidics chips have been used to study kinetics of biochemical systems[93–95] on various time scales (from seconds to minutes), mainly using optical detection techniques.[88,96,97] Recently, Song and Ismagilov[98] presented a microfluidic chip to perform kinetic measurements with millisecond resolution, which relies on chaotic mixing inside moving droplets. Due to the ease of fabrication and their compatibility with aqueous phase biological systems, polymers are the material of choice for the fabrication of microfluidics devices in most of the studies reported in the literature. However, polymer-based microchips might be less suitable for carrying out reactions that require the use of organic solvents. To the best of our knowledge there are no studies in which the use of glass microfluidics devices is reported to study the kinetics of organic reactions and to monitor reactions by MS.

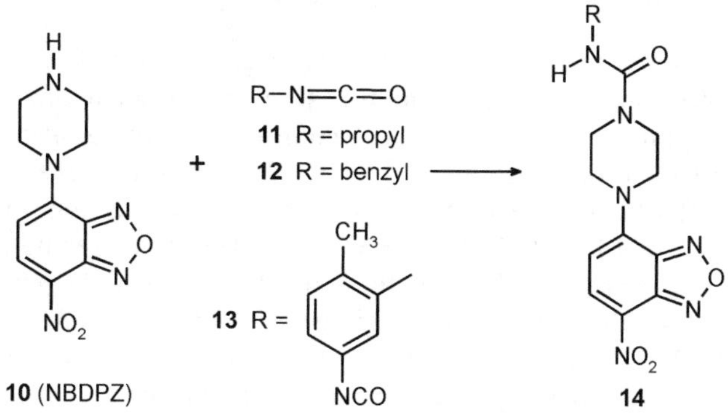

R−N=C=O
11 R = propyl
12 R = benzyl
13 R = (toluene with CH_3 and NCO)
10 (NBDPZ)
14

Scheme 9.3 Reaction of NBDPZ (**10**) with propyl isocyanate (**11**), benzyl isocyanate (**12**) and toluene-2,4-diisocyanate (**13**) to give the corresponding NBDPZ–urea derivatives (**14**).

In this section the use of the integrated NESI-chip device is described to study the kinetics of the reaction of NBDPZ with isocyanates (Scheme 9.3) as a model system. This reaction has been introduced by Karst and co-workers[99,100] for the derivatization of extremely reactive isocyanates[101] in order to allow their analysis by means of HPLC.

9.5.2 Materials and Methods

9.5.2.1 Chemicals

NBDPZ (**10**; Scheme 9.3) was synthesized and purified according to Vogel and Karst.[99] The different isocyanates were purchased from Sigma Aldrich (Steinheim, Germany). Solutions (10^{-6} M) of **10** and each isocyanate were prepared using dry liquid chromatography-MS grade acetonitrile (Biosolve, Valkenswaard, The Netherlands) in order to minimize a loss of reactant by hydrolysis reactions.

9.5.2.2 ESI-MS Detection

On-chip reactions were studied by directly coupling the microreactor to a Micromass (Manchester, UK) LCT ESI-TOF MS instrument and monitoring the decrease of the NBDPZ signal intensity at m/z 251 with time upon reaction of **10** with each isocyanate. Spectra were acquired in the positive ion mode at a capillary voltage ranging from 1.8 kV to 2 kV, at a sample cone of 35 V, and an extraction cone voltage of 2 V. The source temperature was kept constant at 100 °C and the cone gas flow between 80 and 90 L h^{-1}.

All measurement points were in triplicate and the total signal height was used for quantification.

9.5.2.3 On-chip Reaction Kinetics Studies

All experiments were performed by injecting reagent solutions into the chip by means of 100 μL blunt needle Hamilton syringes mounted on a CMA/102 microdialysis pump. The syringes were connected to fused silica capillaries (i.d. = 20 μm) by using Upchurch™ unions. Derivatization reactions of the isocyanates **11**, **12** and **13** with **10** were performed by injecting on-chip equimolar solutions of 10^{-5} M ($C_{on-chip} = 5 \times 10^{-6}$ M) of both **10** and isocyanates **11**, **12** and **13** in acetonitrile, keeping the ratio (1:1) between the injection speed of the reagent solutions constant. Various reaction times were exploited by varying the total flow rate from 100 to 40 nL min^{-1}. Longer reaction times were obtained by quickly replacing the chip in the holder during the experiments and therefore varying the reaction volumes from 71 to 284 nL (Table 9.3).

9.5.2.3.1 On-chip NBDPZ Calibration. Using "two inlet" chips (Figure 9.15a), the calibration of the NBDPZ signal intensity at m/z 251 was carried out by varying the ratio between the injection speeds of a stock solution of **10** (10^{-5} M) in inlet A of the chip and acetonitrile as a solvent in inlet B. Spectra were recorded at on-chip NBDPZ concentrations ranging from 10^{-5} M to 10^{-7} M. Additional calibrations were done using "three inlet"

Table 9.3 Lengths (L, cm) and volumes (V, nL) of the "two inlet" microchannels (Figure 9.15a) and residence times (min) at various flow rates.

Flow rate (nL min^{-1})	*$L = 7.1$ $V = 71$*	*$L = 10.5$ $V = 105$*	*$L = 19.1$ $V = 191$*	*$L = 28.4$ $V = 284$*
100	0.7	1	1.9	2.8
80	0.9	1.3	2.4	3.5
60	1.2	1.8	3.2	4.7
40	1.8	2.6	4.8	7.1
20	3.6	5.2	9.6	14.2

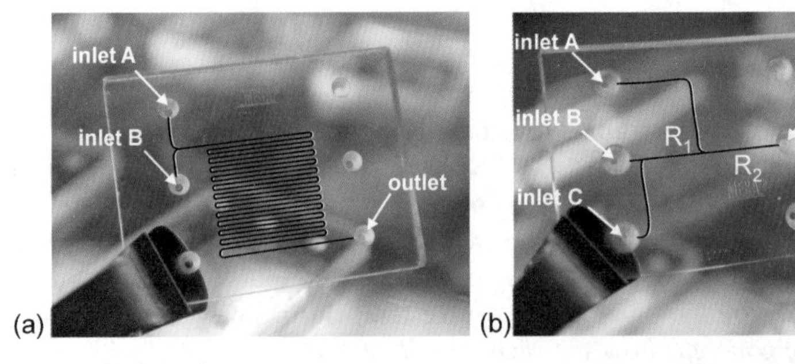

Figure 9.15 Photographs of (a) the "two inlet" and (b) the "three inlet" chips used to study reaction kinetics and ion suppression phenomena, respectively.

chips (Figure 9.15b). Equimolar stock solutions (1.25×10^{-5} M) of **10** and propyl isocyanate **11** in acetonitrile were injected into inlets A and C, respectively. Solvent was injected in inlet B. The injection speed at inlet C was kept constant at 80 nL min^{-1}, while those of inlets B and A were varied from 40 to 100 nL min^{-1} and from 100 to 40 nL min^{-1}, respectively. The total flow rate at the chip outlet was kept constant, resulting in a constant on-chip concentration of **11** (5×10^{-6} M) and a concentration of **10** varying between 5×10^{-6} M and 10^{-6} M. Based on the same principle, calibrations were performed injecting 1.25×10^{-5} M stock solutions of **10** in inlet A at flow rates ranging between 80 and 20 nL min^{-1}, 10^{-5} M stock solutions of either **12** or **13** in inlet C at a constant flow rate of 100 nL min^{-1} and solvent in inlet B at flow rates ranging between 20 and 80 nL min^{-1}.

9.5.2.3.2 On-chip Ion Suppression Studies. On-chip suppression studies were performed for each isocyanate using "three inlet" chips. Experiments were carried out injecting equimolar stock solutions (1×10^{-5} M) of NBDPZ (**10**) and isocyanates **11** and **12** in inlets A and C, respectively. The injection speed of A was kept constant at a value of 100 nL min^{-1}, resulting in an on-chip NBDPZ concentration of 5×10^{-6} M, while that of inlet C was varied from 10 to 90 nL min^{-1} (on-chip concentration of isocyanate varying between 4.5×10^{-6} M and 5×10^{-7} M). Solvent was injected in inlet B at flow rates ranging between 90 and 10 nL min^{-1}, to compensate for the decrease in injection speed of inlet C, resulting in a constant total flow rate at the chip outlet. Variations of the intensity of the NBDPZ signal at *m*/*z* 251 were monitored upon increasing the concentration of isocyanate. In the case of diisocyanate **13** the same method as described for isocyanates **11** and **12** was used, but injecting a 1.25×10^{-5} M NBDPZ stock solution in inlet A at a constant flow rate of 80 nL min^{-1} and varying the injection speeds at inlets B and C from 100 to 20 and from 20 to 100 nL min^{-1}, respectively.

9.5.2.4 Laboratory-scale Control Experiments

For the laboratory-scale experiments, a flow injection mass spectrometry (FIA-MS) setup was used. A G1312a model HPLC pump, which delivered the carrier stream of acetonitrile, and a G1313a model autosampler (both HP1100 series by Agilent, Waldbronn, Germany) for sample introduction were coupled to the mass spectrometric detector. All laboratory-scale MS experiments were performed on an Esquire 3000+ ion trap mass spectrometer from Bruker Daltonics (Bremen, Germany) equipped with a standard ESI source. For the best possible ionization and detection of the analyte, the conditions were optimized with respect to the intensity of the NBDPZ signal resulting in the following parameters: nebulizer gas (N_2) pressure: 40 psi; dry gas (N_2) flow: 9.0 L min^{-1}; dry gas temperature: 365 °C; capillary high voltage: 5000 V; capillary exit: 49.8 V; skimmer: 16.4 V; octopole 1 DC: 13.36 V; octopole 2 DC: 0.00 V; octopole RF: 152.5 Vpp; lens 1: −9.8 V; lens 2: −88.2 V; trap drive level: 35.0.

The FIA-MS system was controlled by a PC using Esquire Control 5.1 and HyStar 2.3; data evaluation was done with DataAnalysis 3.1 (all Bruker Daltonics, Bremen, Germany).

The reaction was started by combining 750 μL of the NBDPZ (**10**) solution (10^{-5} M) and 750 μL of the respective isocyanate (**11–13**) solution of equimolar concentration in a standard HPLC vial, thus resulting in starting concentrations of 5×10^{-6} M for both reactants. In order to follow the reactions progress, 5 μL aliquots were sampled from the reaction mixture every three minutes and injected into the FIA-MS system, where the decrease of the NBDPZ signal intensity at m/z 251 was monitored. Sampling cycles can be shortened by using faster instrumentation.

All measurement points were in triplicate and the peak area was used for quantification.

9.5.3 Results and Discussion

9.5.3.1 *Continuous Flow Microreactors*

The two types of continuous flow microreactors used are depicted in Figure 9.15. All isocyanate reactions were studied in "two inlet" chips (Figure 9.15a). A list of channel lengths (L), channel volumes (V) and respective residence times at flow rates ranging from 20 to 100 nL min^{-1} is given in Table 9.3.

Two "three inlet" microreactors, having a total reaction channel length of 28 and 10 mm, respectively, were used to perform control experiments and to study ion suppression phenomena. "Three inlet" chips differ from the "two inlet" ones in that they consist of two distinct mixing zones R_1 and R_2 (Figure 9.15b) instead of only one. The lengths of the two reaction channels R_1 and R_2 are 13 and 15 mm, respectively, in the 28 mm long microreactor and 4 and 6 mm, respectively, in the 10 mm long microreactor. Based on the channel length and on the flow rates at which all on-chip control experiments were carried out (200 nL min^{-1}), residence times of 4 and 5 s and 1 and 2 s were calculated, respectively, for the two mixing zones (R_1 and R_2) of the 28 mm and the 10 mm long chip. Since no significant conversion is observed for any of the investigated reactions at this time scale (Figure 9.16; see below), this type of microreactor is well suitable to perform control experiments.

An additional advantage of the "three inlet" chips over the "two inlet" ones is that reagents can be mixed in different ratios, keeping the concentration of one of the two reagent solutions and the total flow rate constant. This allows, for example, the study of the influence of co-solutes on the analyte ionization, minimizing side effects on the MS signal due to changes in flow rates.

9.5.3.2 *Reaction of NBDPZ with Isocyanates*

Mono- and diisocyanates are an important class of compounds, being widely used in many industrial processes such as the production of pesticides,

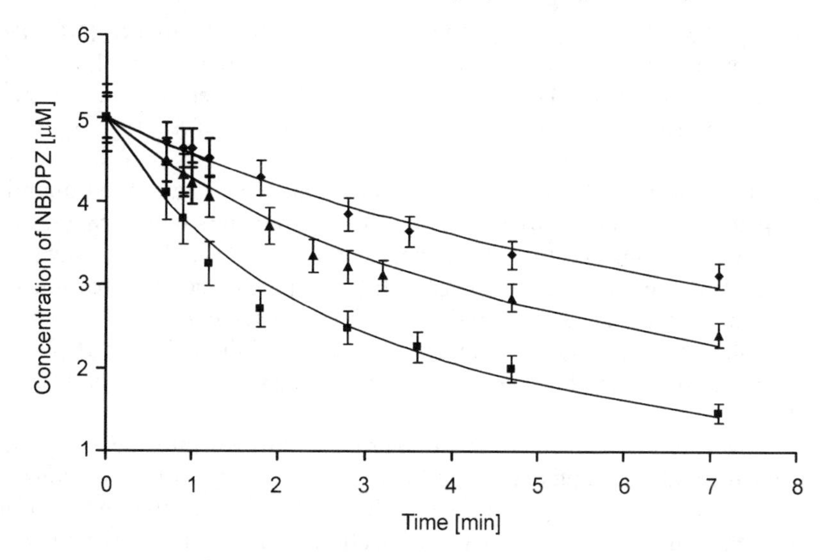

Figure 9.16 Reaction profile (points) of the on-chip derivatization of propyl isocyanate (**11**) (◆),benzyl isocyanate (**12**) (▲) and toluene-2,4-disocyanate (**13**) (■) with NBDPZ and corresponding fits to a second-order kinetics model (lines).

pharmaceuticals and polyurethanes.[102] Due to their high reactivity and toxicity, mono- and diisocyanates are relatively difficult to handle and consequently to study. Strategies based on the use of various derivatizing agents for the analysis of isocyanates have been widely used in analytical chemistry. Derivatization with amine-based reagents is favorable for the analysis of isocyanates compared with other methodologies exploited in the past.[103–105] The use of NBDPZ as a derivatizing agent to quantify the amount of mono- and diisocyanates in air samples has recently been proposed by Vogel and Karst.[99,100] The derivatization involves the reaction of NBDPZ with (di)isocyanates to give the corresponding urea derivatives (**14**; Scheme 9.3). Recently, Henneken *et al.*[101] developed a method using tandem mass spectrometry (MS-MS) to monitor NBDPZ and the corresponding methyl urea derivative (**14**; R = methyl).

Since NBDPZ gives a high-intensity stable singly charged MS signal, it is suitable to study its reaction with (di)isocyanates in the chip-based nanoelectrospray interface. In contrast, the urea derivatives (**14**) gave a very low intensity and unstable MS signal in our experimental setup. The selected reactions were therefore studied by monitoring the decrease of **10** upon reaction with the isocyanates by ESI-MS.

9.5.3.3 *On-chip Control Experiments*

Mass spectrometric calibration of the NBDPZ singly charged signal (m/z 251) revealed a linear relationship between [NBDPZ] and the signal intensity from

[NBDPZ] = 5×10^{-7} M up to [NBDPZ] = 5×10^{-6} M. At higher [NBDPZ], a deviation from linearity was observed, due to suppression phenomena. Reactions of isocyanates **11**, **12** and **13** with NBDPZ were therefore carried out in acetonitrile using equimolar concentration ratios of 5×10^{-6} M for all reagents.

ESI-MS is often used for analyte quantification. However, a good quantification method must take into account all those parameters that during sample ionization may influence the MS signal of the analyte under investigation. The nature of the analyte, the solvent, the instrument operative parameters, the capillary tip distance from the mass spectrometer cone and the sample infusion flow rates are some examples.[64] A study of the influence that some of those parameters might have on the NBDPZ signal at *m*/*z* 251 is therefore required.

The flow rate at which the analyte solution is infused into the mass spectrometer through the capillary can influence the efficiency with which ions are formed, transferred to the gas phase and detected.[106] The chip-based interface relies on the quick variation of reagent injection flow rates to allow a range of reaction times. In order to quantify the influence of the decrease of the total flow rate on the NBDPZ MS signal, a number of control experiments were carried out by diluting on-chip a solution of 10^{-5} M of NBDPZ with solvent. The ratio between the flow rates at the two inlets was kept constant (1:1). The total flow rate was decreased from 100 to 20 nL min^{-1} by steps of 20 nL min^{-1}. No significant changes in the NBDPZ signal at *m*/*z* 251 were observed upon decreasing the total flow rate. Standard deviation values of about 1% were calculated for measurements at each speed, indicating a remarkable stability of the NBDPZ signal. Overall, taking into account the NBDPZ signal intensity at different flow rates, the contribution to experimental error of the method is about 5%.

The composition of the sample and the presence of co-solute in the analyte solution are known to have an influence on the analyte ionization.[64] For the reactions of NBDPZ with isocyanates **11**, **12** and **13**, control experiments were performed in order to study the possible influence of the isocyanates (co-solutes) on the NBDPZ signal intensity at *m*/*z* 251. Calibrations of NBDPZ solutions containing the highest concentration of the corresponding isocyanate present in the mixture during the reaction (5×10^{-6} M) were carried out in "three inlet" chips (Figure 9.15b). The presence of isocyanates in the NBDPZ calibrating solution does not markedly influence the intensity of the NBDPZ signal, as confirmed by the fact that the averaged value of the slopes of the three calibration curves, $(7.94 \pm 0.24) \times 10^9$, coincides with the slope of the calibration curve of pure NBDPZ (7.94×10^9).

Mass spectrometric quantification of analytes in complex samples such as reaction mixtures often suffers from problems related to interference, especially through ion suppression effects.[107,108] Ion suppression is the result of the presence of a co-solute in the sample that has a volatility different from the analyte, which may significantly affect the ionization efficiency. This unwanted, though often observed phenomenon mainly occurs in biological samples rather than in organic systems.[34] By using three inlet chips on-line coupled with ESI-MS, the possible suppression of the NBDPZ signal at *m/z* 251 due to the presence of isocyanates in the reaction mixtures was investigated. For each of the isocyanates **11–13**, the intensity of the signal of a 5×10^{-6} M NBDPZ solution at *m/z* 251 was monitored, upon mixing with increasing isocyanate concentrations. The very short microreactor (*i.e.* 10 cm) ensures that no reaction takes place in the few seconds on-chip residence time, at 200 nL min^{-1} total flow rate (see above). No significant changes in the intensity of the NBDPZ signal were observed in any of the three cases, indicating that under the conditions used ion suppression phenomena are negligible. Standard deviation values of 1.3%, 0.8% and 1.1% ($n = 3$) were calculated for the measurements at increasing concentrations of isocyanates **11**, **12** and **13**, respectively.

9.5.3.4 On-chip Reaction Kinetics

Figure 9.16 shows the experimental reaction profiles for the on-chip reaction of NBDPZ with monoisocyanates **11** and **12** and diisocyanate **13**, determined by ESI-MS by simply varying the total injection speed of the reagent solutions. All reactions were carried out injecting on-chip equimolar (5×10^{-6} M) solutions of **10** and the corresponding isocyanate in acetonitrile. For each reaction, the concentration of **10** at different reaction times was calculated from MS data, using the equations of the calibration curves. Since all reactions were performed under the same conditions, the three decay curves give a quantitative indication of the reactivity of the three reagents. As expected from the nature of their R groups (see Scheme 9.3),[101] propyl isocyanate (**11**) shows the lowest reactivity, while benzyl isocyanate (**12**) is the most reactive of the three isocyanates.

Second-order kinetics is assumed as the most probable mechanism for the investigated reactions, because the concentration of **10** is equal to the concentration of the isocyanates (5×10^{-6} M). The same assumption is valid for diisocyanate **13**, since the formation of the monoderivatized product is much faster than that of the bis-derivatized product, as confirmed by ESI-MS data. The experimental data were fitted to a second-order kinetic model (Figure 9.16) using the equation

$$\frac{1}{[A]} - \frac{1}{[A]_0} = kt \qquad (9.6)$$

where [A] is the concentration of NBDPZ at a given time, $[A]_0$ is the initial concentration of NBDPZ, t [min] is the reaction time and k is the reaction rate constant. Rate constants (Table 9.4) were determined by applying a least squares optimization routine. To verify the assumption that the investigated reactions follow second-order kinetics, the initial rate constants (k_0) were also

Table 9.4 Slope of the first part of the reaction profiles (a) used to calculate the initial rate constant (see Equation (9.7)), second-order rate constant (k) and initial rate constant (k_0) of the reaction of NBDPZ with isocyanates **11–13**.

Isocyanate	a ($M\,min^{-1}$)	k ($M^{-1}\,min^{-1}$)	k_0 ($M^{-1}\,min^{-1}$)
11	4.1×10^{-7}	1.9×10^{4}	1.6×10^{4}
12	1.31×10^{-6}	7.0×10^{4}	5.2×10^{4}
13	6.36×10^{-7}	3.3×10^{4}	2.5×10^{4}

calculated from the slope (a) of the linear initial part of the reaction profiles according to the following equation, which is valid when the concentration of NBDPZ is equal to that of the isocyanates:

$$k_0 = \frac{a}{[\text{NBDPZ}]_0^2} \tag{9.7}$$

where $[\text{NBDPZ}]_0$ is the initial concentration of NBDPZ.

In Table 9.4 the slopes of the linear part of the reaction profiles (a), the second-order kinetic rate constants (k) and the initial rate constants (k_0) are summarized. From Table 9.4 it is clear that for the reactions of NBDPZ with monoisocyanates **11** and **12** and diisocyanate **13** the rate constants calculated by fitting the data to second-order kinetics are in good agreement with those calculated based on initial rates only, supporting the second-order kinetics of the reactions. The experimental data were also fitted to a first-order kinetics model. From the fits, which are not shown here, and the residual values, it was clear that only the second-order reactions kinetics could describe the data well.

9.5.3.5 Laboratory-scale Reaction Kinetics

For comparison reasons, the kinetics of the reaction of NBDPZ with isocyanates **11–13** were studied using conventional laboratory-scale procedures. The relative reactivity of the isocyanates **11–13** is in good agreement with that determined on-chip by continuous flow experiments. However, in all cases the laboratory-scale reactions show slower kinetics compared to the on-chip continuous flow experiments. The laboratory-scale reaction rate constants (k_{lab}) and those determined by on-chip experiments (k_{chip}) are summarized in Table 9.5. The somewhat larger on-chip reaction rates may be attributed to the distinctive conditions under which continuous flow microfluidic chips operate, such as diffusive mixing under laminar flow regime in nanolitre reaction volumes. However, since the differences in rate constants are not significantly large, they might also be attributed to errors in both methods.

Table 9.5 Reaction rates[a] determined by on-chip experiments (k_{chip}), laboratory-scale experiments (k_{lab}) and their ratio (k_{chip}/k_{lab}).

Isocyanate	k_{chip} (M^{-1} min^{-1})	k_{lab} (M^{-1} min^{-1})	k_{chip}/k_{lab}
11	1.6×10^4	4.2×10^3	3.8
12	5.2×10^4	1.6×10^4	3.3
13	2.5×10^4	5.6×10^3	4.3

[a]Initial reaction rates calculated according to Equation (9.7).

9.6 Concluding Remarks

In contrast to many sophisticated interfacing designs, reported in the recent literature, two simple microfluidic integrated devices for coupling microreactors to a nanoflow ESI mass spectrometer were designed and studied. A monolithic and a modular integration approaches were developed and their MS performances were evaluated by infusing a solution of reserpine in a mixture of water and acetonitrile (1:1 v/v) as a standard solution. Sample infusion flow rates and nebulizing gas pressure turned out to be the most important parameters that determine the ionization efficiency of the monolithic design. Both interfaces give a high-quality Taylor cone and signal stability, as indicated by the standard deviations of the TIC over several minutes (standard deviation approximately 8% and 1% for the monolithic and modular interfaces, respectively). In the monolithically integrated system active nebulization of the sample enhances the transfer of sample from solution to gas phase, independently of the sample flow rate, whereas spray tips can be chosen in the modular design to fit the right flow regime.

A simple though efficient mixing concept based on shorter diffusion distances by merging together incoming reagents streams in the vertical direction was integrated in the chip design. Reagent mixing was demonstrated by computational and experimental means to be completed within a few tens of milliseconds, which makes the setup suitable to study reaction kinetics.[42]

All integrated functions are optimized for standard glass fabrication technologies and therefore inexpensive to produce. Together with microreactor and mixer functions, the designs form a complete system to carry out real-time studies of chemical reactions by MS.

The more versatile and stable modular interface was successfully used to investigate supramolecular complexations based on the metal–ligand interaction between Zn–porphyrin **1** and the nitrogen-containing guest molecules **2**, **3**, **4**, **5** and **6**; or the encapsulation of guests **8** and **9** in the cavity of β-CD (**7**).

The titration method, based on rapidly varying the injection speed of reagent stock solutions into the mass spectrometer via a microreactor, offers a valuable alternative to more conventional laboratory-scale methodologies, provided that the selected supramolecular system is suitable to be studied by ESI-MS.[60] The most important advantages of this approach are the limited sample handling, the extremely short delay time between reaction and analysis and the versatility of the integrated system that combines reaction and analytical unit in a very efficient interface.

Furthermore, the use of the continuous flow microchip-based NESI interface for the study of reaction kinetics was successfully demonstrated using the reaction of NBDPZ with propyl isocyanate (**11**), benzyl isocyanate (**12**) and toluene-2,4-diisocyanate (**13**).

Using the NESI-chip interface, in all cases, the reaction rate constants are 3 to 4 times larger compared to that of laboratory-scale experiments. The somewhat larger rate constants determined by using the lab-on-a-chip may be attributed to the distinctive properties of lab-on-a-chip devices. This demonstrates that the continuous flow microfluidics-based nanospray interface provides a unique environment to carry out reactions in a very efficient way, using only tiny amounts of reagents, increasing process safety and reducing sample handling. The limited volume of the reaction vessel (only a few nanolitres), which implies reagent mixing under laminar flow conditions at very short diffusion length scales, offers a high degree of reaction control as compared to macroscale turbulent systems.

A novel type of "three inlet" microreactors offers the possibility to rapidly and systematically study ion suppression phenomena and the effect of different parameters on the ionization of the analyte, all required to study reaction kinetics by ESI-MS in an appropriate way.

The work reported in this chapter clearly demonstrates that the combination of a microreactor with ESI-MS offers a powerful tool to study chemical reactions in a dynamic way.

In conclusion, the NESI-chip system combines the advantages of microfluidics, such as fast, low-volume sample handling, with high sensitivity and specificity of nanoflow ESI-MS analysis, and rapid screening offered by on-line detection.

References

1. S. J. Gaskell, *J. Mass Spectrom.*, 1997, **32**, 677.
2. K. P. Bateman, R. L. White and P. Thibault, *Rapid Commun. Mass Spectrom.*, 1997, **11**, 1253.
3. D. Figeys, A. Ducret and R. Abersold, *Chromatogr. A*, 1997, **763**, 295.
4. M. T. Davis, D. C. Stahl, S. A. Hefta and T. D. Lee, *Anal. Chem.*, 1995, **67**, 4549.
5. C. Henry, *Anal. Chem.*, 1997, **69**, 359A.
6. A. J. De Mello, *Lab Chip*, 2001, **1**, 7N.
7. J. B. Z. Fenn, M. Mann, C. K. Meng, S. F. Wong and C. M. Whitehouse, *Science*, 1989, **246**, 64.
8. C. K. Meng, M. Mann and J. B. Z. Fenn, *J. Phys. D*, 1988, **10**, 361.
9. J. B. Z. Fenn, M. Mann, C. K. Meng, S. F. Wong and C. M. Whitehouse, *Mass Spectrom. Rev.*, 1990, **9**, 37.

10. R. D. Smith, J. A. Loo, C. G. Edmonds, C. J. Barinaga and J. R. Udseth, *Anal. Chem.*, 1990, **62**, 882.
11. M. Wilm and M. Mann, *Anal. Chem.*, 1996, **68**, 1.
12. J. Khandurina and A. Guttman, *J. Chromatogr. A*, 2002, **943**, 159.
13. D. Figeys and D. Pinto, *Electrophoresis*, 2001, **22**, 208.
14. S. Mouradian, *Curr. Opin. Chem. Biol.*, 2001, **6**, 51.
15. Q. Xue, Y. M. Dunayevskiy, F. Foret and B. L. Karger, *Rapid Commun. Mass Spectrom.*, 1997, **11**, 1253.
16. Q. Xue, F. Foret, Y. M. Dunayevskiy, P. M. Zavrachy, N. E. McGruer and B. L. Karger, *Anal. Chem.*, 1997, **69**, 426.
17. R. S. Ramsey and J. M. Ramsey, *Anal. Chem.*, 1997, **69**, 1174.
18. D. Figeys, Y. B. Ning and R. Abersold, *Anal. Chem.*, 1997, **16**, 3153.
19. R. D. Oleschuk and D. J. Harrison, *Trends Anal. Chem.*, 2000, **19**, 379.
20. A. Desai, Y.-C. Tai, M. T. Davis and T. D. Lee, *Proc. Transducers '97*, Chicago, IL, 1997, Vol. **2**, p. 927.
21. J. Wen, *et al., Electrophoresis*, 2000, **21**, 191.
22. C. H. Yuan and J. Shiea, *Anal. Chem.*, 2001, **73**, 1080.
23. J.-S. Kim and D. R. Knapp, *J. Chromatogr. A*, 2001, **924**, 137.
24. T. C. Rohner, J. S. Rossier and H. H. Girault, *Anal. Chem.*, 2001, **72**, 5353.
25. L. Licklider, X. Q. Wang, A. Desai, Y. C. Tai and T. D. Lee, *Anal. Chem.*, 2000, **72**, 367.
26. I. M. Lazar, R. S. Ramsey, S. C. Jacobson, R. S. Foote and J. M. Ramsey, *J. Chromatogr. A*, 2000, **892**, 195.
27. P. Griss, J. Melin, J. Sjödahl, J. Roeraade and G. Stemme, *J. Micromech. Microeng.*, 2002, **12**, 682.
28. C.-H. Chiu, G. B. Lee, H.-T. Hsu, P.-W. Chen and P.-C. Liao, *Sens. Actuators B*, 2002, **86**, 280.
29. M. Schilling, W. Nigge, A. Rudzinski, A. Neyer and R. Hergenröder, *Lab Chip*, 2004, **4**, 220.
30. M. Svedberg, M. Veszelei, J. Axelsson, M. Vangbo and F. Nikolajeff, *Lab Chip*, 2004, **4**, 322.
31. Y.-X. Wang, J. W. Cooper, C. S. Lee and D. L. De Voe, *Lab Chip*, 2004, **4**, 363.
32. J. Kameoka, *et al., Anal. Chem.*, 2002, **74**, 5897.
33. T. P. White and T. D. Wood, *Anal. Chem.*, 2003, **75**, 3660.
34. N. Lion, J.-O. Gellon, H. Jensen and H. H. Girault, *J. Chromatogr. A*, 2003, **1003**, 11.
35. J. Cavanagh, L. M. Benson, R. Thompson and S. Naylor, *Anal. Chem.*, 2003, **75**, 3281.
36. M. C. Mitchell, V. Spikmans and A. J. De Mello, *Analyst*, 2001, **126**, 24.
37. J.-M. Dethy, B. L. Ackerman, C. Delatour, J. D. Henion and G. A. Shultz, *Anal. Chem.*, 2003, **75**, 805.
38. C. A. Keetch, *et al., Anal. Chem.*, 2003, **75**, 4937.
39. S. Benetton, *et al., Anal. Chem.*, 2003, **75**, 6430.
40. A. Zamfir, *et al., Anal. Chem.*, 2004, **76**, 2046.

41. M. Brivio, R. E. Oosterbroek, W. Verboom, A. van den Berg and D. N. Reinhoudt, *Lab Chip*, 2005, **5**, 1111.
42. M. Brivio, *et al., Anal. Chem.*, 2005, **77**, 6852.
43. CFD Research Corporation, Huntsville, Alabama, USA.
44. S. Madhavan-Reese, D. Lim, J. Mazumder and E. F. Hasselbrink Jr., *Proc. μTAS '2002*, 2002, p. 900.
45. M. Brivio, *et al., Chem. Commun.*, 2003, 1924.
46. Upchurch Scientific, Oak Harbor, USA.
47. J. G. E. Gardeniers, R. E. Oosterbroek and A. van den Berg, *Lab-on-a-Chip, Miniaturized Systems for (Bio) Chemical Analysis and Synthesis*, ed. R. E. Oosterbroek and van den Berg, 2003, p. 37.
48. R. E. Oosterbroek, M. H. Goedbloed and A. van den Berg, *Proc. μTAS '2001*, Kluwer Academic, Dordrecht, 2001, p. 627.
49. M. S. Munson and P. Yager, *Anal. Chim. Acta*, 2004, **507**, 63.
50. K. Jensen, *Nature*, 1998, **393**, 735.
51. A. D. Stroock, *et al., Science*, 2002, **25**, 647.
52. B. Zhang, H. Liu, B. L. Karger and F. Foret, *Anal. Chem.*, 1999, **71**, 3258.
53. G. Jenkins and A. Manz, *J. Micromech. Microeng.*, 2002, **12**, N19.
54. F. Vögtle, *Supramolecular Chemistry*, Wiley, New York, 1993.
55. J.-M. Lehn, *Supramolecular Chemistry*, VCH, Weinheim, 1995.
56. M. Vincenti, *J. Mass Spectrom.*, 1995, **30**, 925.
57. B. N. Pramanik, P. L. Bartner, U. A. Mirza, Y.-H. Liu and A. K. Ganguly, *J. Mass Spectrom.*, 1998, **33**, 911.
58. C. A. Schalley, *Int. J. Mass Spectrom.*, 2000, **194**, 11.
59. B. Ganem, Y. T. Li and J. D. Henion, *J. Am. Chem. Soc.*, 1991, **113**, 6294.
60. For a review, see J. M. Daniel, S. D. Friess, S. Rajagopalan, S. Wendt and R. Zenobi, *Int. J. Mass Spectrom.*, 2002, **216**, 1.
61. G. V. Oshovsky, W. Verboom, R. H. Fokkens and D. N. Reinhoudt, *Chem. Eur. J.*, 2004, **10**, 2739.
62. B. D. Berezin, *Coordination Compounds of Porphyrins and Phthalocyanines*, Wiley, Chichester, 1981.
63. D. M. Rudkevich, W. Verboom and D. N. Reinhoudt, *J. Org. Chem.*, 1995, **60**, 6585.
64. For a more detailed discussion on the mechanism of ion formation in ESI-MS and the effect of ionization on solution equilibria, see (a) P. Kebarle, L. Tang, *Anal. Chem.*, 1993, **65**, 972A; (b) P. Kebarle and M. Peschke, *Anal. Chim. Acta*, 2000, **406**, 11; (c) P. Kebarle, *J. Mass. Spectrom.*, 2000, **35**, 804; (d) N. B. Cech and G. E. Enke, *J. Mass Spectrom. Rev.*, 2001, **20**, 362.
65. E. Leiz, A. Jaffrezic and A. van Dorsselear, *J. Mass Spectrom.*, 1996, **31**, 537.
66. W. Blokzijl and J. B. F. N. Engberts, *Angew. Chem. Int. Ed. Engl.*, 1993, **32**, 1545.
67. For reviews on cyclodextrin chemistry, *see Chem. Rev.*, 1998, 98, issue 5.
68. K. Uekama, F. Hirayama and T. Irie, *Chem. Rev.*, 1998, **98**, 2013.
69. J. Szejtli, *Chem. Rev.*, 1998, **98**, 1743.

70. A. Hedges, *Chem. Rev.*, 1998, **98**, 2035.
71. Th. Hofler and G. Wenz, *J. Incl. Phenom.*, 1996, **25**, 81.
72. M. R. de Jong, J. Huskens and D. N. Reinhoudt, *Chem. Eur. J.*, 2001, **7**, 4164.
73. C. B. Lebrilla, *Acc. Chem. Res.*, 2001, **34**, 653.
74. Y. Cay, M. A. Tarr, G. Xu, T. Yalcin and R. B. Cole, *J. Am. Soc. Mass Spectrom.*, 2003, **14**, 449.
75. J. B. Cunniff and P. Vouros, *J. Am. Soc. Mass Spectrom.*, 1995, **6**, 437.
76. J. Ramirez, S. Ahn, G. Grigorean and C. B. Lebrilla, *J. Am. Chem. Soc.*, 2000, **122**, 6884.
77. S. Ahn, J. Ramirez, G. Grigorean and C. B. Lebrilla, *J. Am. Soc. Mass Spectrom.*, 2001, **12**, 278.
78. A. J. Ali and T. M. Lohman, *Science*, 1997, **275**, 377.
79. L. S. Luo, M. D. Burkart, D. Stachelhaus and C. T. Walsh, *J. Am. Chem. Soc.*, 2001, **123**, 11208.
80. B. J. Brazeau, R. N. Austin, C. Tarr, J. T. Groves and J. D. Lipscomb, *J. Am. Chem. Soc.*, 2001, **123**, 11831.
81. A. M. Valentine, S. S. Stahl and S. J. Lippard, *J. Am. Chem. Soc.*, 1999, **121**, 3876.
82. B. A. Krantz and T. R. Sosnick, *Biochemistry*, 2000, **39**, 11696.
83. P. Halonen, A. A. Baykov, A. Goldman, R. Lahti and B. S. Cooperman, *Biochemistry*, 2002, **41**, 12025.
84. C. K. Chan, *et al., Proc. Natl Acad. Sci. USA*, 1997, **94**, 1779.
85. S. F. Nelsen, *et al., J. Am. Chem. Soc.*, 1997, **119**, 5900.
86. J. P. F. Cherry, *et al., J. Am. Chem. Soc.*, 2001, **123**, 7271.
87. M. R. Bringer, *et al., Philos. Trans. R. Soc. London, Ser. A*, 2004, **362**, 1087.
88. X. Zhou, R. Medhekar and M. D. Toney, *Anal. Chem.*, 2003, **75**, 3681.
89. G. Taylor, *Proc. R. Soc. London*, 1953, **A219**, 186.
90. L. Konermann, *J. Phys. Chem.*, 1999, **103**, 7210.
91. M. Schwarz and P. C. Hauser, *Lab Chip*, 2000, **1**, 1.
92. A. J. Mello de, *Lab Chip*, 2001, **1**, 7N.
93. M. Kalkuta, D. A. Jayawickrama, A. M. Wolters, A. Manz and J. V. Sweedler, *Anal. Chem.*, 2003, **75**, 956.
94. H. B. Mao, T. L. Yang and P. S. Cremer, *Anal. Chem.*, 2002, **74**, 379.
95. M. Kerby and R. L. Chien, *Electrophoresis*, 2001, **22**, 3916.
96. A. G. Hadd, D. E. Raymond, J. W. Halliwell, S. C. Jacobson and J. M. Ramsey, *Anal. Chem.*, 1997, **69**, 3407.
97. G. H. Seong, J. Heo and R. M. Crooks, *Anal. Chem.*, 2003, **75**, 3161.
98. H. Song and R. F. Ismagilov, *J. Am. Chem. Soc.*, 2003, **125**, 14613.
99. M. Vogel and U. Karst, *German Pat.*, DE 199 26 731, 1999.
100. M. Vogel and U. Karst, *Anal. Chem.*, 2002, **74**, 6418.
101. H. Henneken, *et al., J. Environ. Monit.*, 2003, **5**, 100.
102. H. Ulrich, *Chemistry and Technology of Isocyanates*, Wiley, Chichester, 1996.
103. K. L. Dunlap, R. L. Sandrige and J. Keller, *Anal. Chem.*, 1976, **48**, 497.

104. D. Karlsson, J. Dahlin, G. Skarping and M. Dalene, *J. Environ. Monit.*, 2002, **4**, 216.
105. Y. Nordqvist, J. Melin, U. Nilsson, R. Johansson and A. Colmsjö, *Fresenius J. Anal. Chem.*, 2001, **371**, 39.
106. A. Schmidt and M. Karas, *J. Am. Soc. Mass Spectrom.*, 2003, **14**, 492.
107. T. M. Annesley, *Clin. Chem.*, 2003, **49**, 1041.
108. R. King, R. Bonfiglio, C. Fernandez-Metzler, C. Miller-Stein and T. Olah, *J. Am. Soc. Mass Spectom.*, 2000, **11**, 942.

SECTION 2
MALDI-MS

CHAPTER 10

On-line and Off-line MALDI from a Microfluidic Device

HARRISON K. MUSYIMI, STEVEN A. SOPER AND KERMIT K. MURRAY

Department of Chemistry and Center for BioModular Multi-Scale Systems, Louisiana State University, Baton Rouge LA 70803, USA

10.1 Introduction

Micro total analysis systems (μTAS) are novel platforms that address the need for high-throughput and automated analysis achieved through miniaturization, integration and multiplexing of processing and analysis elements.[1–4] Integration of sample processing steps reduces analysis time and prevents sample loss, contamination and false positives, while miniaturization affords the assembly of dedicated modular devices that are portable. Fast and parallel analyses of small sample volumes[5,6] using low-cost disposable polymeric microfluidic devices[7–9] have the potential to revolutionize point-of-care testing for clinical and diagnostic screening. The majority of emerging microfluidic devices are capable of integrated and highly multiplexed sample processing.[10–13] Chemical and biochemical analyses often probe qualitative, structural and quantitative information and many efforts continue to progress towards realizing the full potential of these devices by integration with existing analytical detectors.[14–18] A long-term challenge lies in scaling down traditional analytical techniques into miniaturized platforms without compromising their utility and system performance.

A variety of detection methods have been reported for microdevices, including fluorescence, conductivity and mass spectrometry (MS).[19] Fluorescence

Miniaturization and Mass Spectrometry
Edited by Séverine Le Gac and Albert van den Berg

Published by the Royal Society of Chemistry, www.rsc.org

detection is a good choice because it can be performed directly on the chip. Although fluorescence detection is sensitive, it requires sample derivatization or reliance on native fluorescence. Conductivity detection is excellent in terms of size and simplicity, but lacks specificity and the detector can interfere with electrophoretic separations.[20] MS is a good detector in cases where spectroscopic or electrochemical methods are not practical or where a more general detector is desired. The instrument development challenge is to interface the chip to the mass spectrometer while maintaining the performance of both the chip and spectrometer.

Electrospray ionization (ESI) is a versatile ionization technique that produces highly charged ions continuously from solutions.[21] The high ionic charge allows high molecular weight molecules to be analyzed using conventional mass spectrometers due to their lower mass-to-charge ratio. For microfluidic chip detection, a glass chip can be scored and broken across a channel, or a spray tip can be attached to or formed in the chip.[19]

Despite the inherent advantages of ESI for on-line MS, there have been a number of approaches to matrix-assisted laser desorption ionization (MALDI)-MS that have applications to chip separations. The general advantage of MALDI for biomolecule analysis is a higher tolerance for impurities and the singly charged ions that give less complex mass spectra. The MALDI approach has additional multiplexing advantages over ESI for microfluidic chip MS. The directional control of the spray can be difficult for a high-density array of spray tips, potentially leading to crosstalk. Microfluidic chip interfaces with ESI suffer from problems when sprays are started or stopped and this limits the speed of moving from one sample to another and therefore limits high-throughput applications. The MALDI interface has more options for rapid switching, such as a fast x–y translation stage or a galvanometer-driven laser mirror.

The development of MALDI-MS for on-line and off-line analysis from microfluidic devices is influenced by a number of considerations. The low sample volumes typical of analyses in a microfluidic device are readily compatible with low sample consumption in MALDI-MS.[22,23] Further, sample preparation often involves a series of steps to address sample complexity and, for this reason, microfluidic devices are attractive platforms for integrated sample processing prior to MALDI-MS analysis. The fast data acquisition rate of MALDI time-of-flight (TOF) MS readout allows for integration with rapid analysis in microfluidic devices. MALDI-MS provides high-resolution and accurate mass measurements that lead to high specificity and excellent identification capabilities. These figures of merit coupled with the sensitivity and high mass range makes MALDI-MS a versatile detector and increases the scope of microfluidic assays. Integration of microfluidic devices to biological MS is therefore a logical step towards developing platforms for automated and high-throughput processing of biologically complex samples available in limited volumes or in relatively low concentrations.

10.2 Instrumentation

10.2.1 Microfluidics

A microfluidic device consists of a network of channels that are a few micrometers wide and deep fabricated in planar substrates. Microfluidic devices have been fabricated in various materials such as glass, silicon and polymers.[1,3,9,24] The material physical and chemical properties play a significant role in microfabrication, sample handling and processing within a microfluidic device. Recently, in a departure from materials such as glass and silicon, there has been a shift to polymer-based microfluidic devices.[25] Polymers are easy to machine and the microfabrication process is cost effective for mass production. A typical microfabrication process for polymer-based devices requires a molding master with the desired pattern prepared using methods such as photolithography or micromilling techniques.[8,25–27] The molding master is then used to replicate polymer micro-parts by hot embossing or injection molding.[7,8] Once stamped the devices are assembled by thermally annealing a cover plate to enclose the microstructures patterned onto the polymer substrate. Polymers exhibit a wide range of surface properties such as wettability, adsorption and zeta potential that depend on the chemical composition of the surface.[28,29] Such physicochemical properties affect sample processing, experimental conditions and quality of data in diverse ways.[30] Often, various surface modification chemistries are required on polymer substrates for tailoring the surface to accommodate sample processing needs. Samples are usually flowed through a network of channels for processing such as separation, sorting, capture, lysis, digestion, pre-concentrating, desalting and other clean-up procedures.

10.2.2 MALDI-TOF-MS

A typical MALDI-TOF-MS system consists of a sample target, ionization chamber, laser, mass analyzer and detector. A sample mixed with the appropriate matrix is loaded on a target and placed into the ionization chamber. Once the ionization chamber is evacuated, an ultraviolet (UV) or infrared (IR) laser is then used to irradiate the sample for desorption and ionization. Inside the ionization chamber, a series of ion extraction grids are used to accelerate the ions into a flight tube, which serves as a mass-to-charge analyzer. Ions arriving at the end of the flight tube are detected using a particle-multiplier detector. The signal generated from the detector is digitized and converted into a mass spectrum before further computer processing. The ionization chamber and the flight tube are under high vacuum to remove residual gases with which ions might collide before arriving at the detector.

The transfer of material from a microfluidic chip to the mass spectrometer can be performed on-line or off-line. Typically, a MALDI target is under high vacuum while microchip-based electrophoresis is carried out at atmospheric pressure. This difference in pressure makes on-line transfer a challenge.

With the on-line approach, the sample is continuously delivered to the vacuum in real time. Off-line sample introduction entails the running of microfluidic processes on the chip with later MS analysis. Therefore, the introduction of samples into the ion source generally requires breaking the ion source vacuum.

10.3 Off-line Analysis

Off-line analysis from a microfluidic device using MALDI-MS is rather straightforward. Common approaches entail either fraction collection from a microfluidic device and spotting onto a MALDI sample target or placing the entire device inside a MALDI-MS ionization chamber for analysis.[31] In the first approach, bottlenecks are often encountered in the deposition technique when sampling low volumes.[32] However, automation of the deposition step can increase the throughput of this off-line approach.[33,34] A second approach entails sample processing on an open substrate or in enclosed fluidic networks then removing the cover plate of a microfluidic device before analysis with MALDI-MS. Usually, no sample transfer steps are involved with this approach, thereby limiting sample losses. Limitations of off-line approaches are encountered when the size of the device is not compatible with the ion source and therefore cannot fit into the ionization chamber MALDI-MS readout. In some cases, the device can be split into smaller component sizes,[35] or, if the device is a smaller size, slight modifications can be made to the sample target to allow MALDI-MS readout.[31,36] When the cover of a microfluidic device must be removed after sample processing, caution is required to preserve the integrity of the sample.[37]

Off-line analysis from a microfluidic device has recently been demonstrated in our laboratory.[37] In this particular approach, sample processing was performed inside an enclosed polymeric microfluidic device. The device was fabricated by directly micromilling the fluidic channels onto a poly(methyl methacrylate) (PMMA) polymer sheet using a high-precision micromilling machine. The device was assembled by thermally annealing a polydimethylsiloxane (PDMS) cover to enclose the fluidic channels. After the device assembly, a polyacrylamide sieving gel was introduced into the fluidic channel. The PMMA/PDMS hybrid device was used to perform sodium dodecyl sulfate polyacrylamide gel electrophoresis (SDS-PAGE) in the enclosed fluidic channel. After a period of time, the separation was stopped and the PDMS cover used to enclose the fluidic channel was carefully peeled off in order to access the gel.

The entire PMMA device or the PDMS cover was mounted onto a modified MALDI target holder for analysis. A schematic representation of the ion source configuration used with this approach is shown in Figure 10.1 and photographs of the setup from front and behind are shown in Figure 10.2. The target holder consisted of a circular stainless steel piece with a recess milled across the center to accommodate the size of the device. A similar piece of stainless steel was used to hold the device from the rear. A slit was cut across the center of the front piece to allow a laser beam to interrogate the device channel

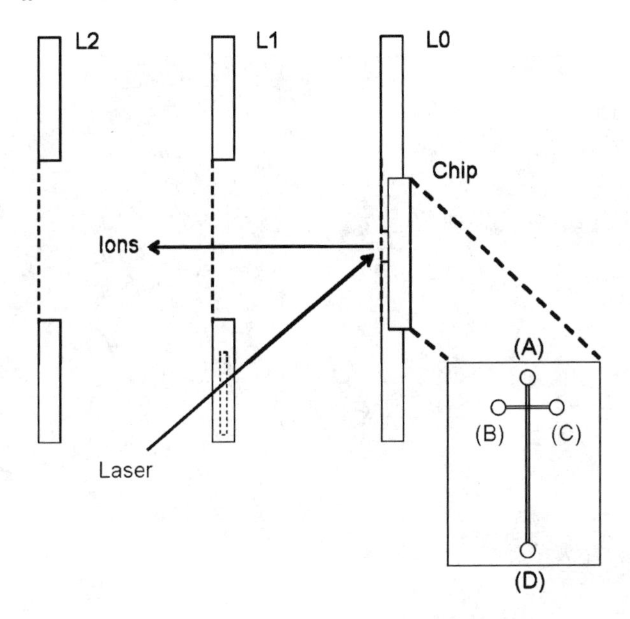

Figure 10.1 Schematic representation of a microfluidic chip ion source. The chip is shown in plan view as an inset with buffer reservoir (A), sample reservoir (B), sample waste reservoir (C) and buffer waste reservoir (D). Ions desorbed from the chip held by target L0 are accelerated into the flight tube by extraction grids L1 and L2.

and also to allow acceleration of the ions into the flight tube. Once the device was mounted on the holder, it was placed inside the MS ion source vacuum chamber for interrogation with an IR laser. No matrix addition step was required under these conditions since compounds present in the crosslinked gel served as matrix for IR laser desorption ionization. The IR laser wavelength was tuned to 2.95 μm and 2 Hz repetition rate. The device was moved across the path of a focused laser line inside the ionization chamber manually or using a micrometer on a linear motion feed-through that was driven by a variable speed motor. The laser beam was aligned with the center of the gel line in the fluidic channel by observing the laser ablation using a video camera.

Figure 10.3 shows the IR-LDI mass spectrum of the peptide bradykinin ($M_r = 1060.2$) desorbed and ionized from the PDMS cover. The bradykinin was loaded into the chip at a concentration of 2.5 mM and injected into the chip channel using a field of 150 V cm^{-1} for 9 min. The spectrum is the result of 10 laser shots at 2.95 μm wavelength. The off-line capillary gel microfluidic mass spectra are obtained after electrophoretic transport through a closed chip channel. Operation with a closed channel requires that the cover be removed from the chip prior to analysis. When removing the cover, sections of the gel tended to adhere to the surface of the PDMS. With some amount of care the entire gel lane could be extracted intact from the chip and, in many cases, irradiation of the PDMS cover resulted in the best mass spectra. The base peak

A

B

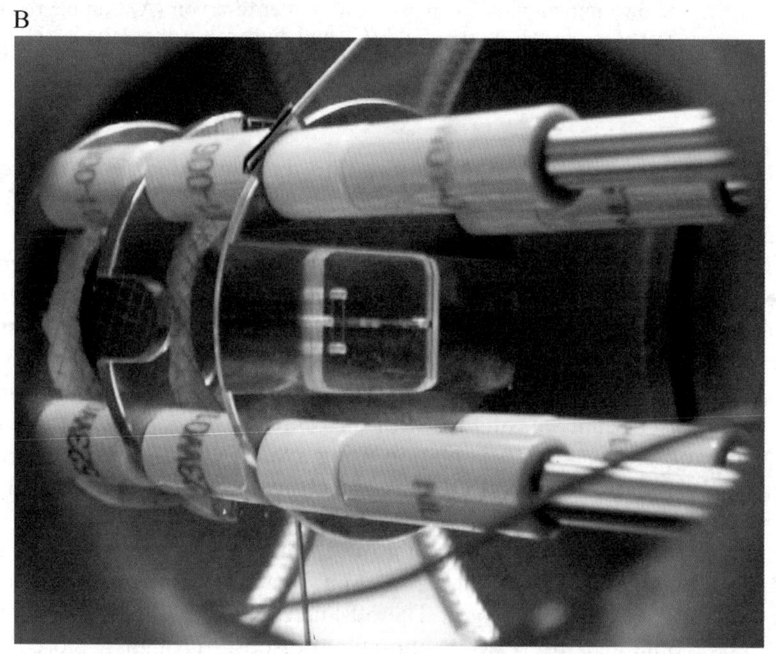

Figure 10.2 Photograph of a PMMA microfluidic chip mounted in a MALDI ion source viewed from (A) the front and (B) the back of the chip.

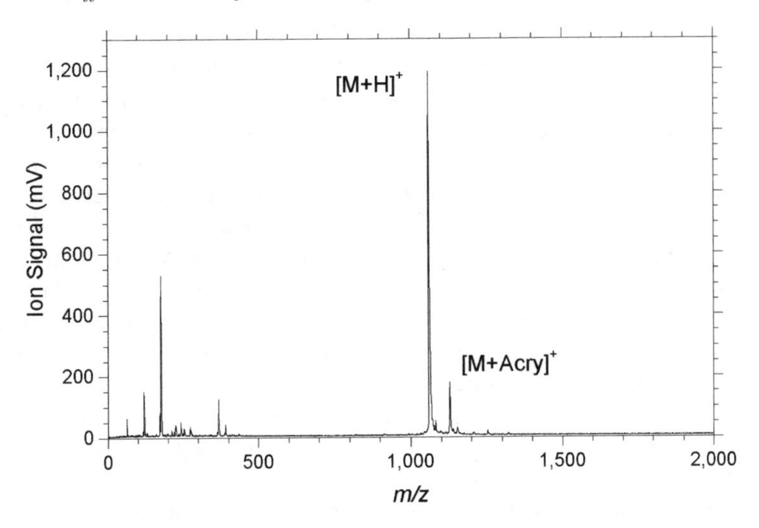

Figure 10.3 Laser desorption/ionization mass spectra from a PDMS microfluidic chip cover from a 750 μM solution of bradykinin run through a polyacrylamide gel chip (70 pmol loaded) with no additional matrix at 2.95 μm.

of the spectrum is assigned to the singly protonated bradykinin, $[M + H]^+$. Interfering peaks due to sodium and acrylamide adducts and ionized gel components are observed, but are relatively small compared to the base peak.

10.4 On-line Analysis

On-line analyses are highly desired for their high sample throughput, automation and ability to acquire real-time data. One obvious caveat with on-line MALDI-MS is that most instruments are designed for off-line sample introduction. On-line interfaces for MALDI-MS can be grouped into those using (1) aerosol particles, (2) capillaries or (3) mechanical means to deliver the matrix and analyte into the mass spectrometer.[38] The aerosol MALDI method uses a milliliter per minute spray of particles to deliver the sample to a TOF mass spectrometer.[39] The particles are dried in a heated tube and irradiated by a pulsed UV laser to form ions. Although aerosol MALDI is compatible with solid matrices and the system can be operated for weeks at a time with no cleaning, it suffers from excessively high sample consumption. The capillary-based continuous flow (CF) MALDI interfaces use a narrow-bore capillary to deliver analyte in a liquid matrix at a few microlitres per minute to the point of laser desorption and ionization.[40] Widespread use of CF-MALDI has been limited due to the lack of suitable liquid matrix materials for use with UV desorption lasers. A continuous flow microfluidic chip interface for MALDI that uses a standard crystalline matrix has been reported recently.[41,42]

Two mechanical interfaces for on-line MALDI have recently been developed; both of them are compatible with more widely used solid matrices. With one of these approaches, liquid samples are transported into the mass spectrometer through a capillary at a flow rate of hundreds of nanolitres per minute and deposited on a rotating quartz wheel.[43] Solvent evaporation results in a thin sample trace on the wheel, which is rotated into position for laser desorption. The most significant drawback to the continuous vacuum deposition method is the current requirement for manual cleaning of the wheel. A second mechanical approach is a modification of the rotating ball inlet[44] that was adapted for MALDI. For rotating ball MALDI analysis, a solution containing matrix and analyte is delivered to a gasket pressed tightly against the surface of a ball that rotates several times each minute. As the ball rotates, the solution is exposed to vacuum where the volatile solvent evaporates, leaving the matrix and analyte. When the ball has rotated one-half turn, it is in position just behind the MALDI target in the ion source.

The ball inlet has been further modified to allow for coupling of microscale separations to on-line MALDI-MS.[45–47] To integrate microfluidic platforms to on-line MALDI-TOF-MS, a configuration that allows on-line transfer of sample and matrix from atmospheric pressure into the high vacuum of a MS ionization chamber was developed. The interface consists of a stainless steel ball with a drive shaft press-fit through the center of the ball. The ball assembly is mounted on an ISO-100 flange that has a circular hole machined on-center. A circular Teflon gasket with a central hole fits between the ball and an ISO-100 flange face. Vacuum grease is applied between the gasket and the flange before the gasket is sandwiched between the ball and the flange to form a vacuum seal. The ball is held in place by two stainless steel blocks with sintered bronze bearings press fit to allow free rotation of the shaft. The stainless steel bearing blocks are held against the ISO-100 flange by adjustment screws. The majority of the ball is at atmospheric pressure and exposes only a small portion of the ball to the vacuum. Centering the ball on the shaft is critical, since slight wobble in the shaft causes a vacuum breach as the ball rotates. The ball–gasket seal is sufficiently vacuum-tight to maintain acceptably low pressures ($>10^{-5}$ torr) when the ball is rotating.

The ISO-100 flange with the ball assembly is mounted on a Delrin flange that allows the entire ball assembly to be raised to the mass spectrometer acceleration voltage while the remainder of the mass spectrometer remains at ground potential. Rotation of the ball is accomplished using a multi-speed transmission motor outside the ionization chamber. The rotating ball acts as a mechanical transport mechanism that introduces samples deposited on its surface past the gasket into the ionization chamber. To eliminate sample crosstalk, a cleaning system made from a solvent-saturated felt pad is mounted on-line for surface regeneration. A clean surface is always presented for subsequent deposits after the previous deposition region passes out of the ionization chamber, thereby making this inlet self replenishing. The ion source chamber is modified so that the ionization laser is directed at the center of the rotating ball flange through a quartz window at a 45° angle with respect to the

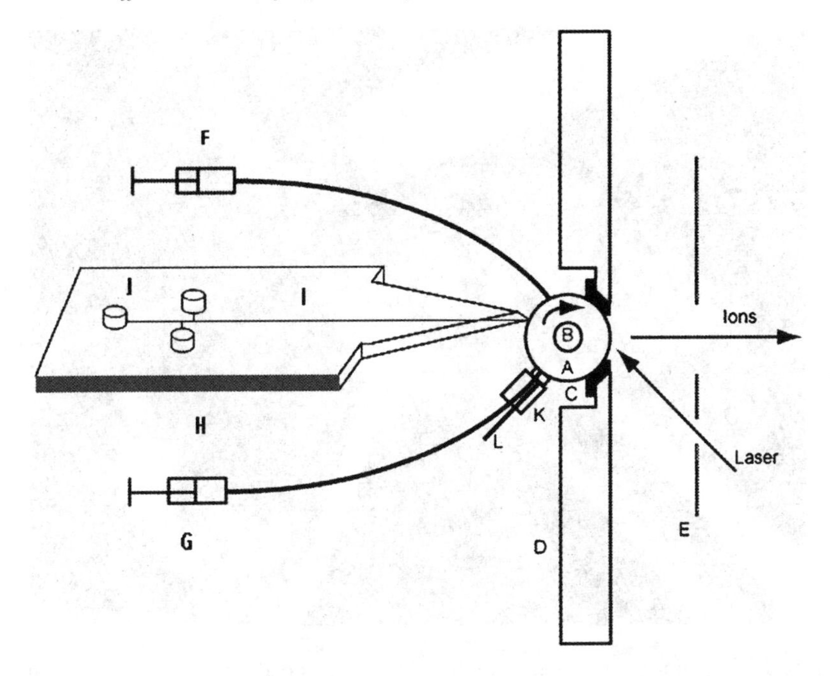

Figure 10.4 Schematic representation of a rotating ball inlet: A, stainless steel ball; B, drive shaft; C, gasket; D, ISO-100 flange; E, ion extraction grid; F, matrix syringe pump; G, cleaning solvent pump; H, microfluidic chip; I, separation channel; J, reservoirs; K, cleaning system; L, waste solvent drain.

flange. A video camera and macro lens is directed through a second port on the opposite side of the chamber and is used to view the ball in operation. A 355 nm Nd:YAG laser was used for the MALDI experiment and the laser repetition rate set at 10 Hz. A schematic representation of the interface coupled to a microfluidic device for on-line analyses is shown in Figure 10.4. A photograph of the interface in operation is shown in Figure 10.5.

Analysis in a microfluidic device with electrophoresis-based separation was coupled to on-line MALDI-MS using a rotating ball inlet.[48] The driving force in microchip electrophoresis is electrokinetic and sample constituents are separated based on differences in their electrophoretic mobilities. Positively charged species migrate towards the cathode, while negatively charged species migrate towards the anode. However, in most cases, the magnitude of the electroosmotic flow (EOF), which is substrate material dependent, is higher than the electrophoretic mobility of negatively charged species. In such cases, the bulk solution flow and therefore the apparent electrophoretic mobility of negatively charged species are towards the cathode. Polymers in general support a low EOF compared to materials like glass. PMMA, for example, has a fairly constant EOF over a wide pH range of separation media. Microchips are integrated to on-line MALDI-MS by adapting the rotating ball inlet to

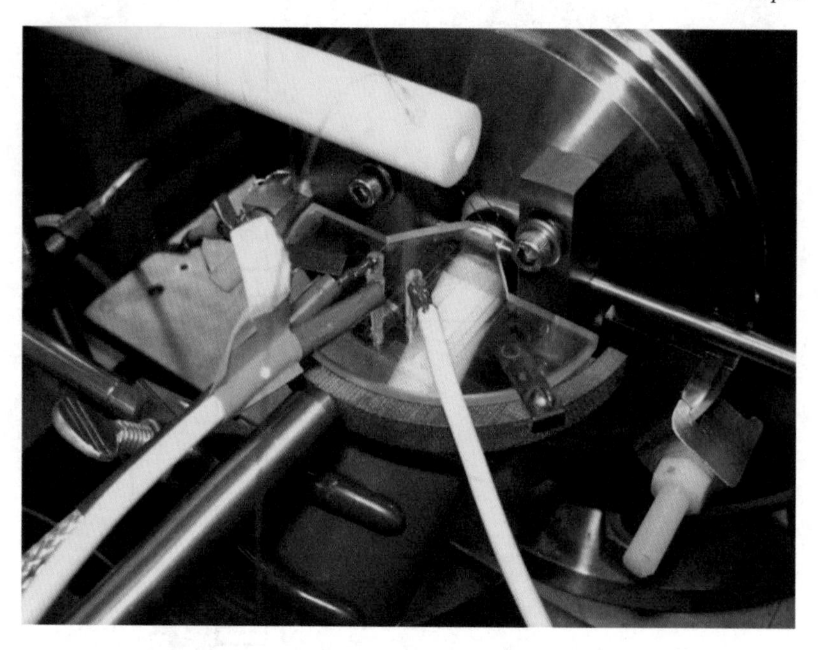

Figure 10.5 Photograph of the rotating ball inlet and PMMA microfluidic chip in operation.

function as an electrophoresis electrode in addition to sample transfer and MALDI target function.[46,48] The voltage applied on the rotating ball acts as a reference to operate microchip electrophoresis in a reversed or normal mode and also a reference for MALDI-MS operation in either positive or negative mode. As a note of caution, MALDI-MS and electrophoresis systems use high voltages and must be operated with care.

Samples separated in a microfluidic channel migrate toward the rotating ball and are deposited onto the ball surface. A matrix is subsequently added on-line using a capillary in contact with the sample track, then the sample and matrix deposited on the ball are transported into the ionization chamber for desorption with a laser. Integration of the microchip electrophoresis to MALDI-MS is done directly without the use of sample transfer interconnects between the microchip and rotating ball inlet as shown in Figure 10.2. This eliminates any broadening of the electrophoresis peaks associated with extra column effects and material mismatch, and it automates MALDI sample preparation. Rotation of the ball is tuned to the speed of the separation in order to minimize contribution to peak variance from interface geometry. It is also important to note that the speed of ball rotation is critical in allowing time for the sample and matrix trace to dry by evaporation before entering the ionization source. The rotating ball inlet further decouples the ionization process from the separation process thereby allowing for optimization of both processes independently without compromising the overall performance. Although

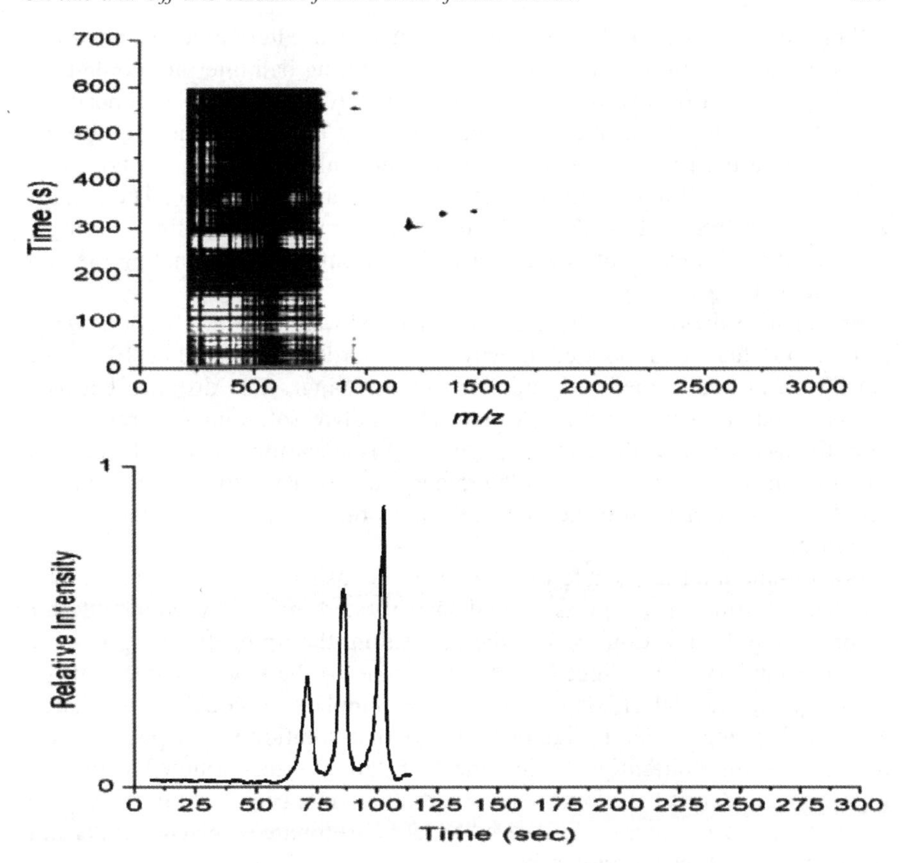

Figure 10.6 Upper panel shows a two-dimensional separation contour plot showing the separation of peptides bradykinin, substance p and bombesin on a PMMA chip. The *x* axis represents *m/z* values, the *y* axis represents the separation and readout times and the gray shade shows the ion intensities. The panel below is a representative electropherogram generated by integrating ion intensities from a two-dimensional plot.

MALDI-MS is credited for a high tolerance of impurities, the sensitivity of low-abundance species is often compromised in complex samples due to ion suppression effects. Coupling microchip electrophoresis to MALDI-MS is one way of simplifying complex mixtures before analysis, and thereby improving the sensitivity of constituents present in low abundance.

The electropherogram shown in Figure 10.6 represent a separation of three peptides with corrected migration times on the *x* axis. The separation efficiencies averaged 10^5 plates per meter. Compared to capillary electrophoresis, lower field strength is employed in PMMA microchip to minimize the Joule heating effect. Nonetheless, separations still occurred in far less time on the chip because of the short channel and generated comparable separation efficiencies.

Single drop deposition has been investigated as an alternative means of on-line sample deposition.[47] A schematic of the rotating ball interface configured for single droplet deposition is shown in Figure 10.7. For droplet deposition, the rotation rate of the ball was approximately 0.12 rpm. Prior to droplet deposition, a matrix spot was deposited on the ball using a 50 mm i.d. fused silica capillary held horizontally using an *x*–*y*–*z* translation stage. The analyte was delivered from a 100 mM solution at a flow rate of 1.5 μL min^{-1} using a syringe pump. Matrix spots were created by intermittently contacting the ball with the matrix capillary.

For analyte deposition, the droplet generator was held in a second *x*–*y*–*z* translation stage and pointed downward towards the ball at a 10° angle from vertical and immediately adjacent to the flange. The distance between the ball and capillary tip was 2.5 mm. The analyte solution was introduced into the generator with a syringe pump. Visualization of sample droplet was obtained using a second video camera and macro zoom lens. Sample droplets were ejected onto the matrix spots as they rotated below the droplet generator.

For on-line analysis, single droplets were deposited onto the rotating ball interface at atmospheric pressure and then rotated into the vacuum side for desorption and ionization. Before the deposition, the tip of droplet generator and the capillary were aligned to the centerline of the stainless steel ball by adjusting the translation stages so that the sample spot could be irradiated without changing the laser alignment. The matrix solution was deposited onto the ball by intermittently moving the translation stage-mounted capillary. Figure 10.8a depicts the mass spectrum of angiotensin I obtained in this manner. The spectrum is characterized by large protonated molecule peaks and less intense alkali metal adduct peaks.

A mass spectrum of a single droplet containing 8 fmol of bovine insulin deposited on the ball coated with CHCA is shown in Figure 10.8b. The protonated insulin peak $[M + H]^+$ and the doubly protonated $[M + 2H]^{2+}$ are observed. Figure 10.8c shows a mass spectrum obtained from an 8 fmol single droplet deposit of cytochrome *c* on the rotating ball interface coated with CHCA. The doubly protonated peak $[M + 2H]^{2+}$ and singly protonated $[M + H]^+$ cytochrome *c* peaks were observed at a mass resolution of 40. The lower resolution for the ball deposition as compared to the off-line single droplet deposit may result from field inhomogeneities in the ion extraction region. Desorption from the round surface of the ball may have an adverse effect on the spatial focusing of the ions and result in a broad distribution in flight times in the accelerated packet. Furthermore, the rotating ball instrument does not utilize delayed ion extraction, which is used in the off-line instrument to improve the mass resolution. Another possible reason for the lower mass resolution and signal is that when the sample spot passes through the gasket, the mechanical force between the gasket and the ball changes the surface morphology or could add contaminants to the deposit.

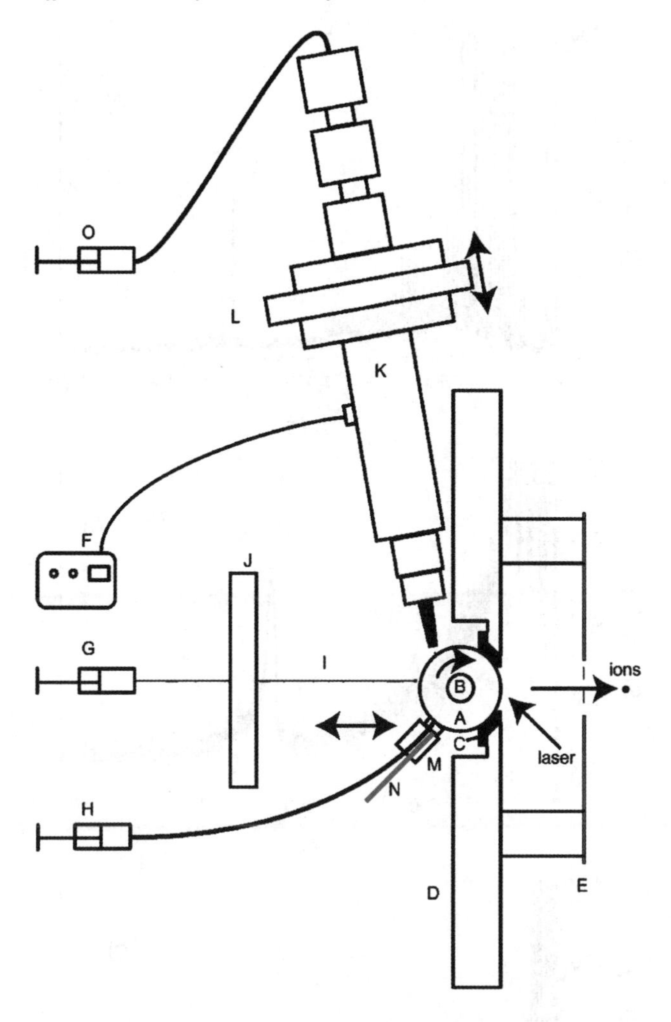

Figure 10.7 Schematic of the single droplet deposition device: A, 19 mm diameter stainless steel ball; B, drive shaft; C, gasket; D, ISO 100 flange; E, ground grid; F, droplet generation electronics; G, syringe pump for matrix solution; H, syringe pump for solvent; I, capillary; J, translation stage for capillary; K, droplet generator; L, translation stage for droplet generator; M, cleaning system (Teflon holder and felt); N, peek plastic tube for waste; O, syringe pump for analyte solution.

10.5 Off-line versus On-line Analysis

Generally, off-line and on-line analysis approaches from microfluidic devices are complementary. Despite the strengths and weaknesses associated with either approach, a comparison can be made with regard to performance, ease of implementation and the robustness of the approach. Off-line approaches are

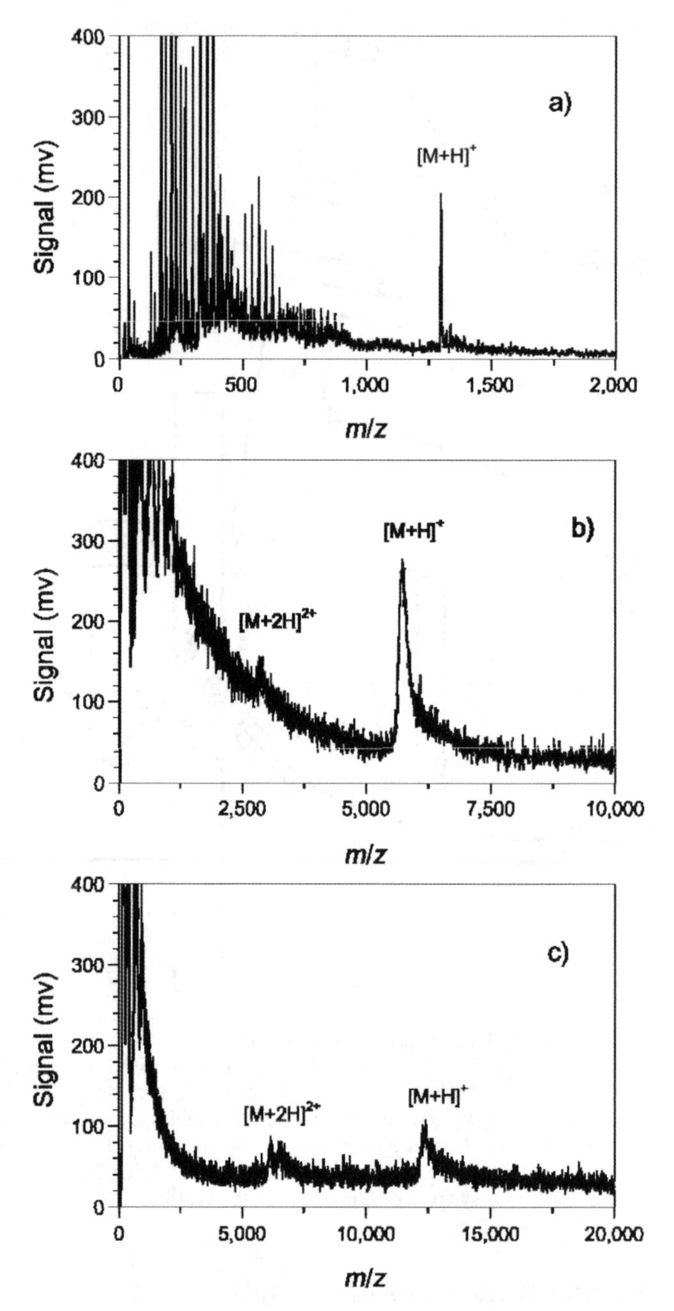

Figure 10.8 On-line MALDI mass spectra from single droplet deposition on the rotating ball interface: (a) angiotensin I; (b) bovine insulin; (c) cytochrome *c*.

performed without major modifications of either the microfluidic device or the mass spectrometer and the system performance is not significantly affected. However, in the off-line approach, during readout the laser or device is moved and depending on the modifications done to the ion source, the mass spectrometer performance can be affected. Mass resolution is often compromised in cases where the ion focusing field is affected by device movements and position within the ion source. Careful consideration of these issues during design and modification of the ion source can tremendously reduce the effects on mass resolution. Some off-line approaches generally suffer from sample losses associated with sample transfer steps and limited sample throughput. However, sample losses arising from sample transfer steps can be avoided by the direct analysis of samples within the microfluidic device. On the other hand, on-line approaches, such as the rotating ball inlet, require major adjustments to a typical MALDI-MS ion source that can affect mass resolution and sensitivity. On-line analysis reduces instrument down time and provides a high sample throughput compared to off-line approaches. Modifications required for a typical MALDI-MS ion source for on-line analysis are demanding and difficult to optimize but avails the instrument to various front end sample processing techniques.

References

1. T. Vilkner, D. Janasek and A. Manz, *Anal. Chem.*, 2004, **76**, 3373.
2. A. J. Tudos, G. A. J. Besselink and R. B. M. Schasfoort, *Lab Chip*, 2001, **1**, 83.
3. D. R. Reyes, D. Iossifidis, P. A. Auroux and A. Manz, *Anal. Chem.*, 2002, **74**, 2623.
4. P. A. Auroux, D. Iossifidis, D. R. Reyes and A. Manz, *Anal. Chem.*, 2002, **74**, 2637.
5. T. Laurell, L. Wallman and J. Nilsson, *J. Micromech. Microeng.*, 1999, **9**, 369.
6. R. G. H. Lammertink, S. Schlautmann, G. A. J. Besselink and R. B. M. Schasfoort, *Anal. Chem.*, 2004, **76**, 3018.
7. O. Rotting, W. Ropke, H. Becker and C. Gartner, *Microsyst. Technol.*, 2002, **8**, 32.
8. L. J. Kricka, P. Fortina, N. J. Panaro, P. Wilding, G. Alonso-Amigo and H. Becker, *Lab Chip*, 2002, **2**, 1.
9. S. M. Ford, J. Davies, B. Kar, S. D. Qi, S. McWhorter, S. A. Soper and C. K. Malek, *J. Biomech. Eng. Trans. ASME*, 1999, **121**, 13.
10. D. Erickson and D. Q. Li, *Anal. Chim. Acta*, 2004, **507**, 11.
11. T. Thorsen, S. J. Maerkl and S. R. Quake, *Science*, 2002, **298**, 580.
12. J. W. Hong and S. R. Quake, *Nat. Biotechnol.*, 2003, **21**, 1179.
13. R. Pal, M. Yang, R. Lin, B. N. Johnson, N. Srivastava, S. Z. Razzacki, K. J. Chomistek, D. C. Heldsinger, R. M. Haque, V. M. Ugaz, P. K.

Thwar, Z. Chen, K. Alfano, M. B. Yim, M. Krishnan, A. O. Fuller, R. G. Larson, D. T. Burke and M. A. Burns, *Lab Chip*, 2005, **5**, 1024.
14. C. L. Xu, B. X. Li and Z. J. Zhang, *Chin. J. Anal. Chem.*, 2003, **31**, 1520.
15. F. von Eggeling, K. Junker, W. Fiedler, V. Wollscheid, M. Durst, U. Claussen and G. Ernst, *Electrophoresis*, 2001, **22**, 2898.
16. K. Uchiyama, H. Nakajima and T. Hobo, *Anal. Bioanal. Chem.*, 2004, **379**, 375.
17. M. A. Schwarz and P. C. Hauser, *Lab Chip*, 2001, **1**, 1.
18. A. J. de Mello, *Lab Chip*, 2001, **1**, 7N.
19. K. Uchiyama, H. Nakajima and T. Hobo, *Anal. Bioanal. Chem.*, 2004, **379**, 375.
20. M. Galloway, W. Stryjewski, A. Henry, S. M. Ford, S. Llopis, R. L. McCarley and S. A. Soper, *Anal. Chem.*, 2002, **74**, 2407.
21. J. B. Fenn, M. Mann, C. K. Meng, S. F. Wong and C. M. Whitehouse, *Mass Spectrom. Rev.*, 1990, **9**, 37.
22. J. S. Page and J. V. Sweedler, *Anal. Chem.*, 2002, **74**, 6200.
23. J. S. Page, S. S. Rubakhin, S. S. and J. V. Sweedler, *Anal. Chem.*, 2002, **74**, 497.
24. H. Becker and L. E. Locascio, *Talanta*, 2002, **56**, 267.
25. A. J. de Mello, *Lab Chip*, 2002, **2**, 31N.
26. S. A. Soper, S. M. Ford, S. Qi, R. L. McCarley, K. Kelly and M. C. Murphy, *Anal. Chem.*, 2000, **72**, 642A.
27. S. M. Ford, B. Kar, S. McWhorter, J. Davies, S. A. Soper, M. Klopf, G. Calderon and V. Saile, *J. Microcolumn Sep.*, 1998, **10**, 413.
28. B. J. Kirby and E. F. Hasselbrink, *Electrophoresis*, 2004, **25**, 187.
29. B. J. Kirby and E. F. Hasselbrink, *Electrophoresis*, 2004, **25**, 203.
30. H. Shadpour, H. Musyimi, J. Chen and S. A. Soper, *J. Chromatogr. A*, 2006, **1111**, 238.
31. D. Finnskog, A. Ressine, T. Laurell and G. Marko-Varga, *J. Proteome Res.*, 2004, **3**, 988.
32. S. Ekstrom, D. Ericsson, P. Onnerfjord, M. Bengtsson, J. Nilsson, G. Marko-Varga and T. Laurell, *Anal. Chem.*, 2001, **73**, 214.
33. S. Ekstrom, P. Onnerfjord, J. Nilsson, M. Bengtsson, T. Laurell and G. Marko-Varga, *Anal. Chem.*, 2000, **72**, 286.
34. A. R. Wheeler, H. Moon, C. J. Kim, J. A. Loo and R. L. Garrell, *Anal. Chem.*, 2004, **76**, 4833.
35. M. Gustafsson, D. Hirschberg, C. Palmberg, H. Jornvall and T. Bergman, *Anal. Chem.*, 2004, **76**, 345.
36. D. J. Rousell, S. M. Dutta, M. W. Little and K. K. Murray, *J. Mass Spectrom.*, 2004, **39**, 1182.
37. Y. Xu, M. W. Little and K. K. Murray, *J. Am. Soc. Mass Spectrom.*, 2006, **17**, 469.
38. K. K. Murray, *Mass Spectrom. Rev.*, 1997, **16**, 283.
39. K. K. Murray and D. H. Russell, *Anal. Chem.*, 1993, **65**, 2534.
40. L. Li, A. P. L. Wang and L. D. Coulson, *Anal. Chem.*, 1993, **65**, 4935.

41. M. Brivio, R. H. Fokkens, W. Verboom, D. N. Reinhoudt, N. R. Tas, M. H. Goedbloed and A. van den Berg, *Anal. Chem.*, 2002, **74**, 3972.
42. M. Brivio, N. R. Tas, M. H. Goedbloed, H. Gardeniers, W. Verboom, A. van den Berg and D. N. Reinhoudt, *Lab Chip*, 2005, **5**, 378.
43. J. Preisler, F. Foret and B. L. Karger, *Anal. Chem.*, 1998, **70**, 5278.
44. H. Orsnes, T. Graf, S. Bohatka and H. Degn, *Rapid Commun. Mass Spectrom.*, 1998, **12**, 11.
45. H. Orsnes, T. Graf, H. Degn and K. K. Murray, *Anal. Chem.*, 2000, **72**, 251.
46. H. K. Musyimi, D. A. Narcisse, X. Zhang, W. Stryjewski, S. A. Soper and K. K. Murray, *Anal. Chem.*, 2004, **76**, 5968.
47. X. Zhang, X. D. A. Narcisse and K. K. Murray, *J. Am. Soc. Mass Spectrom.*, 2004, **15**, 1471.
48. H. K. Musyimi, J. Guy, D. A. Narcisse, S. A. Soper and K. K. Murray, *Electrophoresis*, 2005, **26**, 4703.

CHAPTER 11

Lab-on-a-Chip Devices Enabling (Bio)chemical Reactions with On-line Analysis by MALDI-TOF Mass Spectrometry

MONICA BRIVIO,† WILLEM VERBOOM AND DAVID N. REINHOUDT

Laboratory of Supramolecular Chemistry and Technology, University of Twente, PO Box 217, 7500 AE Enschede, The Netherlands

11.1 Introduction

The introduction of matrix-assisted laser desorption ionization (MALDI)[1,2] more than a decade ago has revolutionized not only mass spectrometry (MS) but also biomolecular analysis. This soft ionization technique provides a powerful means to ionize a large variety of fragile and nonvolatile analytes including peptides and proteins,[3–6] oligonucleotides,[7] oligosaccharides,[8] synthetic polymers[9,10] and small molecules.[11] MALDI sources can be coupled to time-of-flight (TOF) equipment, extending the capability of the method to the analysis of molecules with molecular masses much higher than those detectable with other MS methods. Using this ionization technique Schriemer and

† Present address: MIC-Department of Micro and Nanotechnology, Technical University of Denmark-building 345 east, 2800 Lyngby, Denmark.

Miniaturization and Mass Spectrometry
Edited by Séverine Le Gac and Albert van den Berg

Published by the Royal Society of Chemistry, www.rsc.org

Li[12] have demonstrated the detection of singly charged ions up to 1.5×10^6 Da. MALDI offers the advantage of generating relatively simple spectra because mainly singly charged ions are formed. Other advantages of MALDI-MS include the tolerance for buffers and contaminants, a very high sensitivity[13] and faster analysis compared to atmospheric pressure ionization techniques.

Several approaches have been reported in the literature to increase the high-throughput capability of MALDI-MS. These include off-line analysis of fractions collected from a separation column and deposited on the MALDI target[14] as well as on-line coupling of MALDI with column separations[15–19] or planar separations such as thin-layer chromatography (TLC)[20–24] by means of various types of interfaces. However, on-line chip-MS[25,26] has mainly been restricted to the electrospray ionization method due to the easy interfacing between the electrospray capillary and microfluidics chips, and because this ionization technique operates at atmospheric pressure. The ionization process in MALDI-TOF-MS is mainly carried out in vacuum, which requires special precautions for integration with a microfluidics chip.

Recently, we described the first example of the coupling of a microfluidics device to a MALDI-TOF mass spectrometer, by integrating an on-chip microreaction unit with a MALDI-TOF standard sample plate.[27] This allows (bio)chemical reactions to take place in the MALDI-TOF instrument. Since it is based on a pressure-driven fluidics handling method, using the vacuum in the ionization chamber of the MALDI-TOF-MS system, this approach avoids wires and tubes for feed and flow control.

A variety of examples ranging from simple biochemistry to organic and polymer chemistry will illustrate the effectiveness and versatility of this chip-MS device.

In the first generation of chip design, due to the channel geometry, very high flow rates can be generated using a vacuum-driven actuation, which results in very short residence times. Furthermore, the reaction mixture can only be analyzed at one sampling spot (the outlet) where the reaction mixture is collected. As a consequence no time-dependent experiments can be done. A proper design of the fluidic circuit and the integration of multiple outlet ports would allow a quasi-continuous monitoring of the reaction mixture in time, thus extending the use of the MALDI-chip to kinetic studies.

A second generation of microreactors has therefore been designed,[28] which allows longer reaction and measurement times. In addition, a novel concept is described, consisting of the integration of a "monitoring window" into the chip design for monitoring reactions by using the MALDI-chip setup. As a model reaction the Schiff base reaction[29] between 4-*tert*-butylaniline (**2**) and 4-*tert*-butylbenzaldehyde (**4**) in ethanol to give the corresponding imine **6** (Scheme 11.1) was carried out in the MALDI-chip and the reaction mixture was analyzed through the "monitoring window".

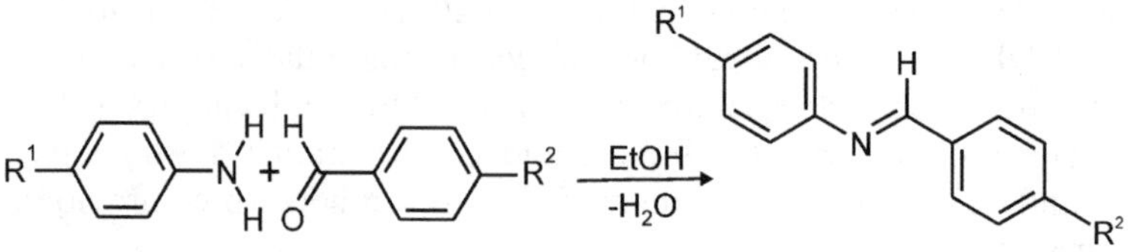

EtOH
$-H_2O$

1 $R^1 = H$ **3** $R^2 = H$ **5** $R^1 = R^2 = H$
2 $R^1 = t$-Bu **4** $R^2 = t$-Bu **6** $R^1 = R^2 = t$-Bu

Scheme 11.1 Schiff Base Formation Reaction.

11.2 First-Generation MALDI-Based Lab-on-a-Chip

11.2.1 Materials and Methods

11.2.1.1 Microreactor

The microfluidics device used in the experiments (Figure 11.1) was fabricated in the cleanroom facilities of the $MESA^+$ Institute for Nanotechnology.[27] Microchannels and inlet/outlet holes were realized in a top borofloat wafer (1.1 mm thick) by high-resolution powder-blast micromachining.[30] This relatively new microfabrication technique[31] is faster but less precise than isotropic HF etching[32] and has the advantage of avoiding underetching. A great advantage of this technique is the possibility to realize an unlimited number of vertical feed-throughs in one process step of about 30 min. In order to avoid contact between the chip and the voltage grid of the MALDI instrument (Figure 11.2b), the top borofloat wafer was anodically bonded to a thinner silicon wafer (500 μm thick).

11.2.1.2 Detection

The reaction products formed in the microreactor channel and collected in the chip outlet were, in real time, identified by MALDI-TOF-MS using a Voyager-DE-RP MALDI-TOF mass spectrometer (Applied Biosystems/PerSeptive Biosystems, Inc., Framingham, USA) equipped with delayed extraction[33,34] and a 337 nm ultraviolet (UV) nitrogen laser producing 3 ns pulses. The mass spectra were obtained both in the linear and reflection mode. To avoid fast solidification of the reaction products at the outlet reservoirs of the microchip, the vacuum pressure in the ionization chamber of the mass spectrometer was gradually reduced while introducing the chip-MALDI sample plate. The pressure inside the TOF analyzer was kept at 10^{-8} torr (1.33×10^{-6} Pa).

11.2.1.3 Testing the Chip Via Schiff Base Formation

Reagents were obtained from Aldrich Chemicals, The Netherlands. Samples were used as supplied commercially without further purification. No matrix was

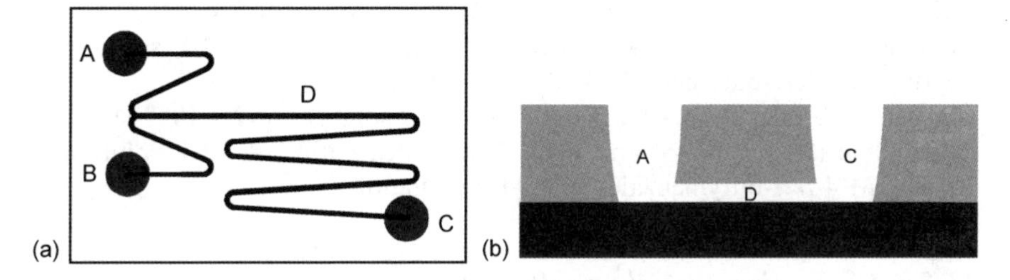

Figure 11.1 (a) Schematic top view of the microfluidics chip with the inlet (A and B) and the outlet (C) cups and the reaction microchannel (D); (b) side view of the micromachined top borofloat wafer with one of the two inlets (A), the outlet (C) and the microchannel (D), anodically bonded with the bottom silicon substrate.

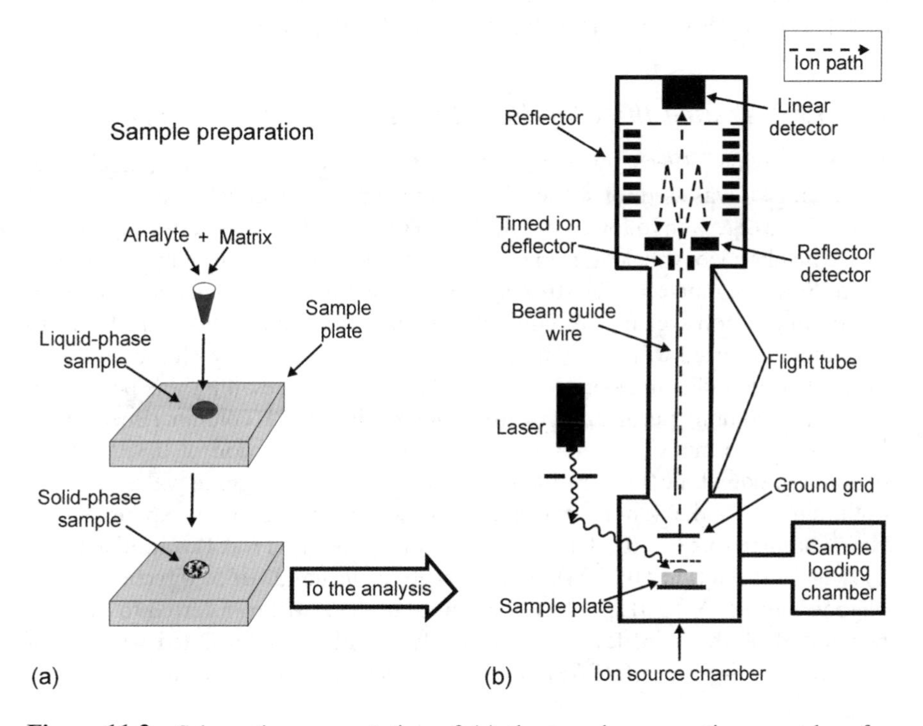

Figure 11.2 Schematic representation of (a) the sample preparation procedure for MALDI-MS analysis and (b) the MALDI-TOF-MS instrument.

added for MALDI detection of the aromatic products of these systems, because of their strong absorption at the laser wavelength ($\lambda = 337$ nm). Three Schiff base reactions were performed on-chip and the products were analyzed on-line.

Reaction 1. An amount of 3 μL of a solution of 2×10^{-4} M aniline (**1**) in ethanol was placed in inlet A; inlet B was filled with 3 μL of a solution of 2×10^{-4} M benzaldehyde (**3**) in ethanol.

Reaction 2. Inlet A was filled with 3 μL of a solution of 2×10^{-4} M 4-*tert*-butylaniline (**2**) in ethanol and inlet B was filled with 3 μL of a solution of 2×10^{-4} M 4-*tert*-butylbenzaldehyde (**4**) in ethanol.

Reaction 3. Inlet A was filled with 3 μL of a solution of 2×10^{-4} M 4-*tert*-butylaniline (**2**) in ethanol and inlet B was filled with 3 μL of a solution of 4×10^{-4} M 4-*tert*-butylbenzaldehyde (**4**) in ethanol.

11.2.1.4 Testing the Chip with Polymers

Reagents were obtained from Aldrich Chemicals, The Netherlands, and were used as supplied commercially without further purification. Dihydroxybenzoic acid (DHB) was used as the matrix. In inlet A, 0.3 μL of polystyrene (PS; $MW_{av} = 3500$ Da) in tetrahydrofuran (THF)/water was injected with DHB as the matrix. In inlet B, 0.3 μL of poly(methyl methacrylate) (PMMA; $MW_{av} =$ 3500 Da) in THF/water was injected with DHB as matrix.

11.2.1.5 Testing the Chip for Oligonucleotide Sequencing

The Sequazyme™ (PerSeptive Biosystems, Framingham, USA) oligonucleotide sequencing kit was used for sequencing synthetic deoxyoligonucleotides by partial exonuclease digestion followed by MALDI-TOF-MS. The method was applied to a 41-base oligodeoxynucleotide in which a nucleobase was replaced by a hydrogen atom. Since oligonucleotides strongly bind trace amounts of salts, sample deionization *via* cation exchange is required. To this end the sample was transferred to Parafilm and mixed with the cation exchange beads provided with the kit. Since the investigated oligonucleotide is longer than 10 bases, desalting was performed using spin column purification with a QuickSpin™ C-25 column (Boehringer, Mannheim, Germany). Inlet A was used for the introduction of 0.3 μL of the oligonucleotide at a concentration of 2×10^{-4} M in HPLC grade water. The pH optimum for snake venom phosphodiesterase (SVP) digestion experiments is basic and therefore 0.3 μL of ammonium citrate buffer (pH = 6.1) was mixed with 1 μL of SVP dilution (10^{-4} M). SVP hydrolyzes in the 3′ to 5′ direction of the oligonucleotide. A 0.3 μL portion of supernatant SVP was transferred to the inlet position B of the chip device. Additionally, 0.2 μL of a saturated solution of 3-hydroxypicolinic acid (HPA; matrix) in water was injected into inlet B.

11.2.1.6 Testing the Chip for Peptide Sequencing

The Sequazyme™ C-peptide sequencing kit enables peptide digestion followed by analysis of sequentially truncated peptides using MALDI-TOF-MS. Before digesting and analyzing unknown samples of the peptide (adrenocorticotropic hormone), the activity of the enzyme carboxypeptidase Y (CPY) was checked by dilution experiments with ammonium citrate buffer (pH = 6.1), CPY dilution and the peptide standard angiotensin I. The adrenocorticotropin ACTH (18–39) fragment (0.3 μL) was transferred as a solution of 5×10^{-5} M in a 1:1

(v/v%) mixture of water and acetonitrile to inlet A. Inlet B contained 0.3 μL of a solution of 0.1 mM CPY in HPLC grade water and ammonium citrate (pH = 6.1) and 0.2 μL of the matrix α-cyano-4-hydroxycinnamic acid in a 1:1 (v/v%) mixture of water and acetonitrile and 0.1% trifluoroacetic acid (TFA). Immediately after the introduction of the ACTH 18–39 peptide the CPY dilution including the matrix was transferred to inlet B.

11.2.2 Results and Discussion

11.2.2.1 MALDI Instrument

A schematic of a MALDI-TOF-MS instrument is depicted in Figure 11.2b. Samples, consisting of a few microlitres of analyte solution (with or without matrix), are deposited on a MALDI target (Figure 11.2a). After the solvent has evaporated the sample plate, carrying the solidified samples, is introduced into the MALDI ionization chamber *via* load-lock. The ionization process takes place in a high-vacuum chamber to which the plate is introduced *via* a pre-chamber kept at a pressure lower than atmospheric. Analyte ions are then accelerated as they are formed and pumped into the TOF analyzer, where they are separated based on their mass-to-charge ratio.

11.2.2.2 Microreactor Design and Fabrication

The microreactor (Figure 11.1) was designed to carry out on-line MALDI-TOF-MS analyses of products of simple chemical and biochemical reactions (A + B = C type). Therefore, the microreactor consists of two inlet reservoirs (A and B) for the injection of a maximum volume of 5 μL of reagent solutions and a 5 μL outlet pocket (C) where the analyte solution is collected and analyzed. The reaction microchannel in which the reagents react is 80 mm long, 200 μm wide and 100 μm deep. Due to the shape of the powder-blasted channel, the cross-section is approximated to a 150 μm wide by 100 μm deep rectangle. The resulting channel volume is 1.2 μL.

11.2.2.3 Inlet Filling Procedure

The inlet reservoirs were filled one by one using a micropipette and taking care that both the second inlet and the outlet holes were closed. Filling the inlet pockets without carefully following the procedure would cause the capillary force to drive the liquid out from the inlet reservoir into the channel. Closing the outlet and one of the inlet cups while loading the liquid in the other one causes a counter-pressure in the channel that keeps the liquid within the inlet pocket that is being filled. During the filling procedure both the outlet and the inlet holes not being filled were closed by means of adhesive tape. After loading of each inlet was completed, the inlet cups were sealed by applying a piece of glass fixed to the surface of the chip by means of adhesive tape.

11.2.2.4 Chip-MALDI Sample Plate Interaction

To perform on-line experiments the glass–silicon chip was integrated in the MALDI-TOF sample plate (Figure 11.3). The integration of the microreaction and detection unit was achieved by placing the chip in a hole (1.35 mm deep) made in the sample plate by milling. Subsequently, the system (chip and sample plate) was introduced into the MALDI-TOF instrument by the standard loading procedure.

11.2.2.5 Activation Method

Just before loading the system the tape that closed the outlet pocket was removed. At this stage the liquid is kept in the inlet cups by the atmospheric pressure at the opened outlet (Figure 11.4a). Once the MALDI sample plate with the integrated microchip enters the vacuum ionization chamber, the counter-pressure from the outlet disappears and the air in both inlet pockets pushes the liquid through the microchannel towards the outlet pocket (Figure 11.4b).

A second driving force that might affect the fluidic flow is the capillary pressure $p_{cap} = 2\gamma/r$ (where r is the radius of the tube and γ is the surface tension

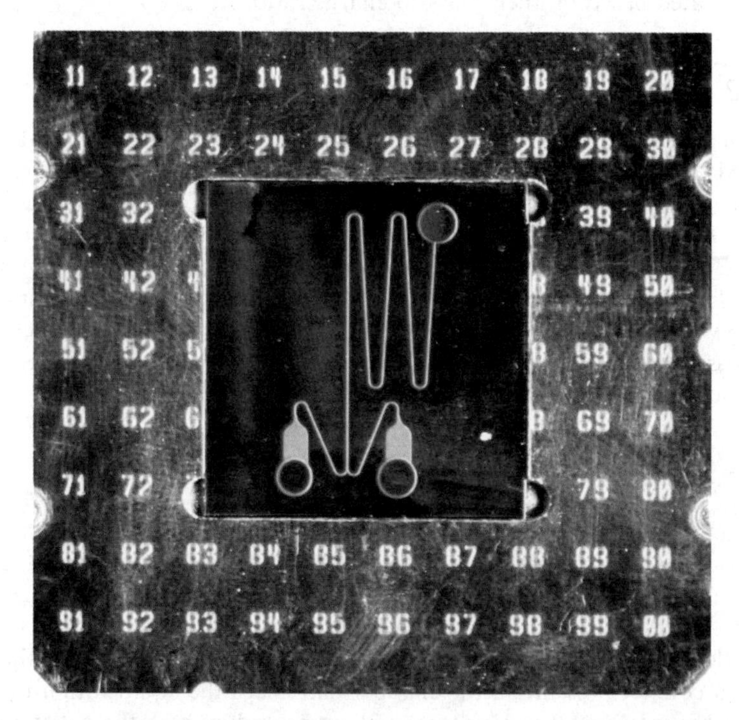

Figure 11.3 Photograph of the microreactor integrated in a standard MALDI-TOF sample plate. Because of the self-activating character of the microfluidics device, the system can be introduced into the MALDI ionization chamber without any wire or tube for the feeding and flow control.

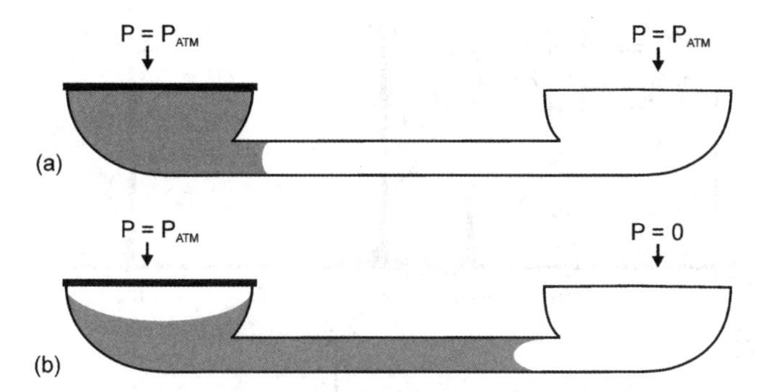

Figure 11.4 Passive pressure-driven mechanism of fluidics handling. (a) The liquid is kept in the inlet reservoirs by air ($p = p_{atm}$) present in the channel. (b) Inside the MS instrument the decrease in pressure at the outlet causes the liquid to flow to the outlet through the microchannel.

of the liquid in the channel).[35] However, in the current chip, in which the channel diameter is of the order of 100 μm, the capillary pressure (1.44×10^3 Pa for water) is small compared to the pneumatic driving force (10^5 Pa).

11.2.2.6 On-Chip Reactions

To demonstrate the effectiveness of the MALDI-based integrated microfluidics device, a few simple chemical and biochemical reactions were carried out in the MALDI-chip and their products analyzed on-line at the chip outlet (see above).

11.2.2.6.1 Schiff Base Formation. The Schiff base reaction (Scheme 11.1), in which primary amines (**1** and **2**) react with aldehydes (**3** and **4**) to give imines (**5** and **6**), was chosen as a primary study of the microreaction unit and its coupling with MALDI-TOF-MS, because it is a straightforward reaction and products are obtained in high yields. The pH control required for the Schiff base formation[29] turned out to be not necessary under "lab-on-a-chip" conditions as a consequence of the high surface-to-volume ratio that characterizes microreactors.[36]

A first experiment was carried out reacting on-chip equimolar (2×10^{-4} M) solutions of aniline (**1**) and benzaldehyde (**3**) in ethanol. The MALDI mass spectrum (Figure 11.5a) only shows the signal of the reaction product, imine **5** (*m*/*z* 182), proving the quantitative reaction of reagents **1** and **3** in the microchannel. Post-source decay experiments[37] were performed on the same sample collected at the chip outlet, in order to confirm the structure of the on-chip-formed imine **5**. By *in situ* fragmentation of the reaction product ions at *m*/*z* 182, fragments F1 (*m*/*z* 104) and F2 (*m*/*z* 77) were generated and detected, validating the structure of imine **5** (Figure 11.5a). Figure 11.5b shows the MALDI mass spectrum of imine **6** (*m*/*z* 294) formed on-chip by reacting equimolar (2×10^{-4} M) solutions of reagents **2** and **4** in ethanol. The same

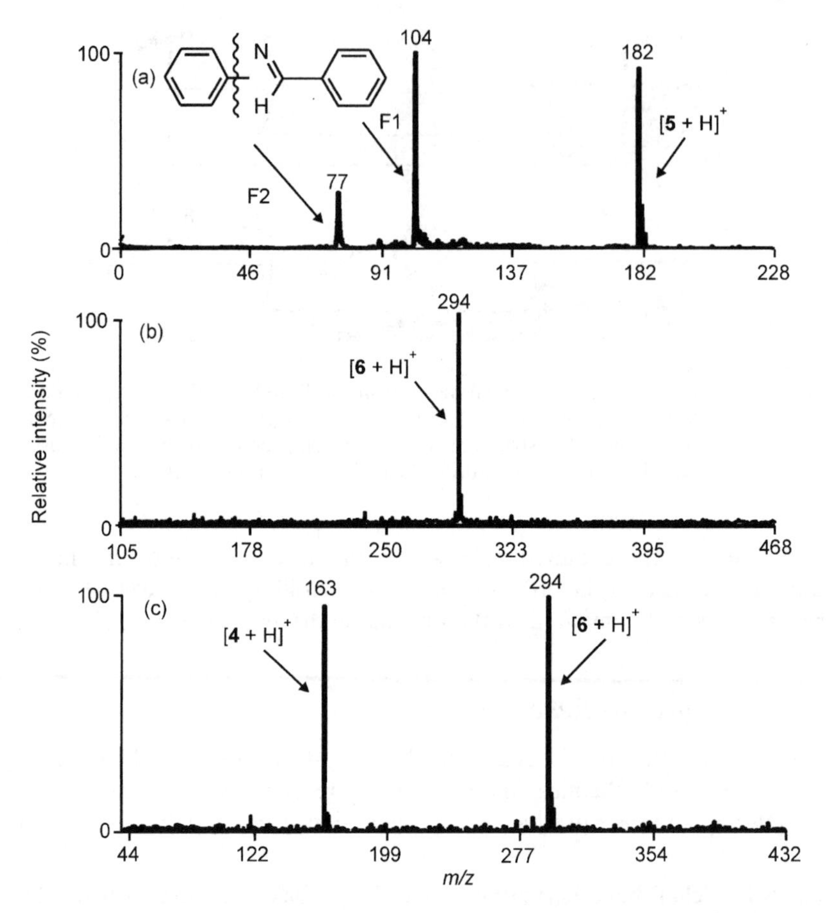

Figure 11.5 MALDI-TOF mass spectra of the Schiff base reaction products: (a) imine **5** at *m*/*z* 182 formed in the reaction microchannel in experiment 1 and its MALDI-TOF PSD mass spectrum fragments F1 and F2; (b) imine **6** at *m*/*z* 294 formed on-chip during experiment 2; (c) imine **6** at *m*/*z* 294 formed on-chip during experiment 3 and the excess of reagent **4** at *m*/*z* 163.

reaction was carried out on-chip by reacting two equivalents of aldehyde **4** with one equivalent of amine **2**. In this case signals of both product and excess of reagent were detected by MALDI-MS (Figure 11.5c).

11.2.2.6.2 Polymers. In order to prove the versatility of the microfluidics system as a platform for different applications, its ability to separate mixtures of polymers was tested. MALDI mass spectra show two different polymer distributions separated in time (Figure 11.6). The two different polymer distributions can be easily assigned to the injected polymers by the peak-to-peak separation due to the different molecular weight of the monomers. The first pattern recorded in time was assigned to PS (Figure 11.6a) and the

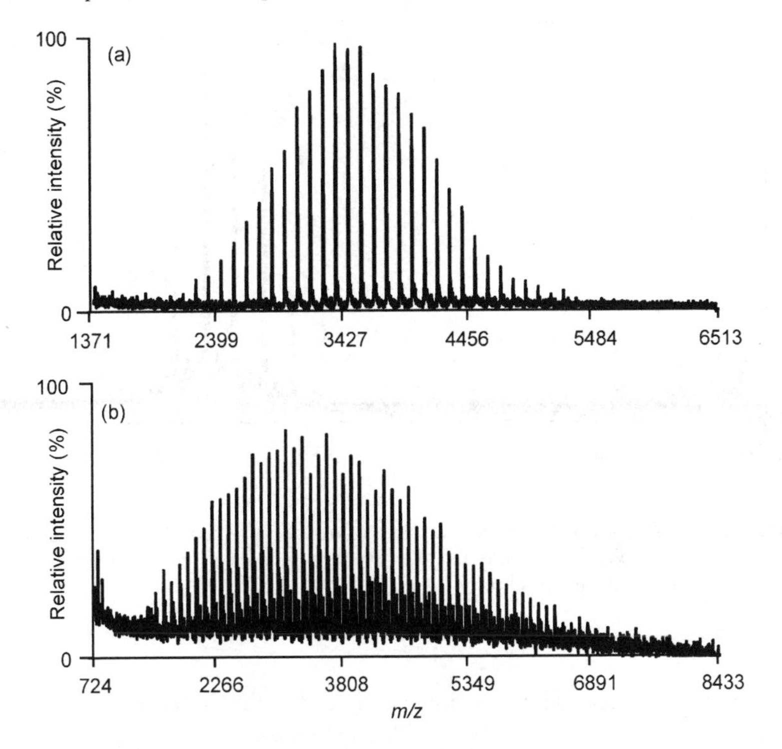

Figure 11.6 MALDI-TOF mass spectra of the two different polymer distributions separated in arrival time at the outlet reservoir and assigned to (a) PS and (b) PMMA.

second to PMMA (Figure 11.6b); the difference in mass between adjacent peaks being 104 and 100 mass units, respectively.

This result shows that the chip is capable of separating the two polymers, on the basis of their different polarities. Due to its polar side chains, PMMA may interact with the channel walls, resulting in longer on-chip residence times than for the apolar PS. These results suggest the possible application of the lab-on-a-chip presented here as a separation tool, comparable to gel permeation chromatography.[38]

11.2.2.6.3 Oligonucleotides. Mass spectrometry is an outstanding technique for the characterization and sequence determination of oligonucleotides generated via solution-phase chemical reactions.[39] The DNA sequence can be determined by the mass difference between the oligonucleotide fragments of (partial) digests from either end of the DNA molecule. Oligonucleotide digestion was performed directly on the microfluidics chip by mixing the oligonucleotide (substrate) with SVP that hydrolyzes DNA in the $3'$ to $5'$ direction. MALDI-MS analysis of the hydrolysis products was performed at

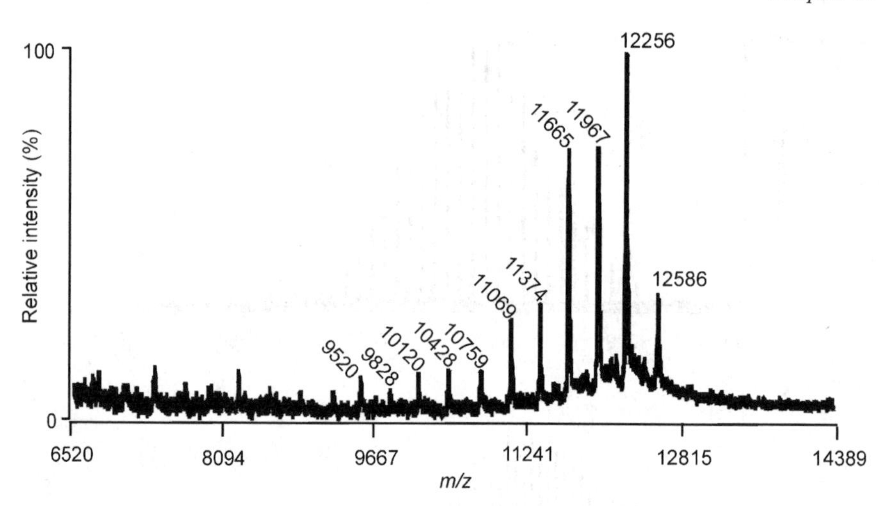

Figure 11.7 MALDI-TOF mass spectrum of an oligonucleotide at m/z 12 586 and the oligonucleotide residue average masses after enzymatic digestion carried out on-chip.

Table 11.1 Average masses of the individual nucleotide bases.

Name	*Abbreviation*	*Average mass (Da)*
Adenine	A	313.2
Cytosine	C	289.2
Guanine	G	329.2
Thymine	T	304.2

the chip outlet where the sample was collected. The mass difference between adjacent pairs of peaks in the MALDI-TOF mass spectrum identifies each nucleotide base in the sequence. In Figure 11.7 the peak of the oligonucleotide (at m/z 12 586) and those of the oligonucleotide residues are shown. From the average masses of the individual bases (Table 11.1) the oligonucleotide sequence can be derived (GCTCTAGACT).

This approach to analyze exonuclease ladders seems to be a particularly promising tool to determine rapidly the sequence of oligonucleotides. Compared to conventional methods, based on the laboratory-scale preparation of a large number of samples at different enzyme and substrate concentrations followed by MALDI-MS analysis, the microfluidics approach offers the advantage of saving time and material. Furthermore, due to the limited sample handling, the risks encountered when manipulating biomolecules are also reduced as well as that of sample contamination.

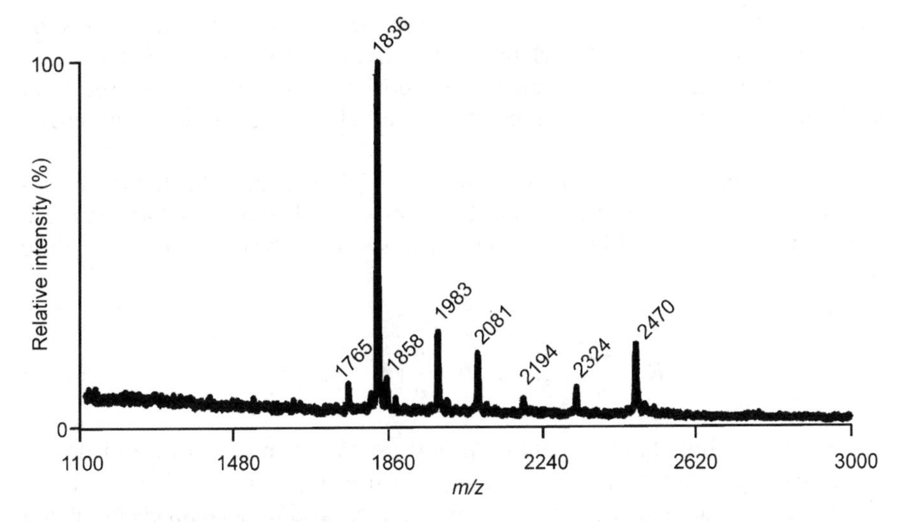

Figure 11.8 MALDI-TOF mass spectrum of the ACTH peptide *m/z* 2470 and the amino acid residues after enzymatic digestion. The sequence resulting from the partial on-chip digestion is: Phe-Glu-Leu-Pro-Phe-Ala.

11.2.2.6.4 Peptide Sequencing. Carboxypeptidase Y sequentially hydrolyzes the C-terminal residues of peptides,[40] which can be subsequently analyzed directly by MS. The sequence of the peptide fragments can be determined from the mass differences between the adjacent peaks in the mass spectrum. In the lab-on-a-chip approach peptide digestion and product analysis are performed directly on the chip, thus minimizing sample handling, sample loss and method development time. In the lab-on-a-chip methods, indeed, only a total of a few picomoles of peptides are required, and the analysis can be made *in situ* immediately after digestion, thereby avoiding sample transfer to the mass spectrometer. After data acquisition and mass calculation, the mass differences between the adjacent peaks in the ladders were used to determine the C-terminal sequence of the peptide (Figure 11.8).

11.2.2.7 Fluid Dynamics

A high degree of temporal control over the (micro)fluidics is a prerequisite to carry out time-dependent studies of (bio)chemical reactions. In the MALDI-based lab-on-a-chip device described here the fluid is transported through the microchannels due to a pressure difference between the inlet, where a small air bubble is present (atmospheric pressure), and the outlet (vacuum). The relation between the generated flow rate and the applied vacuum pressure is defined by the equation[41,42]

$$\Delta p = R\phi_{\mathrm{v}} \tag{11.1}$$

where Δp [N m^{-2}] is the pressure drop across the channel and ϕ_v [m^3 s^{-1}] is the volumetric flow through the channel. The hydraulic resistance R [N s m^{-5}], which is dependent on the channel geometry and the viscosity of the fluid μ [N s m^{-2}], is calculated by integrating the velocity profile over the cross-sectional area.

Based on the channel geometry a number of design formulas for the calculation of the hydraulic resistance has been derived.[43] For rectangular channels, such as that of the MALDI-chip, the resistance R can be calculated according to

$$R = \frac{4\mu l}{ab^3}\left[\frac{16}{3} - 3.36 \times \frac{b}{a}\left(1 - \frac{b^4}{12a^4}\right)\right]^{-1} \tag{11.2}$$

where l [m] is the channel length, $2a$ and $2b$ the channel width and depth, respectively, and μ [N s m^{-2}] the liquid viscosity. Equation (11.2) is valid for low Reynolds numbers (*i.e.* laminar flow regimes) and long channels, implying that entrance effects can be neglected. Using Equation (11.1) and knowing the channel length l of the MALDI-chip, the residence time t_{res} [s] can be calculated according to

$$t_{res} = \frac{lAR}{\Delta p} \tag{11.3}$$

where A [m^2] is the channel cross-sectional surface area. In the MALDI-chip microchannel ($2a = 1.5 \times 10^{-4}$ m; $2b = 10^{-4}$ m; $l = 8 \times 10^{-2}$ m) the calculated resistance is $R = 1.55 \times 10^{13}$ N s m^{-5} (for water; $\mu = 10^{-3}$ N s m^{-2}). The residence time for an applied pressure difference of 1 bar (10^5 N m^{-2}) is 1.9×10^{-1} s.

It is evident that no time-dependent experiments are possible using the first generation of microfluidic devices (see Section 11.2.2.2), since the residence times are of the order of 200 milliseconds. Even when opening multiple sampling ports along the channel, the time required to move the ionizing laser from one window to the next would be much too long compared to the on-chip residence times. Channels that have a higher hydraulic resistance are a possible solution to the short on-chip residence times. A new microfluidic system was therefore designed to allow a longer residence time.

11.3 Second-Generation MALDI-Based Lab-on-a-Chip

11.3.1 Materials and Methods

11.3.1.1 Microreactors

The microreactors (Figure 11.9) were fabricated in the cleanroom facility of the MESA$^+$ Institute for Nanotechnology at the University of Twente. Inlet and

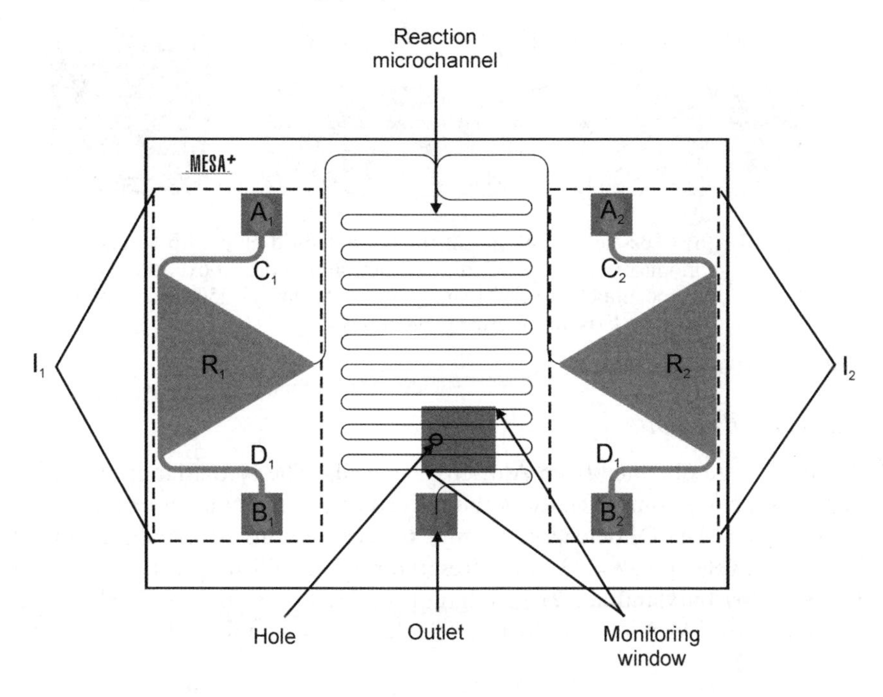

Figure 11.9 Schematic overview of the chip used to prove the "monitoring window" principle. Each inlet loading system (I_1 and I_2) consists of two inlet holes (A_x and B_x) connected to a reservoir (R_x) through channels C_x and D_x (with $x = 1$ and 2).

outlet holes as well as the "monitoring window" were fabricated in the top silicon wafer by KOH wet-etching[44] in combination with a boron etch stop.[45] The reaction microchannel was fabricated in the bottom borofloat wafer by HF etching,[32] while the triangular reservoirs as well as the bigger inlet channels (going from the inlet holes to the reservoirs) were fabricated by powder-blast micromachining.[30,31] Square-shaped holes in the monitoring window were fabricated by focused ion beam (FIB). Various combinations of dimensions and number of holes were exploited. The preliminary experiment reported was carried out using a monitoring window with only one sampling hole of 500 nm by 500 nm.

11.3.1.2 Focused Ion Beam

Holes in the boron-doped silicon membranes (250 nm by 250 nm and 500 nm by 500 nm) were made using an FEI 200 focused ion beam system (see Figure 11.10). The resulting holes were imaged *in situ* using scanning electron microscopy (SEM).

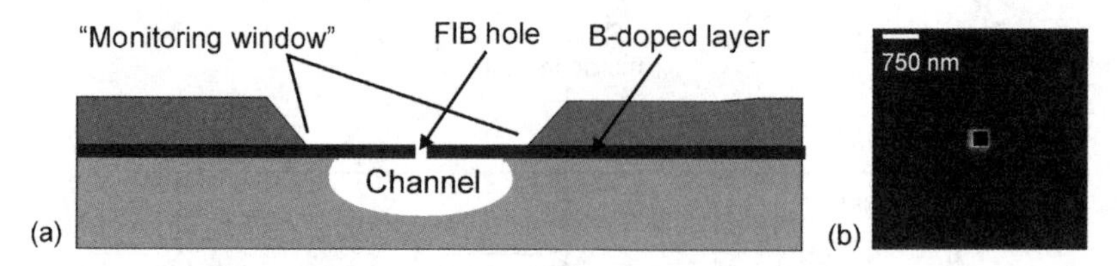

Figure 11.10 (a) Cross-section of the silicon–glass hybrid chip with a KOH-etched "monitoring window" and a sampling hole above the HF-etched microchannel. (b) SEM image of a 250 nm by 250 nm sampling hole made by FIB in the doped silicon layer.

11.3.1.3 Reagents

Chemicals were obtained from Aldrich Chemicals, The Netherlands, and were used as supplied commercially without further purification. No matrix was added for the MALDI detection of imine **6**, because of its strong absorption at the laser wavelength ($\lambda = 337$ nm). Reservoir A was filled with a solution of 10^{-4} M 4-*tert*-butylaniline (**2**) in ethanol and reservoir B was filled with a solution of 10^{-4} M 4-*tert*-butylbenzaldehyde (**4**) in ethanol.

11.3.1.4 Filling Procedure

The reservoirs filling procedure was monitored by mounting the chip, placed on a suitable holder, on an Olympus CK40M inverted microscope.

11.3.2 Results and Discussion

11.3.2.1 Microreactor Design and Fluid Dynamics

A microreactor was designed (Figure 11.9) having 30 μm wide, 10 μm deep and 125 mm long channels, fabricated in a glass substrate by wet chemical etching using HF. The hydraulic resistance as well as the residence time in this new channel design were calculated using Equations (11.2) and (11.3) to give values of $R = 6.3 \times 10^{16}$ N s m^{-5} and $t_{res} = 24$ s.

Due to downscaling of the microchannel, the surface free energy[46,47] ("surface tension") becomes considerable. As a consequence, the contribution of capillary pressure to the MALDI-chip pumping mechanism will be significant. For a flat rectangular channel the capillary pressure P_c [Pa] is given by[35,48]

$$P_c = \frac{2\gamma \cos \theta_c}{h} \tag{11.4}$$

where γ [N m^{-1}] is the surface tension of the liquid in the channel, θ_c is the contact angle between the meniscus and the channel wall (which for water in hydrophilic channels is assumed to be between 0 and 30°) and h [m] is the channel depth. However, the capillary pressure P_c ($1.3 \times 10^4\,\mathrm{N\,m^{-2}}$) estimated for water ($\gamma = 0.07\,\mathrm{N\,m^{-1}}$; cos θ_c ~0.9 with $\theta_c = 25°$) is only about 10% of the pneumatic pressure used to drive the flow in the channel ($10^5\,\mathrm{N\,m^{-2}}$). Therefore it can be concluded that the hydraulic resistance and residence time calculated according to Equations (11.2) and (11.3), without taking into account the contribution of the capillary pressure, are a good estimation of the fluid dynamics in the microchannel depicted in Figure 11.9.

Although the estimated residence times in the new microchannel are about two orders of magnitude higher than those in the previous MALDI-chip design, they are still too short to study reaction kinetics by MALDI-MS. However, a considerably longer total measurement time (~75 min) can be obtained by the introduction of relatively large reservoirs for the reagent solutions (see below). Due to the long measurement time obtainable, this chip was used to demonstrate the "monitoring window" principle.

11.3.2.2 "Monitoring Window" Description

The key element of the new microreactor (Figure 11.9) is the "monitoring window". This consists of a freestanding ~2 μm thick silicon membrane doped with boron, positioned above the channel (Figure 11.10a). Ions can be extracted from the flowing reaction mixture through holes within the silicon membrane. Holes (250 nm by 250 nm and 500 nm by 500 nm), which could be visualized by SEM, were made by means of a FIB. A SEM micrograph of the 250 nm by 250 nm holes is shown in Figure 11.10b. Hole dimensions are important, since a pressure drop along the channel may arise from too large holes in the membrane, thereby affecting the microfluidic actuation mechanism.

The position and the dimensions of the "monitoring window" are also important factors. The dimensions of the window in the top silicon wafer must be large enough to let the laser beam, which hits the chip surface under an angle of about 30°, reach the small hole fabricated in the boron-doped silicon membrane on the bottom side of the same wafer.

11.3.2.3 Inlet Loading System

In the first version of the MALDI-chip reagent solutions were loaded in inlet cups fabricated in the top wafer by powder-blast micromachining. The inlets were then sealed with adhesive tape. A drawback of this inlet system is that the solutions can easily become contaminated by coming into contact with the adhesive tape. A new reagent loading scheme was developed (Figure 11.9) consisting of two inlet systems (I_1 and I_2). Each system comprises two via-holes etched in the top silicon wafer (A_1, B_1 and A_2, B_2, respectively), a triangular reservoir (R_1 and R_2) and two 200 by 100 μm powder-blasted channels (C_1, D_1 and C_2, D_2) that connect each inlet hole with the corresponding reservoir. Flap

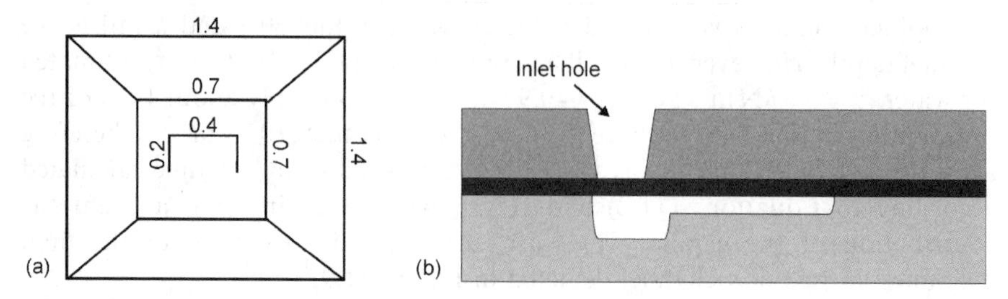

Figure 11.11 (a) Top view of the flap valve at the bottom side of each inlet hole (the dimensions are given in mm) and (b) cross-section of the inlet hole. The microreactor was used to test the feasibility of monitoring on-chip reactions by ionizing liquid-phase samples through the "monitoring window".

valves (Figure 11.11a) were fabricated in the doped silicon layer at the bottom of the inlet holes (Figure 11.11b). These valves keep the loaded solutions in the bottom wafer, thereby eliminating the risk of contamination. The triangular reservoirs (5.4 mm high, 6.2 mm wide and 100 μm deep) were powder-blasted in the bottom glass wafer.

11.3.2.4 "Monitoring Window" Testing

11.3.2.4.1 Inlet filling procedure. The reservoirs were filled by suction of reagent solutions using a syringe. About 3 μL of reagent solutions were placed in holes A_1 and A_2, while the syringe was placed on holes B_1 and B_2. During the filling procedure the chip was placed in a suitable holder. In order to prevent the liquid from filling the reaction channel before entering the MALDI chamber, all other openings (outlet, monitoring window and second inlet) were sealed with adhesive tape during the filling procedure of each inlet. Furthermore, due to the triangular shape of the reservoirs an air bubble was entrapped during the filling procedure (at the reservoir corner where the small channel starts). This prevents filling of the small channel due to capillary forces. After both reagents were placed in the reservoirs, the four inlet holes were closed using UV-curable glue. A glass slide (700 μm by 700 μm) was placed in the inlet holes to prevent the glue from entering the inlet channels, thereby polluting the reactants.

11.3.2.4.2 On-chip Reaction and Analysis. To prove the principle of the "monitoring window", the Schiff base formation reaction between **2** and **4** in ethanol (Scheme 11.1) was carried out using the MALDI-chip device equipped with the chip of Figure 11.9. The chip placed on the MALDI sample plate was introduced into the vacuum chamber by load-lock. The first MALDI-TOF mass spectrum was acquired as soon as the plate reached the right spot in the chamber. The analysis started after about 1 min. Ions were extracted from the

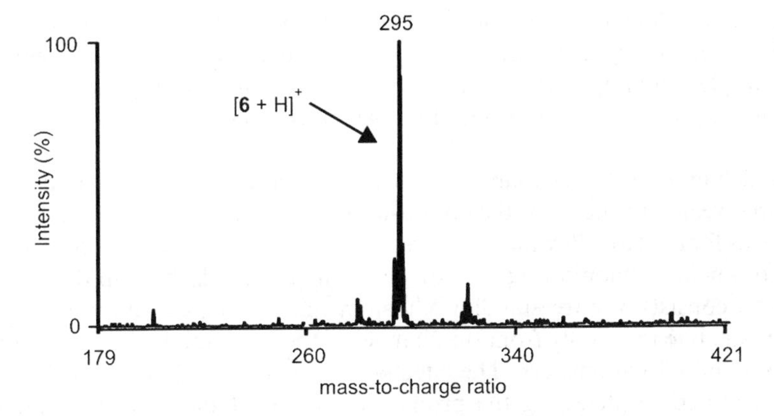

Figure 11.12 MALDI-TOF mass spectrum of imine **6** (*m*/*z* 295) formed on-chip within the ionization chamber of the MALDI instrument. The product was detected by extracting ions from the reaction mixture through the "monitoring window". After the last mass spectrum was recorded ($\sim$1 h), the chip was examined by an optical microscope. The air bubble in the reservoirs expanded into the channel, while liquid was still present in the portion of the channel close to the outlet.

channel through the hole, by hitting the "monitoring window" with the laser. The reaction product, imine **6**, was clearly visible in the mass spectrum at *m*/*z* 295 (Figure 11.12). Even after about 1 h, imine **6** could still be detected, indicating that reagent solutions were still available for the reaction, as predicted by theoretical calculations.

11.4 Concluding Remarks

An integrated microfluidics device that allows chemical syntheses, separations and biochemical processes on a microscale and in real time monitored by MALDI-TOF-MS was developed for the first time. The effectiveness of the lab-on-a-chip device was illustrated for a variety of systems ranging from simple synthetic chemistry to polymer analysis and enzymatic digestion of peptides and oligonucleotides. After the chip is placed in the MALDI-TOF vacuum chamber, and as soon as the reaction mixture (products and unreacted reagents) reaches the outlet, the laser beam ionizes the molecules, which are detected by the TOF analyzer. An important novelty of this approach is the possibility to analyze by MALDI-TOF-MS the reaction products *in situ* as soon as they are formed. Molecules with masses ranging from a few hundred to a few thousand daltons were successfully detected, thus demonstrating the wide range of applicability of the microfluidics system. This result was achieved by realizing a "self-activating" chip, which avoids the need of introducing wires and tubes into the ionization chamber of the MALDI-TOF instrument.

The main limitation of the integrated microfluidics device is the poor fluidics control. By a proper design of the fluidic circuit and the integration of multiple sampling ports, direct, real-time monitoring of product formation during the reaction will become feasible, thus opening the way to kinetics studies.

Modeling of the fluid dynamics in the first-generation MALDI-based lab-on-a-chip revealed the need for the optimization of the microfluidics circuit. A new chip was fabricated, allowing longer reaction and measuring times. As foremost improvement a "monitoring window" was integrated in the chip design. This window consists of a freestanding silicon membrane doped with boron, which allows analyte ions to fly from the channel to the MALDI detector *via* holes of a few hundred nanometers. The effectiveness of the window design has been demonstrated by detecting the product (imine **6**) of the on-chip Schiff base reaction of 4-*tert*-butylaniline (**2**) and 4-*tert*-butylbenzaldehyde (**4**) in ethanol within the MALDI chamber. Though very preliminary, these results clearly demonstrate the possibility of extracting ions from a flowing solution for analysis of the composition of the reaction mixture.

11.5 Outlook

Although improvements in residence and measuring times were achieved by resizing the microfluidics circuit, a poor flow control resulting from the passive fluidics actuation remains the main limitation of the MALDI-chip device. An accurate control of the liquid flow is a prerequisite for the realization of a fully functioning MALDI-based lab-on-a-chip device for the study of reaction kinetics. This can be achieved either by an optimal channel design or by the use of an active pumping mechanism. In particular, when increasing the hydraulic resistance by resizing the reaction channel, contributions to the pumping as a result of surface forces as well as pressure drops due to the integration of multiple openings along the channel must be taken into account. Also, the delay time between the beginning of the vacuum pumping and the first possible analysis contributes to the loss of flow control and, consequently, of the reaction. This limitation might be avoided by closing the chip outlet with a membrane, which would prevent the liquid from starting to flow. The membrane may be opened only when the setup is ready for the first analysis.

References

1. M. Karas and F. Hillenkamp, *Anal. Chem.*, 1988, **60**, 2299.
2. A. Overberg, M. Karas, U. Bahr, R. Kaufmann and F. Hillenkamp, *Rapid Commun. Mass Spectrom.*, 1990, **4**, 293.
3. M. Karas, U. Bahr, A. Ingendoh and F. Hillenkamp, *Angew. Chem. Int. Ed. Engl.*, 1989, **28**, 760.
4. T.-W. D. Chan, A. W. Colburn and P. Derrik, *J. Org. Mass Spectrom.*, 1992, **27**, 53.

5. K. Stupat, M. Karas and F. Hillenkamp, *Int. J. Mass Spectrom. Ion Processes*, 1991, **111**, 89.
6. M. Karas, U. Bahr, A. Ingendoh, E. Nordhoff, B. Stahl, K. Strupat and F. Hillenkamp, *Anal. Chim. Acta*, 1990, **241**, 175.
7. A. Overburg, A. Hassemburger and F. Hillenkamp, in *Mass Spectrometry in the Biological Sciences: A Tutorial*, ed. M. L. Gross, Kluwer Academic, Amsterdam, 1992, p. 181.
8. B. Stahl, M. Steup, M. Karas and F. Hillenkamp, *Anal. Chem.*, 1991, **63**, 1463.
9. U. Bahr, A. Deppe, M. Karas, F. Hillenkamp and U. Giessmann, *Anal. Chem.*, 1992, **64**, 2886.
10. P. O. Danis, D. E. Karr, F. Mayer, A. Holle and C. H. Watson, *Org. Mass Spectrom.*, 1992, **27**, 843.
11. L. H. Cohen and A. I. Gusev, *Anal. Bioanal. Chem.*, 2002, **373**, 571.
12. D. C. Schriemer and L. Li, *Anal. Chem.*, 1996, **68**, 2721.
13. B. O. Keller and L. Li, *J. Am. Soc. Mass Spectrom.*, 2001, **12**, 1055.
14. For reviews on off-line fraction collection-MALDI analysis, see (a) K. K. Murray, *Mass Spectrom. Rev.*, 1997, **16**, 283 (b) A. I. Gusev, *Fresenius J. Anal. Chem.*, 2000, **366**, 691.
15. K. K. Murray and D. H. Russel, *Anal. Chem.*, 1993, **65**, 2534.
16. S. J. Lawson and K. K. Murray, *Rapid Commun. Mass Spectrom.*, 2000, **14**, 129.
17. L. Li, A. O. L. Wang and L. D. Coulson, *Anal. Chem.*, 1993, **65**, 493.
18. Q. Zhan, A. I. Gusev and D. M. Hercules, *Rapid Commun. Mass Spectrom.*, 1999, **13**, 2278.
19. H. Ornes, T. Graf, H. Degn and K. K. Murray, *Anal. Chem.*, 2000, **72**, 251.
20. Y.-C. Chen, J. Shiea and J. Sunner, *Rapid Commun. Mass Spectrom.*, 2000, **14**, 86.
21. Y.-C. Chen, *Rapid Commun. Mass Spectrom.*, 1999, **13**, 821.
22. A. I. Gusev, A. Proctor, Y. I. Rabinovich and D. M. Hercules, *Anal. Chem.*, 1995, **67**, 1805.
23. J. Guittard, X. L. Hronowski and C. E. Costello, *Rapid Commun. Mass Spectrom.*, 1999, **13**, 1838.
24. S. Mowthorpe, M. R. Clench, A. Cricelius, D. S. Richards, V. Parr and R.W. Tetler, *Rapid Commun. Mass Spectrom.*, 1999, **13**, 2.
25. A. J. De Mello, *Lab Chip*, 2001, **1**, 7N.
26. R. D. Oleschuk and D. J. Harrison, *Trends Anal. Chem.*, 2000, **6**, 379.
27. M. Brivio, R. H. Fokkens, W. Verboom, D. N. Reinhoudt, N. R. Tas, M. Goedbloed and A. van den Berg, *Anal. Chem.*, 2002, **74**, 3972.
28. M. Brivio, R. E. Oosterbroek, M. Goedbloed, W. Verboom, D. N. Reinhoudt and A. van den Berg, *Lab Chip*, 2005, **5**, 1111.
29. T. W. G. Solomons, *Fundamentals of Organic Chemistry*, Wiley, New York, 5th edn, 1998, ch. 8.
30. H. Wensink, J. W. Berenschot, H. V. Jansen and M. C. Elwenspoek, *Proc. 13th Int. Workshop on Micro ElectroMechanical Systems (MEMS2000)*, Miyazaki, Japan, 2000, p. 769.

31. P. J. Slikkerveer, P. C. P. Bouten and F. C. M. De Haas, *Sens. Actuators*, 2000, **85**, 296.
32. T. Corman, P. Enoksson and G. J. Stemme, *J. Micromach. Microeng.*, 1998, **8**, 84.
33. M. L. Vestal, P. Juhasz and S. A. Martin, *Rapid Commun. Mass Spectrom.*, 1995, **9**, 1044.
34. P. Juhasz, M. L. Vestal and S. A. Martin, *J. Am. Soc. Mass Spectrom.*, 1997, **8**, 209.
35. P. W. Atkins, *Physical Chemistry*, Oxford University Press, Oxford, 5th edn, 1994, ch. 28.
36. M. Brivio, R. E. Oosterbroek, W. Verboom, M. Goedbloed, A. van den Berg and D. N. Reinhoudt, *Chem. Commun.*, 2003, 1924.
37. J. Stahl-Zeng, F. Hillenkamp and M. Karas, *Eur. J. Mass Spectrom.*, 1996, **2**, 23.
38. M. W. F. Nielen, *Mass Spectrom. Rev.*, 1999, **18**, 309.
39. P. A. Limbach, *Mass Spectrom. Rev.*, 1996, **15**, 297.
40. D. H. Patterson, G. E. Tarr and S. A. Martin, *Anal. Chem.*, 1995, **67**, 3971.
41. F. W. White, *Fluid Mechanics*, McGraw-Hill, New York, 3rd edition, 1994.
42. T. S. J. Lammerink, N. R. Tas, J. W. Berenschot, M. C. Elwenspoek and J. H. J. Fluitman, *8th Int. Workshop on MicroMechanical Systems (MEMS 2000)*, Amsterdam, The Netherlands, 1995, p. 13.
43. R. E. Oosterbroek, PhD thesis, University of Twente, 1999.
44. K. E. Petersen, *IEEE Trans. Electron. Dev.*, 1978, **25**, 1241.
45. H. Seidel, L. Csepregi, A. Heuberger and H. Baumgärtel, *J. Electrochem. Soc.*, 1990, **137**, 3626.
46. A. W. Adamson and A. P. Gast, *Physical Chemistry of Surfaces*, Wiley, New York, 6th edition, 1997.
47. R. F. Probstein, *Physiochemical Hydrodynamics: An Introduction*, Wiley, New York, 2nd edition, 1994.
48. N. R. Tas, J. W. Berenschot, T. S. J. Lammerink, M. C. Elwenspoek and A. van den Berg, *Anal. Chem.*, 2002, **74**, 2224.

CHAPTER 12

MALDI-TOF Mass Spectrometry and Digital Microfluidics for the Investigation of Pre-steady State Enzyme Kinetics

KEVIN P. NICHOLS AND HAN J. G. E. GARDENIERS

MESA+ Institute for Nanotechnology, Mesoscale Chemical Systems, University of Twente, Postbus 217, 7500 AE Enschede, The Netherlands

12.1 Introduction

Lab-on-a-chip systems have been previously demonstrated to have significant utility in the field of analytical chemistry.[1] Lab-on-a-chip systems miniaturize standard bench-scale operations that are normally carried out in containers of the order of millilitres or litres to chip-scale operations in channels and droplets on the order of nanolitres to microlitres. The majority of lab-on-a-chip devices proposed thus far utilize micrometer-scale sealed channels. These sealed channels can then be used to separate, combine and mix chemical reagents in such a way as to recreate traditional wet chemistry protocols on a much smaller scale. When used in such a manner, these sealed channels are referred to as "continuous flow" microfluidic channels.[2–4]

Miniaturization and Mass Spectrometry
Edited by Séverine Le Gac and Albert van den Berg

Published by the Royal Society of Chemistry, www.rsc.org

Recently, digital microfluidics, a method for constructing lab-on-a-chip devices without using continuous flow microfluidics, has emerged. Digital microfluidics utilizes discrete liquid droplets manipulated on arrays of individually addressable electrodes instead of the sealed channels of continuous flow microfluidics. The primary advantages of digital microfluidic systems include greater reconfigurability leading to more "general-purpose" devices, more straightforward translations from bench-scale to chip-scale processes and reduction of dead volumes associated with macroscale to microscale interconnections. The primary disadvantages of digital microfluidic systems are that they require high voltages and polarizable liquids.[5,6]

The most common digital microfluidic fluid actuation techniques utilize a combination of strategically placed electrodes and changes in contact angle induced by one of two principles: electrowetting on dielectric (EWOD) or dielectrophoresis (DEP). EWOD and DEP can be considered as the low- and high-frequency cases, respectively, of the application of a sufficient electric field to polarizable liquids along the correct axes.[7]

EWOD-based digital microfluidic devices typically require a planar electrode covered by a hydrophobic dielectric film, above which droplet actuation occurs. For reversible droplet actuation, a ground plane integrated on a cover above the droplet is required. Upon application of a sufficient voltage, the charge imbalance at the liquid–dielectric interface results in an electrochemical (coulombic) force that causes a change in contact angle, and subsequent droplet movement. A simple digital microfluidic EWOD device can be constructed by arranging two adjacent coplanar electrodes, coated with a hydrophobic dielectric, above which is a polarizable liquid droplet covered by a ground plane. By selectively applying a DC potential to one of the two bottom coplanar electrodes, the apparent hydrophobicities of the surfaces above these two electrodes will vary, causing a droplet placed above one of the two electrodes to move preferentially towards the regions of higher field strength.[8–11]

The work presented in this chapter is an optimized combination of the work of Houston *et al.*, who demonstrated the utility of matrix-assisted laser desorption/ionization time-of-flight mass spectrometry (MALDI-TOF-MS) in studying pre-steady state kinetics using quench-flow techniques,[12] and the work of Wheeler *et al.*, who developed an elegant system for manipulating microlitre-scale droplets on planar surfaces for subsequent analysis using MALDI-TOF-MS.[13] The purpose of this work is improve the pre-steady state kinetic analysis technique of Houston *et al.* by using digital microfluidic techniques to analyze the smallest liquid volume possible with the most rapid quenching possible; it attempts to overcome the "fundamental limitations" previously encountered in pre-steady state kinetic studies. Typically, for pre-steady state kinetic analysis, one combines enzyme with substrate and quenches the reaction after specified time intervals. Pre-steady state kinetic analysis is particularly challenging due to the extremely short time scales typically involved, of the order of milliseconds.[14] The digital microfluidic system previously described by Wheeler *et al.* for use with MALDI-TOF-MS was not suited for this particular application due to unexpected throughput issues, and unavoidable contamination due to protein fouling, thus necessitating a significant rethinking of its design. We discuss how, for certain applications of digital microfluidics to MALDI-TOF-MS, the removal of the on-chip reservoirs and droplet handling systems of the Wheeler *et al.* device allows for higher throughput than can be achieved by their inclusion. Additionally, we describe a novel electrohydrodynamic mixing scheme incorporated in our device.

The purpose of combining these previously described techniques and including the novel mixing scheme described in this chapter was to enhance the throughput of the Houston *et al.* method for pre-steady state kinetic analysis, allowing for the analysis of inherently noisy samples. We sought to create a lab-on-a-chip implementation of the Houston *et al.* method that was capable of overcoming the inherent inaccuracy in certain pre-steady state kinetic measurements. We achieved this by designing a system with a high enough throughput to produce sufficient data to overcome the noise inherent in pre-steady state kinetic analysis. The advantage of this system over that of Houston *et al.* is not necessarily greater accuracy of individual measurements, but higher throughput, allowing for innate inaccuracy to be overcome.

EWOD-based digital microfluidics have already been combined with MALDI-TOF-MS by Wheeler *et al.*[13] However, the Wheeler *et al.* device would have several drawbacks when used in pre-steady state kinetic studies, including lower throughput – for reasons explained in the Section 12.3 – and the absence of a mixing element. Further, while Paik *et al.*[5] have demonstrated droplet mixers for EWOD-based systems, their system is simply not fast enough to be of use in pre-steady state kinetics.

MALDI-TOF-MS has been demonstrated as a useful tool in pre-steady state kinetic research by Houston *et al.*, who combined it off-line with quench-flow methods to follow the appearance of a protein tyrosine phosphatase (PTPase) reaction intermediate.[12] Houston *et al.* were able to measure rate constants up to $30\,s^{-1}$, with k_2/k_3 ratios up to approximately 15. The device described in this chapter extends this technique to measure rate constants approximately 5 times greater, with k_2/k_3 ratios approximately twice previously measurable values. MALDI-TOF-MS is typically conducted on a centimeter-scale conducting plate; the digital microfluidic system employed is a square plate approximately 2 cm on each side, with 16 experimental units per chip, which can be placed directly inside a standard MALDI-TOF-MS plate that has been machined appropriately. By grounding the exposed wires on the otherwise insulated chip, charging is negligible.

There are a variety of other spectroscopy techniques available for investigating reaction kinetics. However, MALDI-TOF-MS allows for the determination of rate constants regardless of the incorporation of a chromophore, and at a wide variety of buffer concentrations. Both of these advantages are useful if one wishes to study enzymatic reactions under naturally occurring conditions. The advantages of MALDI-TOF-MS can be compared to spectrophotometric methods, which require a chromophore, and electrospray ionization mass spectrometry, which is sensitive to buffer concentrations. As MALDI-TOF-MS is an off-line technique requiring quenching for the analysis we performed, a rapid mixer must be incorporated into the system.

As a model system, Yop51 PTPase[15] was analyzed. Deprotonated PTPases act as nucleophilic thiolates during attacks on phosphates. MS analyses of PTPases are possible due to the predictable formation of a covalent phosphoenzyme intermediate.[12] In this study, unphosphorylated enzyme is used as an internal standard to permit quantitation of the phosphorylated to unphosphorylated enzyme ratio, used to determine pre-steady state kinetic values.

12.2 Experimental

12.2.1 Device Fabrication

The digital microfluidic system used to obtain kinetic values consists of individually addressable Cr/Au electrodes (Figure 12.1), covered by 100 nm of silicon dioxide, 500 nm of silicon nitride (which together act as a sufficient dielectric layer) and 1 μm of Teflon AF 601S1-x-6 (DuPont) film. The Teflon significantly improves the repeatability of the measurements, as it increases the hydrophobicity, thus reducing droplet sticking. However, it can cause problems with fouling if the system is reused, and must be cleaned with isopropyl alcohol if multiple measurements on the same chip are to be conducted. The total length and width of each electrode shown is 0.75 mm.

All patterns were designed using CleWin (WieWeb Software, The Netherlands) using a script written in Matlab 7.1 (MathWorks, USA). Cr masks were produced using a Heidelberg Instruments DWL 2.0 (Germany) laser pattern generator. Double side polished Pyrex wafers were metallized using a 20 nm Cr adhesion layer and a 200 nm Au thin film was sputtered using a custom built sputtering tool available at the MESA+ Institute for Nanotechnology (University of Twente, The Netherlands). Pyrex wafers were used only because of their resistivity. Silicon wafers could also be used after passivation. Standard photolithography techniques were employed to produce a photoresist etch mask for the metallization layer. In short, a photosensitive resist is dispensed onto the Pyrex wafer, and spun to a uniform thickness. This is then exposed under a Cr mask and developed. The photoresist etch mask was oxidized for five minutes using a standard ozone reactor, prior to wet etching using standard Cr and Au etching recipes. This decreases the hydrophobicity of the photoresist thus facilitating wet etching. The silicon dioxide and silicon nitride layers were deposited using plasma-enhanced chemical vapor deposition, and patterned using buffered hydrofluoric acid (BHF). BHF is dangerous even at low concentrations, and special handling is required. Calcium gluconate gel should be readily available for first aid whenever working with hydrofluoric acid. Vias for wiring were patterned by leaving holes in the dielectric layer onto which a 200 nm Cr layer was sputtered at 2×10^{-2} bar using the above-mentioned sputtering device. A hydrophobic Teflon AF 600 film (DuPont, USA) was manually dispensed and spun to a thickness of approximately 1 μm. Adhesive tape was applied over the contact pads before pouring and spinning the Teflon layer, thus allowing for very simple patterning. After removing the adhesive tape, the Teflon AF was baked at 150 °C for 20 min.

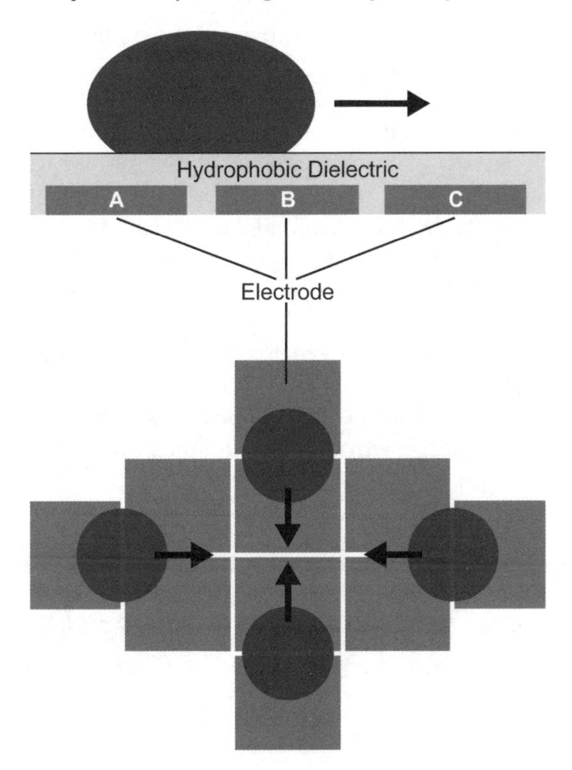

Figure 12.1 Simplified cross-section (top) and top-down (bottom) view of one experimental unit on the chip. Droplets are loaded robotically or manually at each of the four outermost positions shown. During loading, all the central electrodes are set to negative DC voltage, and all the outer electrodes are set to positive DC voltage, to facilitate easier, more accurate droplet placement. This also allows smaller volumes to be accurately dispensed, as capillary forces in the pipette tip can be overcome. The droplets are then sequentially combined using AC voltage in the order described in the text (enzyme with substrate, then quench, then matrix). The cross-section shows a droplet over an electrode gap and a hydrophobic dielectric. Wires are not shown, though during analysis charging is negligible if the system is grounded, since small areas of the wires are in direct (uninsulated) contact with the matrix.

Individual 2 cm chips were diced using a Disco DAD-321 (Disco Corporation, Japan) dicing saw. Dicing lines were included on the electrode layer.

12.2.2 MALDI-TOF-MS Kinetic Analysis on Chip

When a voltage is applied across adjacent electrodes (in the pattern shown in Figure 12.1) at 1000 Hz, 250 V_{RMS}, liquid droplets quickly center themselves over the inter-electrode gap. Electrohydrodynamic forces act as an efficient mixer in such a system.

Experiments were conducted by sequentially combining 0.5 μL buffered 50 μM Yop51 PTPase (Sigma) at pH = 7.2 with its substrate, 20 mM *p*-nitrophenyl phosphate (Acros), then quenching the reaction with 1 M dichloroacetic acid (Acros), and finally forming a matrix using 25 mg mL^{-1} ferulic acid in 2 : 1 H_2O–acetonitrile (Acros). The temperature of the system was controlled at 30 °C using a simple resistive heater and thermistor element. Samples were directly analyzed on the digital microfluidic chip using an Applied Biosystems Voyager MALDI-TOF-MS system. During analysis the electrodes were grounded. This effectively minimized charging due to the uninsulated vias described in Section 12.2.1.

Combination and mixing of droplets at well-defined time intervals was accomplished using a custom-built relay board interfaced with a laptop PC controlled using Matlab. Each 0.5 μL droplet was initially manually loaded across one of the outermost electrode gaps, and then brought into the center for rapid mixing.

Details of the model PTPase reaction are available elsewhere.[12,15] However, in brief, the following equations were used to deduce the rate constants k_1 and k_2:

$$C = \frac{k_2}{(k_2 + k_3)(1 + K_m/[S]_0)} \tag{12.1}$$

$$b = k_3 + \frac{k_2}{1 + (k_2 + k_3)K_m/k_3[S]_0} \tag{12.2}$$

$$\frac{[EP]}{[E]_0} = C(1 - e^{-bt}) \tag{12.3}$$

(where K_m is the Michaelis–Menten constant, $[S]_0$ is the initial substrate concentration, [EP] is the enzyme–product complex concentration and $[E]_0$ is the initial enzyme concentration).

Nonlinear regression using Matlab was used to fit Equation (12.3) to K_m values obtained from the literature,[15] experimentally relevant concentrations and scaled $[EP]/[E]_0$ data calculated as the area under EP divided by the sum of the EP and E areas.

12.2.3 Calibration of Electrohydrodynamic Mixer

Two reactions were analyzed to determine mixing behavior: the mixing of 0.5 μL droplets of 10^{-5} M fluorescein (Sigma, USA) with 0.5 μL deionized water droplets (Figure 12.2), and the mixing of 0.5 μL droplets of 10^{-5} M fluorescein with 0.5 μL droplets of 1 M acetic acid (Sigma, USA).

A standard function generator and a Krohn-Hite 7602M (Krohn-Hite Corporation, USA) amplifier operating at 250 V_{RMS} generated AC voltages with frequencies ranging from 1 Hz to 300 kHz. Mixing efficiency was analyzed using a HI-CAM high-speed video camera (Lambert Instruments, The Netherlands). Videos were subsequently analyzed using a custom script

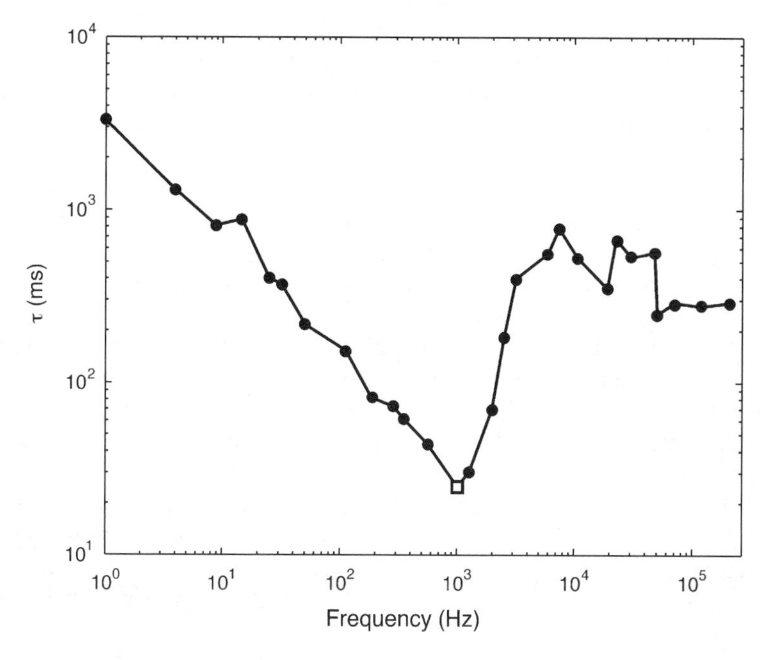

Figure 12.2 Mixing constant τ for the EHD mixer as a function of frequency. Filled circles represent the mean of between 2 and 4 measurements. The white square at 1000 Hz was used to calibrate the device as described in the text. At higher frequencies, operation becomes more complicated as additional phenomena begin to affect the droplet. We have previously presented work on more complicated electrode geometries.[21]

(available on request) for the Mac OS X version of Matlab 7.1 (MathWorks, USA) to determine the mean gradient of the variance as a function of time. It is believed that this value can provide an accurate metric for mixing analysis, as long as images are manually inspected to confirm that the last image in the series is in fact homogenized. For a given frequency, a time constant τ was defined as the point at which 63% of the final gradient was reached. This common technique in circuit analysis provides a useful method to describe exponentially decaying phenomena. High-speed video analysis was also used to measure average droplet speeds during transfer to the center of the device. These two times were then used to define zero points for Equation (12.3).

12.3 Results and Discussion

12.3.1 Primary Results

Houston *et al.*[12] reported the capability to measure pre-steady state kinetics using "rapid chemical quench-flow methods" coupled with MALDI-TOF-MS. However, they reported difficulties in obtaining accurate pre-steady state kinetic measurements from systems with high k_2/k_3 ratios. They reported that above a certain critical value, random error prevents the differentiation of slightly varying progress curves. The highest k_2/k_3 ratio reported by Houston *et al.* was approximately 15 ([29.2 ± 3.8]/[1.89 ± 0.46] for *p*-fluorophenyl phosphate and the enzyme Stp1).[12]

To demonstrate the increased performance possible by performing an equivalent set of measurements in our system, we chose an enzyme–substrate system with a higher k_2/k_3 ratio. Yop51 PTPase at pH = 7.2 with its substrate, 20 mM *p*-nitrophenyl phosphate, has a reported k_2/k_3 ratio of 30.1 (193/6.5).[15] Unphosphorylated enzyme is used as an internal standard to permit quantitation of the phosphorylated to unphosphorylated enzyme ratio, which is used to determine pre-steady state kinetic values (Figure 12.3). As is shown in Figure 12.4, performing a sufficiently large number of measurements will produce enough normally distributed data to overcome even relatively high k_2/k_3 ratios. Our analysis produced a k_2 value of 170 ± 10, and a k_3 value of 8 ± 3. Since previous studies of Yop51 PTPase have not provided standard deviations for the results (presumably due to the difficulty in obtaining them) it is not possible to tell precisely how close our measured values are to previously measured values. However, in relative terms, our results do not appear to significantly deviate from previously published

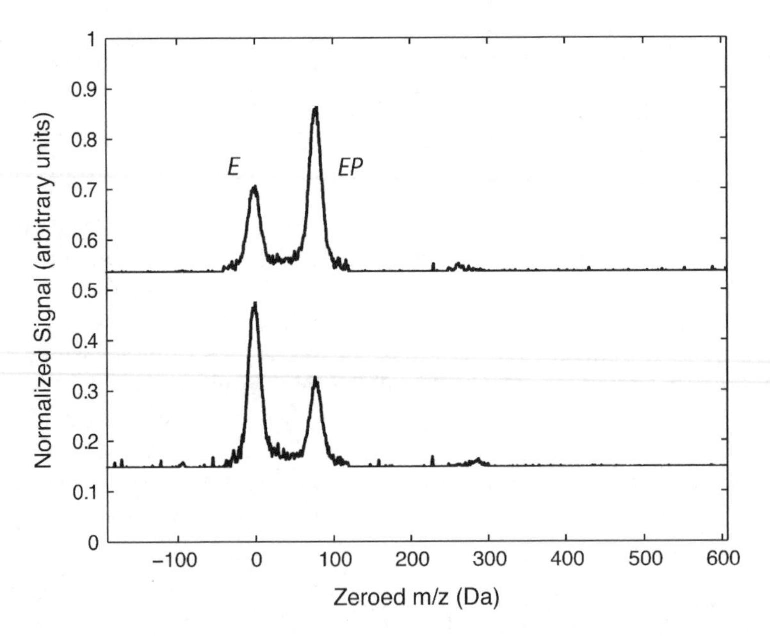

Figure 12.3 Typical MALDI-TOF-MS spectra acquired at two time intervals, after signal processing. The two peaks represent unphosphorylated (E) and phosphorylated (EP) PTPase. The top spectrum is early in the reaction, and the bottom spectrum is late in the reaction. Curves are smoothed and leveled to facilitate proper definition of limits of integration. This image is fundamentally identical to Figure 5 in Houston *et al.*[12]

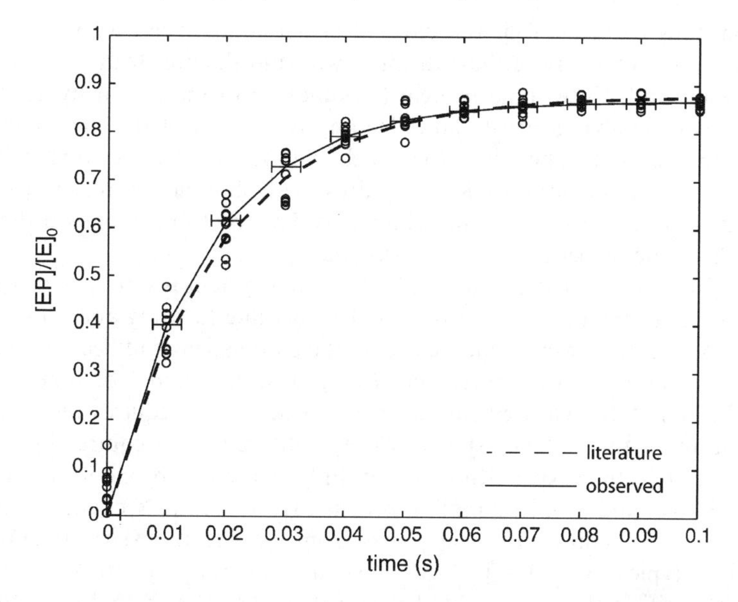

Figure 12.4 Literature data (parameters from Zhang *et al.*;[15] model described in Section 12.2) and observed data. Greater scatter is observed at earlier time points. The horizontal error bars represent two standard deviations from the mean mixing time for these droplet volumes and actuation frequencies, as determined in separate experiments.

results. It should be noted that the rate constants cited above (not our own measurements) were not determined using MALDI-TOF-MS. Our study thus demonstrates the highest rate constants yet measured with this particular analysis technique. The advantages of using MALDI-TOF-MS are described in Section 12.1.

12.3.2 Droplet Handling Systems

As can be seen in Figure 12.1, our device has few of the elegant sample preparation and dispensing systems demonstrated in the Wheeler *et al.* device for MALDI-TOF-MS measurements in a digital microfluidic system. While our initial designs did include analogous systems, we quickly found that, using the best practices currently published, they require too much space on the chip to be of practical use for the particular system we wished to apply it to, for the following reasons.

First is the issue of droplet routing and contamination. The Wheeler *et al.* system performs a relatively simple set of chemical operations; they move an impure sample to a specific location, dry and rinse the sample, and then move a droplet containing matrix-forming solution over the sample, thus requiring two reservoirs. Our system has the additional complications of combining a larger number of droplets: an enzyme, a substrate, a quench and a matrix-forming solution, thus requiring four reservoirs. Further, since even a very slight contamination between any of these droplets will ruin the pre-steady state kinetic experiment, and, since some protein (enzyme) absorption on a hydrophobic surfaces is inevitable, separate lanes leading from each of the four reservoirs would be required. The difficulties with routing such a network without crossing lanes are obvious: to do so requires as much, if not more, space on the chip being devoted to droplet handling as is devoted to positions on the chip where the actual experiments are carried out.

Therefore, the inclusion of a droplet handling system is only justified if the time it saves is greater than the time that is lost due to being able to perform fewer experiments per chip since space on the chip has instead been devoted to droplet handling infrastructure. Performing a large number of time-stepped MALDI-TOF-MS measurements, as we wished to do, requires two primary expenditures of time. The most obvious expenditure is that required by droplet dispensing. This time expenditure can certainly be reduced by incorporating on-chip reservoirs and droplet handling systems. However, the second, and in our case greater expenditure of time comes from loading the MALDI-TOF-MS plate. In a typical MALDI-TOF-MS system, it takes approximately 1 min to properly position the tray in the device, and as many as 5 to 10 min for the system to load the tray, move it into position and pump the system down to a vacuum level sufficient for analysis. Unloading the system takes an additional 1 to 2 min. There is, therefore, a tradeoff between the time that can be saved by using on-chip dispensing systems, and the time that can be saved by using this space for additional measurements per loading cycle. In our case, eliminating droplet dispensers was the better choice.

Although it goes against the general design ethos of lab-on-a-chip systems, an external robotic droplet dispenser is superior to on-chip droplet handling systems for this particular application. Robotic dispensing is already a common tool for MALDI-TOF-MS analysis. Even dispensing by hand takes at most one to two seconds per droplet, and is surprisingly easy if the electrodes are properly energized to "automatically" center the droplet.

We do not provide a detailed numerical analysis of this balance, as the specifics are highly dependent on the exact design, and rather trivial to calculate. However, it is a consideration that future work should take into account.

12.3.3 Calibration of Electrohydrodynamic Mixer

Convective electrohydrodynamic (EHD) flow, utilized in our system for mixing within a droplet, is a consequence of the tangential stress at the interface between a droplet and its surrounding medium, where tangential viscous stress arises to directly balance tangential electric stress.[16] While we believe the use of EHD-based mixing to be novel as applied to a digital microfluidic system, a physical model would be beyond the scope of this analytical chemistry-focused chapter. However, EHD is a well-established field, and numerous papers spanning several decades confirm the effects we describe.[17–19] Therefore, we only discuss the basic characteristics and limitations of the device to the extent necessary that it could be duplicated in a similar digital microfluidic system.

Most digital microfluidic systems described already contain the necessary components for an EHD-based mixer. EHD flow can be initiated by placing a conductive droplet over an insulated electrode gap, and applying a sufficient voltage at the correct frequency across the electrodes. However, care should be taken, as the presence of a lid will decrease the efficiency of EHD mixing. EHD mixing will be most efficient in a suspended droplet not pinned to any surface, and should decrease proportionally to the level of pinning present. Thus, EHD flow in a confined microchannel will not occur under typical conditions, though EHD mixing in a microchannel between liquids with "different electrical properties" has recently been reported.[20] Likewise, EHD mixing will be most efficient in a digital microfluidic device without a lid, and will decrease as the ratio of pinned to unpinned droplet surface increases. This is another factor making the inclusion of the droplet handling systems described by Wheeler *et al.* difficult, as analogous droplet actuation without a grounded lid is significantly more difficult.

Efficient EHD-driven mixing was achieved at 250 V_{RMS} from 150 Hz to 2250 Hz. Above 2250 Hz, mixing efficiency decreases as the decreasing impedance of the dielectric reaches a critical threshold allowing for the pinning of a greater percentage of the droplet to the floor of the device. Above 46 kHz, impedance is low enough to permit sufficient current to cause unacceptable Joule heating. At the chosen frequency of 1000 Hz, mixing time for 0.5 μL droplets was approximately 15 ms (Figure 12.2, white square). This time was used to calibrate the kinetic model as described in Section 12.2.

Analysis of the high-speed videos shows that mixing in this frequency regime appears to be the result of spiraling fluid flow originating at the center of the droplet, moving outwards, curving around the droplet wall and returning to the center of the droplet, as would be expected in an EHD-driven flow field.

Additionally, it was demonstrated that at 1000 Hz, 250 V_{RMS}, no measurable Joule heating occurs. Droplets were left on the device (without a lid) and continuously mixed until no visible liquid remained. No significant difference in evaporation rate was observed between mixing at 1000 Hz, 250 V_{RMS}, and droplets left on the device without the application of any external potential.

12.4 Concluding Remarks

We have demonstrated a lab-on-a-chip system for the study of pre-steady state chemical kinetics utilizing digital microfluidics and MALDI-TOF-MS. The device incorporates an EHD mixing scheme, and is designed with the particularities of pre-steady state kinetics in mind. Future work should focus on an on-chip droplet dispensing system that would meet the requirements discussed in this chapter. If a device could be constructed capable of moving droplets over multiple planes (a three-dimensional digital microfluidic system), and with a removable lid, it would be possible to obtain the densities required for a true high-throughput MALDI-TOF-MS digital microfluidic system.

Acknowledgements

This research was financially supported by the Technology Foundation STW, applied science division of NWO and the technology program of the Ministry of Economic Affairs of The Netherlands (project no. 6626).

References

1. A. Rios, A. Escarpa, M. C. Gonzalez and A. G. Crevillen, *Trac-Trends Anal. Chem.*, 2006, **25**, 467.
2. A. Manz, N. Graber and H. M. Widmer, *Sens. Actuators B*, 1990, **1**, 244.
3. J. G. E. Gardeniers and A. Van Den Berg, *Anal. Bioanal. Chem.*, 2004, **378**, 1700.
4. K. P. Nichols, J. R. Ferullo and A. J. Baeumner, *Lab Chip*, 2006, **6**, 242.
5. P. Paik, V. K. Pamula and R. B. Fair, *Lab Chip*, 2003, **3**, 253.
6. H. Moon, A. R. Wheeler, R. L. Garrell, J. A. Loo and C. J. Kim, *Lab Chip*, 2006, **6**, 1213.
7. T. B. Jones, K. L. Wang and D. J. Yao, *Langmuir*, 2004, **20**, 2813.
8. F. Saeki, J. Baum, H. Moon, J. Y. Yoon, C. J. Kim and R. L. Garrell, *Abstr. Pap. Am. Chem. Soc.*, 2001, **222**, U341.
9. P. Paik, V. K. Pamula, M. G. Pollack and R. B. Fair, *Lab Chip*, 2003, **3**, 28.
10. F. Mugele and J. C. Baret, *J. Phys. Condens. Matter*, 2005, **17**, R705.
11. D. Chatterjee, B. Hetayothin, A. R. Wheeler, D. J. King and R. L. Garrell, *Lab Chip*, 2006, **6**, 199.
12. C. T. Houston, W. P. Taylor, T. S. Widlanski and J. P. Reilly, *Anal. Chem.*, 2000, **72**, 3311.
13. A. R. Wheeler, H. Moon, C. A. Bird, R. R. O. Loo, C. J. Kim, J. A. Loo and R. L. Garrell, *Anal. Chem.*, 2005, **77**, 534.
14. X. Z. Zhou, R. Medhekar and M. D. Toney, *Anal. Chem.*, 2003, **75**, 3681.
15. Z. Y. Zhang, B. A. Palfey, L. Wu and Y. Zhao, *Biochemistry*, 1995, **34**, 16389.
16. J. Q. Feng, *Proc. Math. Phys. Eng. Sci.*, 1999, **455**, 2245.
17. G. Taylor, *Proc. R. Soc. London, Ser. A*, 1966, **291**, 159.
18. J. R. Melcher and G. I. Taylor, *Annu. Rev. Fluid Mech.*, 1969, **1**, 111.
19. D. A. Saville, *Annu. Rev. Fluid Mech.*, 1997, **29**, 27.
20. A. O. El Moctar, N. Aubry and J. Batton, *Houille Blanche-Revue Internationale De L Eau*, 2006, 31 [in French].
21. K. P. Nichols and J. G. E. Gardeniers, *Proceedings of microTAS 2006*, Japan Academic Association Inc., 2006, p. 582.

SECTION 3
TOWARDS THE INTEGRATION OF MASS SPECTROMETERS ON CHIPS

CHAPTER 13

Development of Miniaturized MALDI Time-of-Flight Mass Spectrometers for Homeland Security and Clinical Diagnostics

ROBERT J. COTTER, SARA MCGRATH, CHRISTINE JELINEK AND THERESA EVANS-NGUYEN

Johns Hopkins University School of Medicine, Department of Pharmacology & Molecular Sciences, Middle Atlantic Mass Spectrometry Laboratory, Baltimore MD 21205, USA

13.1 Introduction

Practically every kind of mass analyzer has been subjected to miniaturized design of some form or another. Magnetic sector analyzers were used in the double-focusing and gas chromatography-mass spectrometry (GC-MS) instruments designed by Alfred Nier and Klaus Biemann, respectively, in the two Viking missions to Mars launched in 1975.[1,2] Permanent magnets used in these instruments provided a very low power design with a total mass of 600 kg. In the Nier instrument, mass scanning was achieved by sweeping the accelerating and electrostatic energy analyzer (ESA) voltages, with two collectors recording ions in 1–7 amu and 7–49 amu mass range. The GC-MS instrument had a mass range from 12 to 200 u. More recently, a novel sector instrument based upon a crossed field (ExB) design by Diaz *et al.*[3] has been developed as a residual gas analyzer

Miniaturization and Mass Spectrometry
Edited by Séverine Le Gac and Albert van den Berg

Published by the Royal Society of Chemistry, www.rsc.org

with applications ranging from space to the monitoring of evolved gases on the Kilauea volcano in Hawaii.

Miniaturized quadrupoles were used in the ion/neutral mass spectrometer (INMS) in the orbiter launched in 1997 as part of the Cassini mission to Saturn/Titan, and as well in the Huygens probe aerosol pyrolyzer GC-MS. In this case, the mass range was 1–99 amu and the analyzer carried a mass of 9.25 kg.[4] Microelectromechanical systems (MEMS) technology has been utilized in the fabrication of quadrupole mass analyzers,[5] and multiple quadrupole mass analyzers have been assembled in bundles, or as arrays.[6] Quadrupole ion traps have perhaps been the most miniaturized instruments, in terms of both the number of reports on the subject and the ultimate reduction in size. Cooks and co-workers[7] and Ramsey and co-workers[8] describe in detail the design of multiple cylindrical trap arrays on a chip. Ion traps are in general nicely *scalable*, with reductions in size accompanied by changes in RF frequency and amplitude to accommodate a target mass range and/or operation at elevated pressures.[8]

Time-of-flight (TOF) mass spectrometers have been miniaturized as well, and a number of compact instruments have been commercialized. These include the *miniTOF* by Comstock (Oak Ridge, TN, USA), the *orthoTOF* by IonWerks (Houston, TX, USA) and the portable *MS-200* volatile organic compound (VOC) analyzer in a briefcase by Kore Technology (Cambridgeshire, UK). The Kore instrument has a membrane inlet system and the vacuum is sustained entirely by a self-contained, battery-powered getter pump. The instrument has a mass range of 1000 amu and a battery life of $2\frac{1}{2}$ days between pumpdowns. Finally, a microfabricated TOF instrument has also been described,[9] with a mass range around 60 amu.

Our own interest in the TOF mass analyzer comes from its theoretically unlimited mass range. This means of course that TOF analyzers are nicely compatible with *laser desorption* (LD) and *matrix-assisted laser desorption/ionization* (MALDI). In fact, because these ionization techniques generally make singly charged ions, it is likely that the very high mass ions observed by the two very different methods employed by Karas and Hillenkamp[10] and Tanaka *et al.*,[11] and which led to the now popular MALDI technique, could only have been possible with the TOF mass analyzer. In any case, the TOF mass analyzer is unique in that the mass range should not (in theory) be reduced, along with the reduction in size of the instrument. Given our current dependence on particle detectors (primarily multipliers and multichannel plates), whose detection efficiency depends upon ion velocities, this is at least true within a size range that enables the instrument to sustain high voltages. While this limitation does not lead easily to a TOF-on-a-chip with very high mass range, it does lead to some very interesting, compact mass spectrometer designs that, when one includes all the accompanying pumping, electronics and data acquisition hardware, should be as portable as any other mass analyzer.

Our own efforts to miniaturize these instruments are driven by two major interests: biological agent detection and clinical diagnostics. The development of a biological threat sensor (rather than a chemical sensor) has been addressed in the past by instruments using pattern recognition techniques for spectra containing volatile low-mass species, degradation or pyrolysis products. However, our intent has been to utilize a *proteomics* approach that takes advantage of high-mass biological molecules (primarily peptides and small proteins) that are specific for a given organism and can be used in reference to a growing number of available databases for identification. In the diagnostics area, the miniaturized TOF mass spectrometer is not likely to be the biomarker discovery tool, or the means for characterizing and validating specific disease markers. However, once these are known and the instrument is used in conjunction with appropriate affinity methods up front, the small TOF mass spectrometer will indeed find an important role as a point-of-care instrument. In both of these applications mass range is of primary importance. The TOF mass spectrometer can provide that and the task then is to retain good performance (sensitivity, resolution and accuracy) in the miniaturization process.

13.2 Time-of-Flight Basics

The 1955 publication by Wiley and McLaren[12] still provides an excellent description of today's TOF instruments using multiple-stage ion acceleration and delayed extraction. The TOF mass spectrometer consists of a source region (the shaded regions in Figure 13.1) and a drift length. The voltage V imposed across the source region accelerates all ions (with a mass m and a charge e) to the same kinetic energy, $\frac{1}{2}mv^2 = eV$, while the drift region D is field-free. Since the drift region is much larger than the source region (in a normal sized

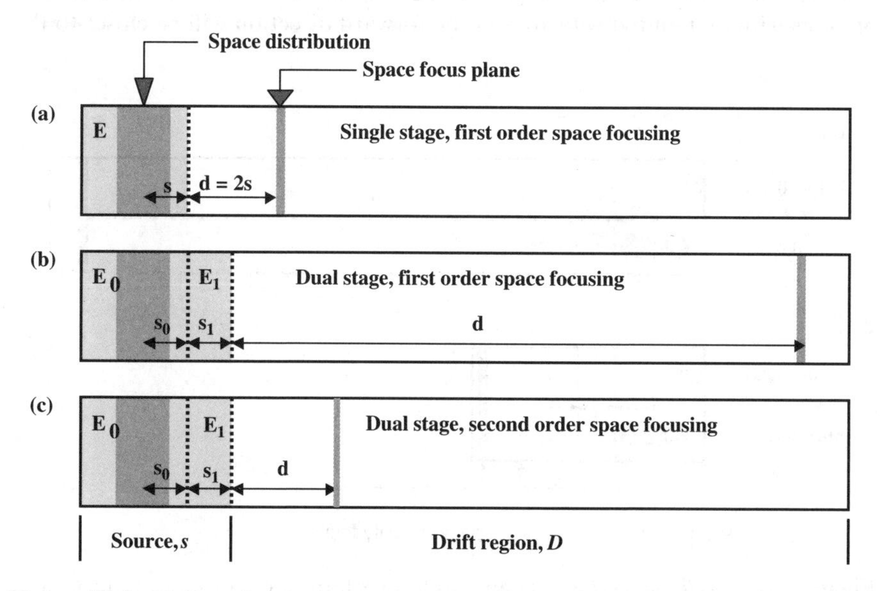

Figure 13.1 Schematic of TOF mass spectrometers having sources with (a) single-stage ion extraction, (b) dual-stage extraction and (c) dual-stage extraction to a second-order focus plane.

instrument) then the flight time t is approximately that needed to cross the drift region, or $t = (m/2eV)^{1/2}D$.

In the TOF mass spectrometer mass resolution is given by $m/\Delta m = t/2\Delta t$. Although the accuracy of measurement of ion flight times accounts for a portion of Δt, the major contributors to the loss in time (and mass) resolution come from the initial spread in ion positions in the source region and their different initial kinetic energies. One can attempt to correct for the initial *spatial distribution*. As shown in Figure 13.1a, ions formed in different regions (dark shading) of a single-stage source will arrive at a distance $d = 2s$, the so-called *space-focus plane* at the same time (within "first order"). This occurs because ions formed towards the back of the source will be accelerated in the field to higher velocity than those formed near the exit, and will "catch up" at this point. Because the space focus occurs only a short distance from the source, placement of a detector at this point would record ions with very short flight times and limit the time (and mass) resolution. Thus, the dual-stage scheme shown in Figure 13.1b is generally used to lengthen t with respect to Δt, by shifting the space focus to 1 m or more.[12,13] Note also (Figure 13.1c) that a particular solution for the parameters s_0, E_0, s_1 and E_1 produces a much better, "second-order" focus. This focal point is much closer to the source.

Wiley and McLaren addressed the initial *kinetic energy distribution* as well by using a technique they called *time-lag focusing*.[12] A short time delay is introduced after the ionization pulse (in their case a pulsed electron beam), and before switching on the extraction field. This allows the ions to drift slightly depending upon their initial kinetic energies, or more precisely their initial velocities along the TOF axis (Figure 13.2a). When the extraction field is turned on, ions with high initial velocities in the forward direction will be closer to the

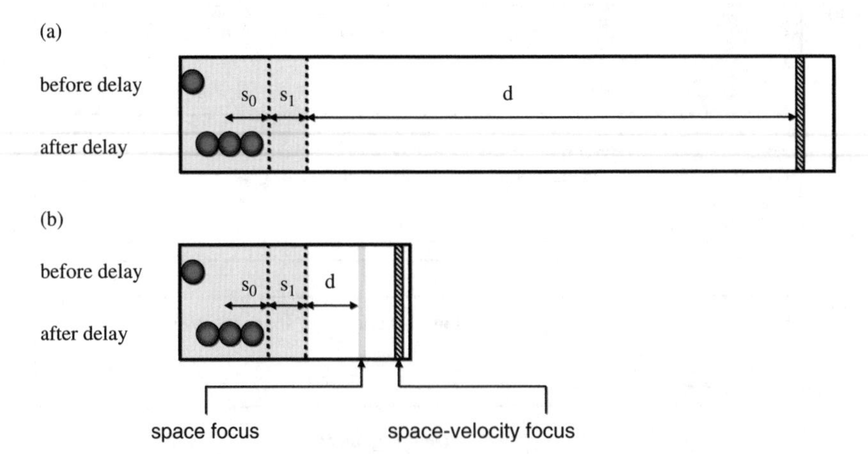

Figure 13.2 (a) *Time-lag focusing* applied to ions formed on a surface to focus ions with initial kinetic energy spread; also known as *delayed extraction*; (b) focusing for a compact TOF mass spectrometer.

source exit and spend less time in the extraction field compared with those with low initial velocities. Ions headed initially in the reverse direction will receive the largest boost from the field. Consequently, ions of the same mass, but different initial kinetic energies, should all arrive at the detector at very nearly the same time, reducing the time peak width and improving resolution. Unlike space focusing, however, *time-lag focusing* is mass dependent. That is, an increasing delay time is required for focusing ions of increasing mass. This means that it is necessary to set the delay times to record mass spectra over a particular mass range. As this is inconvenient it is common in most commercial instruments, used to record peptide and protein mass spectra, to set the delay time to achieve optimal focusing at around m/z 5000.

Today, the modern version of *time-lag focusing*, or *delayed extraction*,[14–16] is used in almost all MALDI-TOF mass spectrometers. It is in fact a special case of the Wiley and McLaren method, in which the initial spatial distribution (usually a very thin sample and matrix mixture dried on a stainless steel plate) is assumed to be zero. Therefore, after the delay time the velocity and spatial distributions are correlated, *i.e.* ions of highest velocity have moved the greatest distance. This *space velocity correlated focusing* has been described by Colby and Riley,[17] and provides considerably better focusing than can be obtained from two independent and uncorrelated initial distributions in space and velocity/energy that comprise the general Wiley–McLaren case.

With that introduction, we note that our design strategy for a compact, linear instrument is shown in Figure 13.2b. We retain source dimensions that are similar in size to those in laboratory instruments in order to maintain the high voltages necessary to accelerate heavy ions to velocities that can be detected by a multichannel plate detector. However, we reduce the size of the drift region substantially and take advantage of higher order space and space-velocity focusing that occurs at this shorter distance. In part this is possible because the very high bandwidth, 8 Gsamples per second digital waveform recorders available today can provide very short time measurement accuracy. Mass resolution will not be as high as that for the larger laboratory instruments, but it can be optimized for this more compact instrument.

13.3 Linear, Pulsed Extraction Instrument

This approach was used to design the compact instrument shown schematically in Figure 13.3, and consisting of a 3-inch drift length. This instrument is an updated version of an instrument that has been described previously[18] and consists of a two-stage ion source bounded by a source plate and two grids, separated by 0.325 cm and 0.55 cm, respectively. The source exit grid and flight tube are floated to the same potential as the front of the dual channel plate detector, approximately −1.9 to 2.3 kV, in order to eliminate any post-acceleration of ions before impacting the detector. The intermediate grid is connected to a power supply set at 6.45 kV, and to the backing plate through a 25 MΩ resistor. A 3.3 kV pulse applied to the backing plate, at the end of the

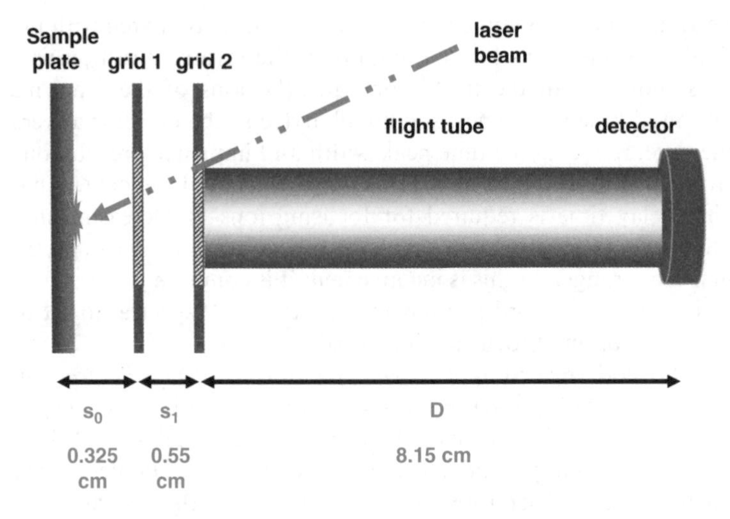

Figure 13.3 Schematic of the source and 3-inch mass analyzer. The flight tube and the front channel plate are at the same potential.

Figure 13.4 Assembly of the ion source, mass analyzer and detector in the vacuum chamber.

delay period, raises the backing plate voltage from 6.45 kV to 9.75 kV. The overall kinetic energy of ions from this scheme is about 11.75 keV, enabling detection of molecular ions up to 300 kDa.

A picture of the ion source, mass analyzer and detector is shown in Figure 13.4, and the entire instrument in Figure 13.5. The instrument uses a 1 in × 1 in sample

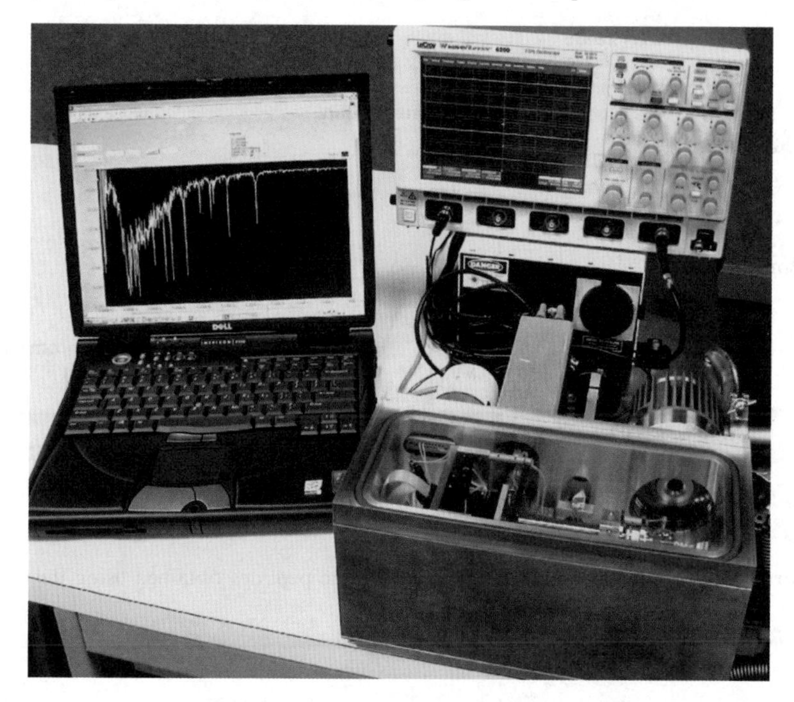

Figure 13.5 Picture of the compact mass spectrometer.

stage that can accommodate 16 MALDI samples, and is mounted on a National Aperture (Salem, NH, USA) XY stage that can be controlled externally. Samples are visualized using a Watec (Las Vegas, NV, USA) WAT-502A CCD camera equipped with a Computar (Commack, NY, USA) MLH 10X macro zoom lens. The laser is a Laser Science (Franklin, MA, USA) 337 nm pulsed nitrogen laser, attenuated with an Oriel Corporation (Stratford, CT, USA) neutral density filter. The extraction pulse is derived from a Eurotek Inc. (Morganville, NJ, USA) HTS series fast high-voltage transistor switch. The detector is a Hamamatsu (Bridgewater, NJ, USA) model 4655-13 microchannel plate, output to a LeCroy (Spring Valley, NY, USA) LC584AM 16 GHz (bandwidth) digital oscilloscope operating with data acquisition rates from 1 to 8 Gsamples per second.

Figure 13.6 shows the mass spectrum of a mixture of nine peptides and small proteins with masses beyond 12 000 Da. We previously reported a mass resolution of one part in 1200 for the molecular ion of the $ACTH_{1-39}$ peak at m/z 4542, corresponding to a peak width of only 2 ns, recorded using a detector with a rise time of 293 ps and FWHM of 455 ps.[18] Because the time delay was optimized for this peak, the widths for other peaks were generally somewhat broader, illustrating the basic mass dependence for delayed extraction. The instrument is capable of detecting even higher masses as is shown in the mass spectrum of bovine serum albumin (BSA) (Figure 13.7), where masses of greater than 132 kDa (dimer of BSA) are recorded.

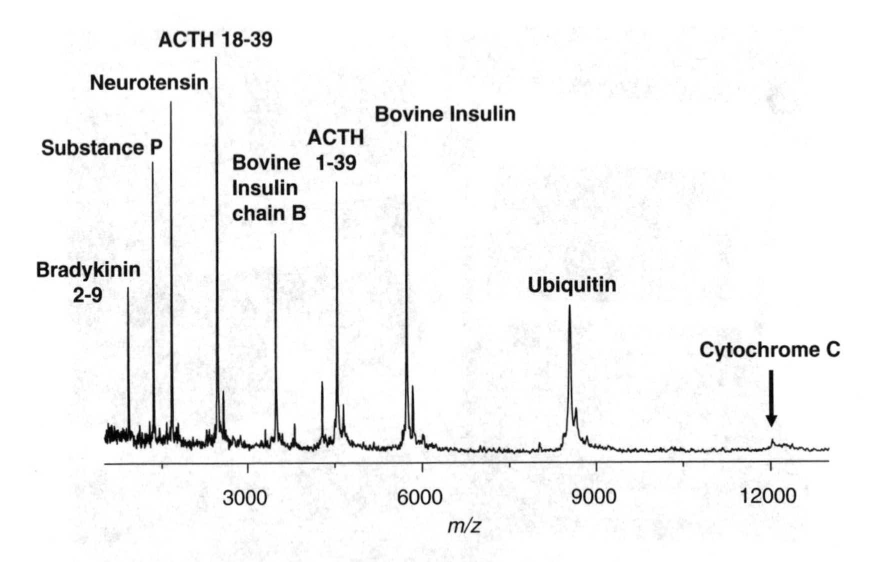

Figure 13.6 Mass spectrum of a mixture of nine peptides obtained using the 3-inch TOF mass analyzer.

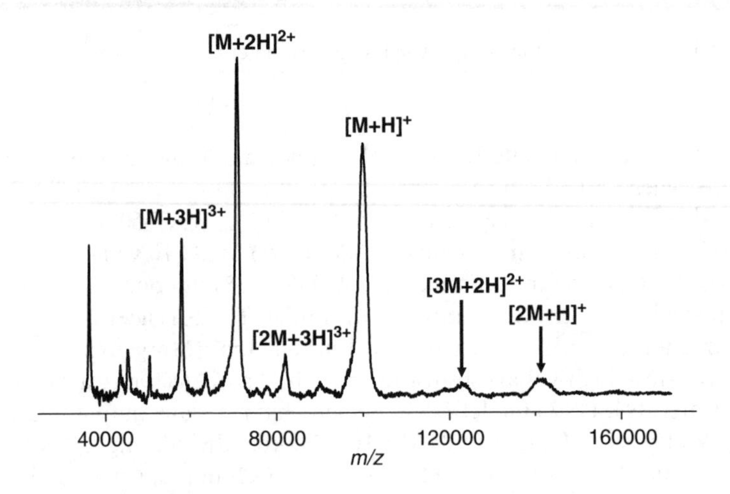

Figure 13.7 Mass spectrum of bovine serum albumin obtained on the 3-inch mass analyzer.

13.4 Compensating for Mass Dependence

We noted above that the space focus plane for a single extraction source is just twice the mean distance s in the source, or $d = 2s$. For a two-stage extraction

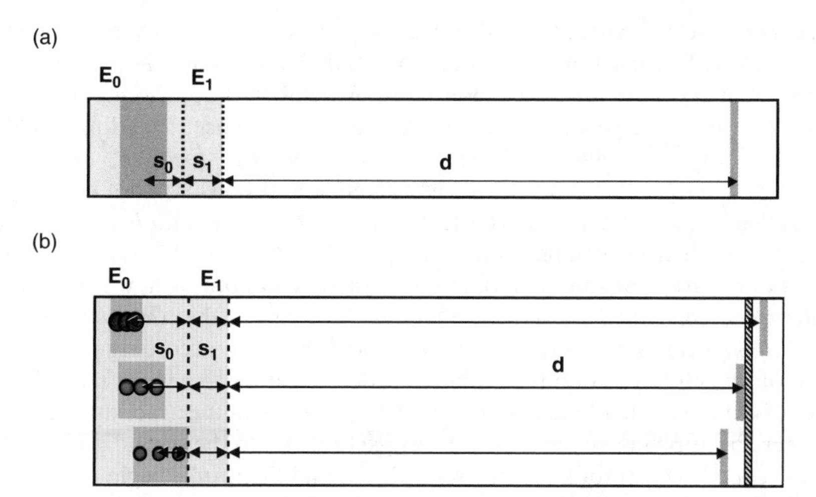

Figure 13.8 Schematic of (a) space focusing on a dual-stage source and (b) space-velocity correlated focusing from a dual-stage source after delayed extraction.

source (shown in Figure 13.8a), the space focus depends upon E_0, s_0, E_1 and s_1 and is given (to first order) by

$$d = 2\sigma^{3/2}\left[\frac{1}{s_0^{1/2}} - \frac{2s_1}{s_0^{1/2}(\sigma^{1/2} + s_0^{1/2})^2}\right]$$

where $\sigma = s_0 + (E_1/E_0)s_1$. This focus condition is correct for all masses, since space focusing is basically independent of mass. Initial kinetic energy focusing is of course much more complex, is mass dependent and depends upon the delay time as well as the dimensions and fields in the source. In the special case of space-velocity correlated focusing, however, it is possible to illustrate the mass dependence using Figure 13.8b. Assuming that all ions are formed directly on the sample surface that forms the back of the source, ions will then drift during the delay time toward the front of the source. Their drift distances will depend upon their initial kinetic energies *and their masses*. That is, the lighter ions will distribute themselves closer to the source exit than heavier ions. While these ions are now distributed spatially in a manner that is correlated with their initial velocities, they do still retain their initial energy (velocity) spreads. While this is not the simple spatial distribution depicted in Figure 13.8a, one can appreciate the fact that s_0 is effectively different for each mass, so that lighter ions will be in focus at a location before the detector and heavier ions will focus beyond the detector.

In the Wiley–McLaren instruments produced for many years by the Bendix Corporation (Rochester, NY, USA) the mass-dependence problem was addressed by *scanning* the delay time, bringing a different mass into focus in each TOF cycle.[13] This so-called *boxcar* approach resulted in very poor duty cycle and is of course not appropriate for MALDI, where it is hoped to record the entire mass range after each laser pulse. A number of approaches have been developed over the years to address this mass dependence, including *impulse-field* focusing,[19,20] *velocity compaction*,[21] *dynamic-field focusing*[22] and *post-source pulse focusing*.[23,24] Since the dimensions s_0 and s_1 also cannot be changed dynamically, the basic approach is to change E_0, E_1 or the ratio E_1/E_0. This can be accomplished by voltage functions applied to the sample (backing) plate, either of the grids, or some combination of these. Our approach, which we have called *mass-correlated acceleration* (MCA),[25] recognizes that the *space-velocity correlation* nicely sorts ions in the first extraction region according to mass, so that ions reach the second region in the order of lighter to heavier mass. Thus, MCA is implemented by changing the field in the second extraction region, in this case by applying an appropriate voltage function to the second grid and the flight tube.[26,27] Operationally, the second grid and flight tube begin the cycle at a negative voltage, and rise to ground as the higher masses reach the second extraction region. As shown in Figure 13.9a, lighter ions will see a higher E_1/E_0 than the heavier ions. The change in the exit grid and flight tube potentials does of course mean that the acceleration voltage, and thereby the final kinetic energies of the ions, change with mass. Thus, the mass calibration is different

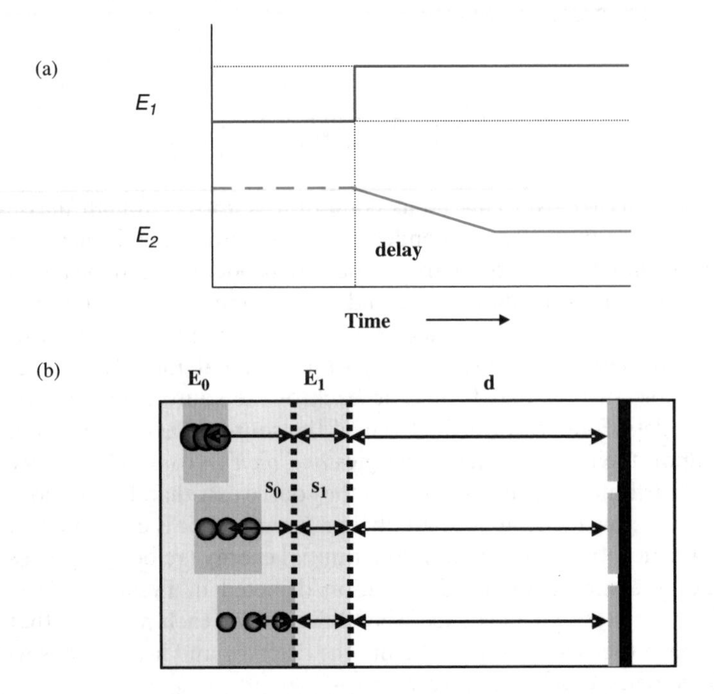

Figure 13.9 (a) Schematic of the time-dependent changes in the electric fields E_0 and E_1 before and after the time-delayed extraction pulse. (b) Implementation of the *mass-correlated acceleration* scheme on a miniaturized instrument.

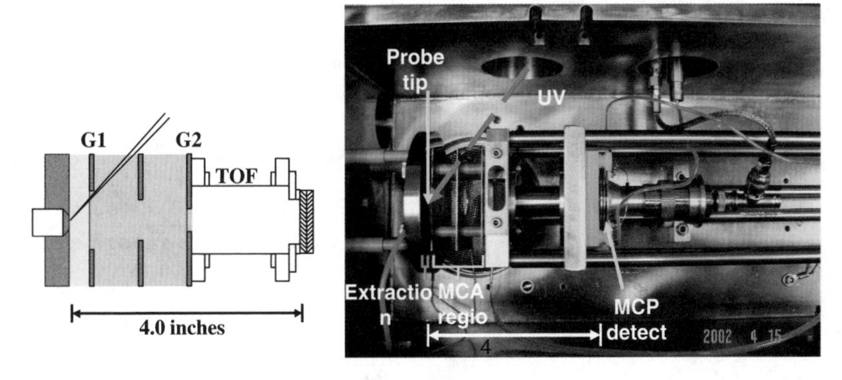

Figure 13.10 Schematic and picture of the 4-inch TOF mass spectrometer using the *mass-correlated acceleration* (MCA) technique.

from the simple square root law described above, but can be determined from two peaks of known mass and the voltage function.

Mass-correlated acceleration has been implemented as well on a miniaturized instrument, shown schematically in Figure 13.9b.[28] The novel 4-inch (total length) TOF mass spectrometer used for this work is shown in Figure 13.10. Some unusual features of this instrument are that the second extraction region is much larger than the first, and that the combined source region is longer than the flight tube. Operationally, the instrument is tuned (with the MCA off) to a delay time that will focus a mass equal to or greater than the upper mass to be recorded, since this will focus the instrument when the flight tube is at ground potential. The MCA is then turned on to record the mass spectrum. Figure 13.11a shows the molecular ion peaks for a mixture of bradykinin, dynorphin and bovine insulin, when the normal delayed extraction is tuned to m/z 5733. The insulin peak is well focused at 4.4 ns, while the peaks for dynorphin and bradykinin have peak widths of 8.8 and 10.6 ns, respectively. When the MCA is turned on, the mass spectrum shows all of these peaks focused between 3 and 4 ns.

This is particularly helpful when analyzing a mixture of tryptic peptides covering a wide mass range. Table 13.1 shows the molecular ion peak widths for a mixture of tryptic peptides from BSA in the normal delayed extraction mode, and in the MCA mode. In both cases, the instrument was tuned optimally for the mass of insulin (m/z 5733) to around 4 ns. In the normal delayed extraction mode, peak widths range from 8.6 ns at m/z 2832 to 11.4 ns at m/z 928. In the MCA mode, peak widths are in the range of 2.8 to 4.5 ns.

13.5 The TinyTOF Project and the Detection of Bacillus Spores

In the late 1980s the *Middle Atlantic Mass Spectrometry* (MAMS) Laboratory began a collaboration with investigators at the Johns Hopkins Applied Physics

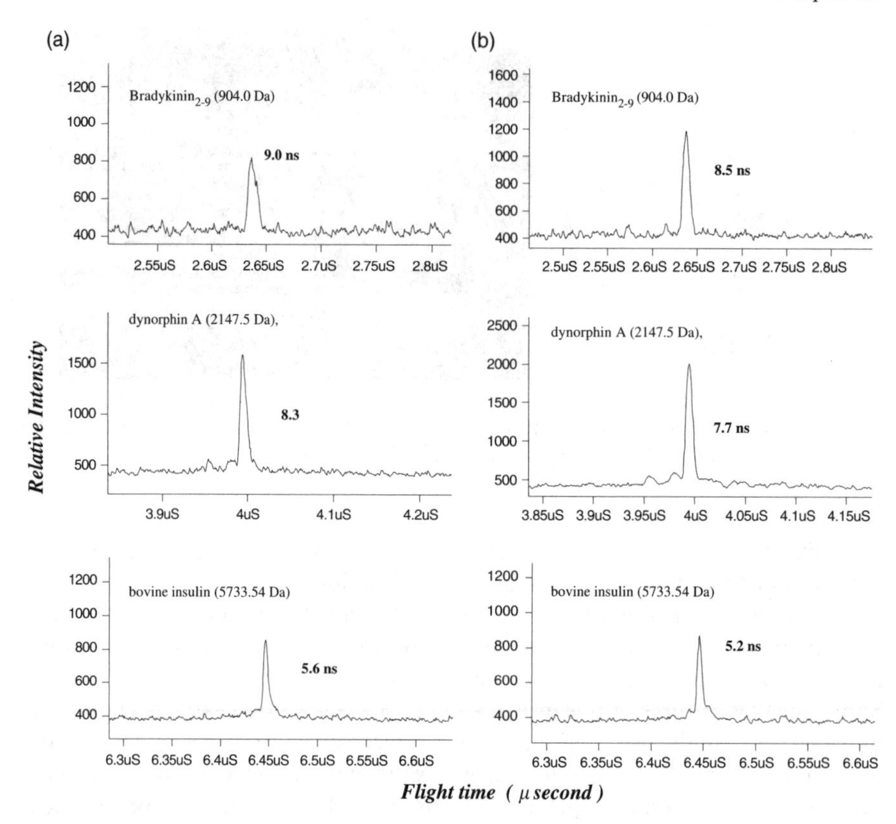

Figure 13.11 Mass spectra of a peptide mixture using (a) time-delayed extraction and (b) mass-correlated acceleration.

Table 13.1 BSA (66 432.96 Da) tryptic digest: comparison of peak widths using delayed extraction and mass-correlated acceleration (MCA).[28] Time delay was optimized for the bovine insulin peak at m/z 5733.

Peptide sequence	*Molecular mass (Da)*	*DE peak width (ns)*	*MCA peak width (ns)*
YLYEIAR	928.075	11.4	3.2
LVNELTEFAK	1164.343	11.0	4.2
HPEYAVSVLLR	1284.500	11.6	3.1
HLVDEPQNLIK	1306.504	8.0	2.8
RHPEYAVSVLLR	1440.688	9.2	3.1
LGEYGFQNALIVR	1480.706	10.6	3.2
DAFLGSFLYEYSR	1568.725	10.8	2.6
KVPQVSTPTLVEVSR	1640.920	9.4	3.1
MPCTEDYLSLILNR	1668.963	10.4	3.4
RPCFSALTPDETYVPK	1825.089	10.3	3.0
DDSPDLPKLKPDPNTLCDEFKADEK	2832.108	8.6	4.4

Laboratory (APL) to develop an approach to the identification of bacteria and other biological agents from the mass spectra of intact biomolecules having high specificity for particular organisms. Intended as an alternative to existing pyrolysis/pattern recognition methods that utilized small molecules, this effort initially involved the use of *fast atom bombardment* (FAB) to exploit polar lipids as potential biomarkers.[29] These were mainly the phospholipids phosphatidylethanolamine (PE), phosphatidylglycerol (PG) and phosphotidylcholine (PC). Linear regression techniques were used to identify bacteria from mixtures,[30] while MS-MS and *constant neutral loss* (CNL) mass spectra were used to deconvolute the polar head group distributions from the less species-specific fatty acyl distributions.[31] In some cases unusual head groups, such as monomethyl PE and dimethyl PE, were unique to only a few species. With the introduction of MALDI,[10,11] it became obvious that proteins could provide considerably greater specificity for organisms, at the species level and possibly at the strain level. In addition, our movement away from the large double-focusing sector instrument that was used in the FAB study, to the TOF mass spectrometer, motivated our efforts to develop compact and field portable instruments. For a number of years, the *tinyTOF* project[32] has been a continuing and fruitful collaboration between investigators from the MAMS laboratory, APL and the University of Maryland. The instruments described here, while designed within our own laboratory, were an integral part of that larger effort. In addition, the completion of the human genome and subsequent development of proteomics for biological research, combined with the availability of completed genomes for an increasing number of organisms, suggested the possibilities for a similar proteomics approach to bioagent identification.[33,34]

Figure 13.12 shows the mass spectra of unfractionated tryptic peptide mixtures derived from *in situ* digestion of a series of five *Bacillus* spore strains. The spectral patterns themselves distinguish these bacilli. More importantly, however, one can show (Table 13.2) that the peaks observed map to known proteins in the genome/proteome databases of these microorganisms. In particular, these are primarily the *small acid-soluble proteins*, or SASPs.[35] The importance of the proteomics approach, compared with simple pattern recognition methods, lies in the ability to identify unknowns, *i.e.* microorganisms whose mass spectra have not been recorded previously for comparison, but whose genomes are known. Pineda *et al.*[36] have developed a *significance test* algorithm that normalizes the number of matches to the relative sizes of the genomes, considers specific post-translational modifications (in particular, nitrogen-terminal methionine cleavage) and restriction to the ribosomal proteins to increase the significance level.

In one example reported from the University of Maryland group,[37] *H. pylori* 26995 was utilized as an unknown microorganism. A mass spectrum of the *in situ* tryptic digest was obtained using a commercial MALDI-TOF mass spectrometer, producing matches to a number of species as shown in Table 13.3. Tested against the entire database (as shown in Table 13.3A), *H. pylori* provided the best match, but with significance level not very different from other candidates. Note, however, that the 14 matches to *H. pylori* provide a better

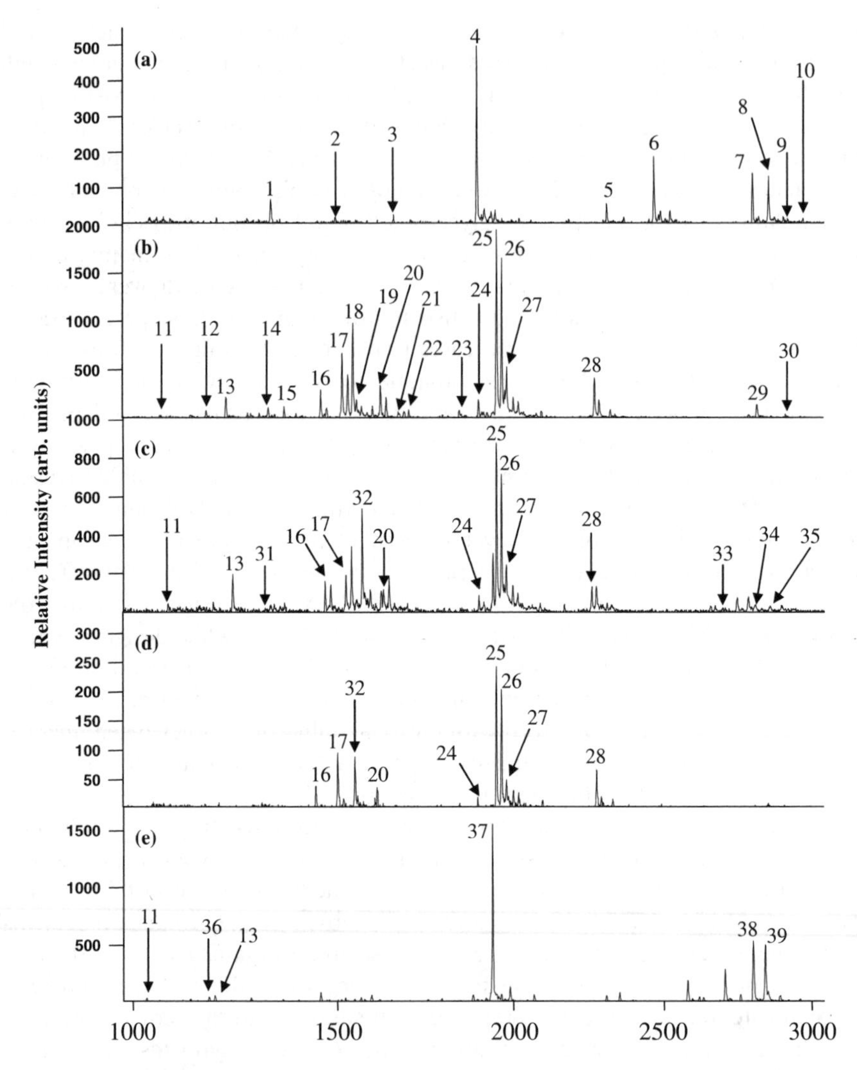

Figure 13.12 MALDI spectra of the tryptic digests generated *in situ* from (a) *Bacillus subtilis* 168, (b) *Bacillus anthracis* Sterne, (c) *Bacillus cereus* T, (d) *Bacillus thuringiensis* subs. *Kurstaki* HD-1 and (e) *Bacillus globigii* spores, and analyzed with the miniaturized TOF mass spectrometer. Peaks that were matched to peptides in the SASP database are numbered 1–39. Peaks that occur in more than one spectrum carry the same number. (Reprinted with permission from ref. 37).

significance level than the 18 matches to *E. coli*, based upon the relative sizes of their genomes. The number of matches increases, for both *H. pylori* and *E. coli*, when one includes the possibility for nitrogen-terminal methionine cleavages (Table 13.3B), while the significance level improves considerably for *H. pylori*. When one restricts the search to only the ribosomal proteins (Table 13.3C), the

Table 13.2 Identification of the peaks found in the mass spectra of the tryptic digests of several *Bacillus* species. (Adapted with permission from ref. 37).

Peak no.	*$[M+H]^+$ obs.*	*$[M+H]^+$ calc.*	*Sequence*	*SASP type (Da)*
Bacillus subtilis 168				
1 [a]	1322.1	1322.5	LVSVAQQQMGGR	−β (6848.6)
2 [a]	1483.9	1484.7	LEIASEFGVNLGADTTSR	−α (6939.6)
3 [a]	1640.3	1640.9	RLVSFAQQNMGGGQF	−α (6939.6)
4 [a]	1880.3	1881.0	LEIASEFGVNLGADTTSR	−α (6939.6)
		1881.0	LEIASEFGVNLGADTTSR	−β (6848.6)
5 [a]	2285.6	2286.5	ANQNSSNDLLVPGAAQAIDQMK	−β (6848.6)
6 [a]	2442.6	2442.7	ANNNSGNSNNLLVPGAAQAIDQMK	−α (6939.6)
7 [a]	2784.2	2783.9	QNQQSAAGQGQFGTEFASETNAQQVR	−γ (9136.5)
8 [a]	2842.5	2841.9	QNQQSAGQQGQFGTEFASETDAQQVR	−γ (9136.5)
9 [a]	2911.5	2912.1	QNQQSAAGQGQFGTEFASETNAQQVRK	−γ (9136.5)
		2912.1	KQNQQSAAGQGQFGTEFASETNAQQVR	−γ (9136.5)
10 [a]	2970.6	2970.1	KQNQQSAGQQGQFGTEFASETDAQQVR	−γ (9136.5)
Bacillus cereus T and *Bacillus thuringensis* subs. Kuristaki HD-1				
11	1032.7	1033.1	ANGSVGGEITK	−α (6939.6)
		1033.1	ANGSVGGEITK	−β (6848.6)
13	1189.3	1189.3	ANGSVGGEITKR	−α/β (6834.6)
		1189.3	ANGSVGGEITKR	−2 (6710.5)
		1189.3	ANGSVGGEITKR	−α/β (7162.9)
31 [a]	1273.1	1273.5	LVAMAEQQLGGR	−1 (7335.1)
16	1430.6	1430.7	LAVPGAESALDQMK	−2 (6710.5)
		1430.7	LAVPGAESALDQMK	−1 (6678.5)
17	1488.3	1488.8	LVAMAEQSLGGFHK	−α/β (6834.6)
32 [a]	1534.6	1534.8	LVSLAEQQLGGYQK	−2 (6710.5)
20	1594.7	1594.8	LVAMAEQQLGGGYTR	−α/β (7080.8)
24	1885.3	1885.1	NSNQLASHGAQAALDQMK	−α/β (7080.8)
25	1940.0	1940.1	YEIAQEFGVQLGADATAR	−2 (6710.5)
		1940.1	YEIAQEFGVQLGADATAR	−1 (6678.5)
		1940.1	YEIAQEFGVQLGADATAR	−α/β (7162.9)
26	1955.3	1956.1	YEIAQEFGVQLGADSTAR	−α/β (6834.6)
27	1971.7	1972.1	YEIAQEFGVQLGADTSSR	−α/β (7080.8)
28	2258.7	2258.5	ANQNSSNQLVVPGATAAIDQMK	−α/β (6834.6)
33 [a]	2728.4	2728.8	AQASGAQSANASYGTEFATETDVHSVK	−γ (9540.1)
34 [a]	2851.7	2852.0	ATSGASIQSTNASYGTEFSTETDVQAVK	−γ (9540.1)
35 [a]	2954.8	2954.2	YEIAQEFGVQLGADATARANGSVGGEITK	−2 (6710.5
Bacillus anthracis Sterne				
11	1033.1	1033.1	ANGSVGGEITK	−α (6939.6)
			ANGSVGGEITK	−β (6848.6)
12 [a]	1140.1	1140.6	LVALAQQQLR	−α/β (6375.2)
13	1189.1	1189.3	ANGSVGGEITKR	−α/β (6834.6)
		1189.3	ANGSVGGEITKR	−2 (6710.5)
		1189.3	ANGSVGGEITKR	−α/β (7162.9)
14 [a]	1296.1	1296.5	GLDGGAVSDMAFR	[b]
15 [a]	1336.4	1336.4	VGDYLANEVEAR	[b]
16	1430.6	1430.7	LAVPGAESALDQMK	−2 (6710.5)
		1430.7	LAVPGAESALDQMK	−1 (6678.5)
17	1488.9	1488.8	LVAMAEQSLGGFHK	−α/β (6834.6)
18 [a]	1518.9	1518.8	LVSLAAEQQLGGFQK	−1 (6678.5)
19 [a]	1529.4	1528.8	LVSLAAEQQLGGGVTR	−α/β (7162.9)
20	1595.3	1594.8	LVAMAEQQLGGGYTR	−α/β (7080.8)
21 [a]	1644.8	1644.9	RLVAMAEQSLGGFHK	−α/β (6834.6)
		1644.9	AIEIAEQQLMKQNQ	−α/β (6668.6)
22 [a]	1675.7	1674.9	RLVSLAEQQLGGFQK	−α/β (6678.5)
23 [a]	1827.7	1827.1	VADEQEQHTIANLMVK	[b]
24	1885.6	1885.1	NSNQLASHGAQAALDQMK	−α/β (7080.8)
25	1940.7	1940.1	YEIAQEFGVQLGADATAR	−2 (6710.5)
		1940.1	YEIAQEFGVQLGADATAR	−1 (6678.5)
		1940.1	YEIAQEFGVQLGADATAR	−α/β (7162.9)
26	1956.5	1956.1	YEIAQEFGVQLGADSTAR	−α/β (6834.6)
27	1972.8	1972.1	YEIAQEFGVQLGADTTSR	
28	1158.6	2258.5	ANQNSSNQLVVPGATAAIDQMK	−α/β (6834.6)
29 [a]	2835.9	2835.0	ATSGASIQSTNASYGTEFATETNVQAVK	[c]
30 [a]	2944.1	2943.1	AQASGASIQSTNASYGTEFATETDVHAVK	[c]

[a]Peptides specific to that species.
[b]Hypothetical protein-1 (9207.3).
[c]Hypothetical protein-2 (9737.3).

Table 13.3 "Best hits" and significance level testing for identification of an "unknown" bacterial spore sample.

Organism	*Partial proteome size*	*Number of matches*	*Significance level*
A. Whole proteome (4–20 kDa)			
***H. pylori* 26995**	**443**	**14**	**0.036**
H. pylori J99	291	10	0.065
M. leprae	656	15	0.198
E. coli	2030	18	0.998
A. aelolicus	353	1	0.999
B. Include N-terminal methionine cleavage			
***H. pylori* 26995**	**443**	**17**	**0.002**
R. prowazekii	207	6	0.268
T. pallidum	251	6	0.427
E. coli	2030	23	0.893
Str. coelicolor	567	5	0.992
C. Restrict to ribosomal proteins			
	Ribosomal proteome size		
***H. pylori* 26995**	**37**	**6**	**0.00001**
P. aeruginosa	39	3	0.018
T. maritime	34	2	0.085
B. subtilis	42	1	0.455
E. coli	44	1	0.470

proteome size and number of matches are reduced, but there is further improvement in the significance level for *H. pylori*. Note that the single match to *E. coli* suggests that the matches in the prior iterations were to tryptic peptides from incorrect proteins, but with the same mass. The rates for such false positives will be highest for the largest genome sizes.

13.6 Point-of-Care Mass Spectrometry

One of our interests is to develop such miniaturized instruments for use in clinical diagnosis, in a point-of-care setting. In many respects, instrument development is a far less challenging task than the development of robust, validated biomarkers along with the means to present these to the instrument in some selective fashion. This is likely to involve some form of affinity or immunoprecipitation step that can be integrated directly with sample introduction system.

While much of the proteomics studies derive proteins from blood, there are indeed other proteome sources that may be exploited, such as urine, bronchoalveolar lavage (BAL), breath condensate and cerebral spinal fluid (CSF). In one particular study, the 3-inch TOF mass spectrometer is being used to monitor hepcidin levels in urine. Decreased hepcidin levels are related to anemia and critical in iron metabolism.[38] Figure 13.13 shows the mass spectrum of 100 nanomolar hepcidin in 10 mM NaCl and 5 mM phosphate buffer used as a

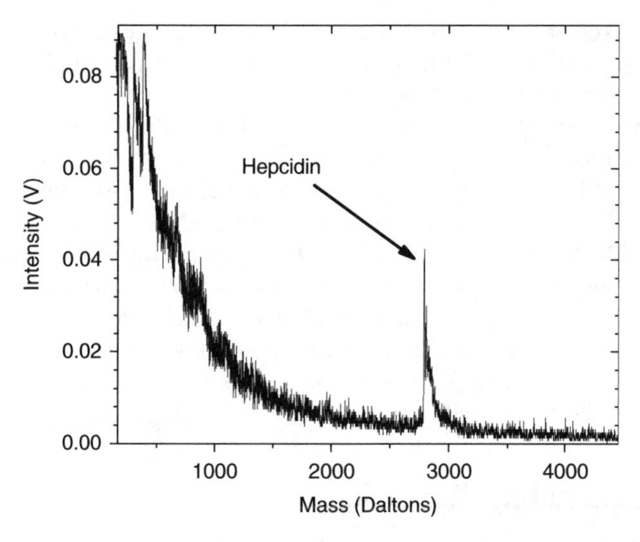

Figure 13.13 Mass spectrum of hepcidin standard obtained using the 3-inch TOF mass spectrometer.

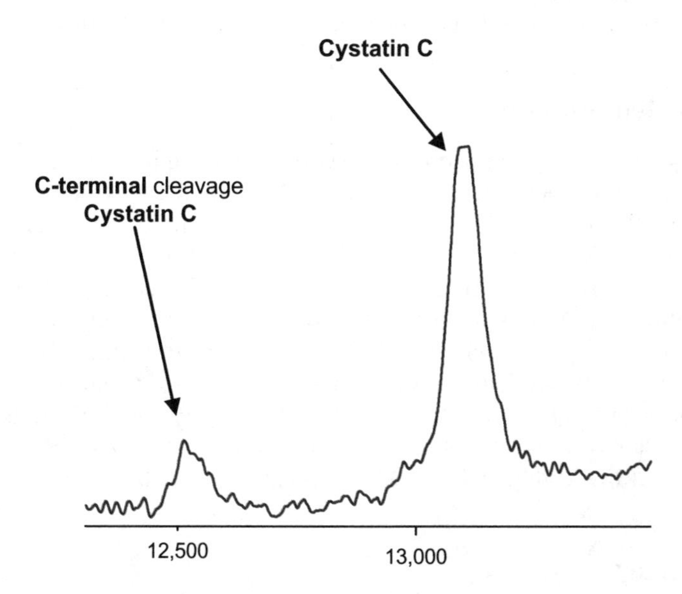

Figure 13.14 Mass spectrum of cystatin sample obtained from CSF and spotted onto a weak cation exchange SELDI chip. Matrix solution consisting of 1 μL acetonitrile, 1 μL 0.1% TFA in 1 mg mL^{-1} water and 1 μL CHCA matrix in 1:1 0.1% TFA–CAN was added to the SELDI chip spots before mass spectrometric analysis.

standard spiked in urine. In this case, samples are spotted on a Ciphergen (Fremont, CA, USA) NP20 hydrophobic SELDI chip placed directly into the instrument using a SELDI chip adapter.

In a separate study we have compared the CSF from a small set of control patients and those with multiple sclerosis.[39] Mass spectra show peaks at *m/z* 12 539 and 13 361 whose relative intensities appear to distinguish these two groups. Using a Kratos (Manchester, UK) AXIMA CFR mass spectrometer, we obtained MALDI-MS-MS spectra of the tryptic peptides from each of these two species, showed that both species map to the same protein, cystatin C, and that the peak at *m/z* 12 539 results from C-terminal cleavage to remove the last eight amino acids. Cystatin C is easily captured using the weak cation exchange CM10 SELDI chip, and the 3-inch TOF mass spectrometer can be used to monitor the relative intensities of these two peaks (Figure 13.14).

13.7 Concluding Remarks

Compact TOF mass spectrometers with mass range, resolution and sensitivity suitable for microorganism detection or point-of-care diagnostics are clearly within reach using currently available technology. The more difficult tasks involve sampling (air, water, aerosols, postal envelopes, *etc.*) in the case of homeland security, or robust diagnostics tests for clinical use. While our laboratory has been involved for some years in the development of the instrumentation and its performance, we are currently pursuing the measurement of several potential diagnostic markers, as described in this chapter.

Acknowledgements

Over the years a number of people have worked on the miniaturized instruments described in this chapter, including Viatcheslav V. Kovtoun, Mari Prieto, Robert English and Ben D. Gardner. The analysis of *Bacillus* spores was a collaboration with Catherine Fenselau and Bettina Warscheid. The hepcidin project is a collaboration with Richard Semba and Luigi Ferruci. The cystatin C project is a collaboration with Avindra Nath and David Wheeler. Machining and fabrication were carried out by Michael Kratfel of Consolidated Instruments Corp (Baltimore, MD, USA). Development of the miniaturized instruments was supported in part through funding from the Defense Advanced Research Projects Agency (MDA972-00-0014) and the National Institutes of Health (GMRR64402), and carried out at the Middle Atlantic Mass Spectrometry Laboratory.

References

1. K. Biemann, *et al., J. Geophys. Res.*, 1977, **82**, 4641.
2. G. A. Soffen, *J. Geophys. Res.*, 1977, **82**, 3959.
3. J. A. Diaz, C. F. Giese and W. R. Gentry, *J. Am. Soc. Mass Spectrom.*, 2001, **12**, 619.

4. Jet Propulsion Laboratory, California Institute of Technology, NASA, Pasadena, CA 91109, USA (http://saturn.jpl.nasa.gov).
5. S. Taylor, R. F. Tindall and R. R. A. Syms, *J. Vac. Sci. Technol.*, 2001, **8**, 557.
6. S. Boumsellek and R. J. Ferran, *Inst. Environ. Sci. Technol.*, 1999, **42**, 27.
7. G. E. Patterson, A. J. Guyman, L. S. Riter, M. Everly, J. Griep-Raming, B. C. Laughlin, Z. Ouyang and R. G. Cooks, *Anal. Chem.*, 2002, **74**, 6145.
8. S. Pau, C. S. Pai, Y. L. Low, J. Moxom, P. T. A. Reilly, W. B. Whitten and J. M. Ramsey, *Phys. Rev. Lett.*, 2006, **96**, 120801.
9. H. Y. Yoon, J. H. Kim, E. S. Choi, S. S. Yang and K. W. Jung, *Sens. Actuators A*, 2002, **97–98**, 441.
10. M. Karas and F. Hillenkamp, *Anal. Chem.*, 1998, **60**, 2299.
11. K. Tanaka, H. Waki, Y. Ido, S. Akita, Y. Yoshida and T. Yoshida, *Rapid Commun. Mass Spectrom.*, 1988, **2**, 151.
12. W. C. Wiley and I. H. McLaren, *Rev. Sci. Instrum.*, 1955, **26**, 1150.
13. R. J. Cotter, *Time-of-Flight Mass Spectrometry: Instrumentation and Applications in Biological Research*, American Chemical Society, Washington, DC, 1997.
14. R. M. Whittal and L. Li, *Anal. Chem.*, 1995, **67**, 1954.
15. R. S. Brown and J. J. Lennon, *Anal. Chem.*, 1995, **67**, 1998.
16. M. L. Vestal, P. Juhasz and S. A. Martin, *Rapid Commun. Mass Spectrom.*, 1995, **9**, 1044.
17. S. M. Colby and J. P. Reilly, *Anal. Chem.*, 1996, **68**, 1419.
18. M. C. Prieto, V. V. Kovtoun and R. J. Cotter, *J. Mass Spectrom.*, 2002, **37**, 1158.
19. N. L. Marable and G. Sanzone, *Int. J. Mass Spectrom. Ion Phys.*, 1974, **13**, 185.
20. J. A. Browder, R. L. Miller, W. A. Thomas and G. Sanzone, *Int. J. Mass Spectrom. Ion Phys.*, 1981, **37**, 99.
21. M. L. Muga, *Anal. Instrum.*, 1987, **16**, 31.
22. G. E. Yefchak, C. G. Enke and J. F. Holland, *Int. J. Mass Spectrom. Ion Processes*, 1989, **87**, 313.
23. G. R. Kinsel and M. V. Johnston, *Int. J. Mass Spectrom. Ion Processes*, 1989, **91**, 157.
24. G. R. Kinsel, J. M. Grundwürmer and J. Grotemeyer, *J. Am. Soc. Mass Spectrom.*, 1993, **4**, 2.
25. S. V. Kovtoun and R. J. Cotter, *J. Am. Soc. Mass Spectrom.*, 2000, **11**, 841.
26. S. V. Kovtoun, R. D. English and R. J. Cotter, *J. Am. Soc. Mass Spectrom.*, 2002, **13**, 135.
27. S. V. Kovtoun and R. J. Cotter, *US Pat.*, 6 518 568, 2003.
28. R. D. English and R. J. Cotter, *J. Mass Spectrom.*, 2003, **38**, 296.
29. D. N. Heller, R. J. Cotter, C. E. Fenselau and O. M. Uy, *Anal. Chem.*, 1987, **59**, 2806.
30. J. A. Platt, O. M. Uy, D. N. Heller, R. J. Cotter and C. E. Fenselau, *Anal. Chem.*, 1988, **60**, 1415.

31. D. N. Heller, C. Murphy, R. J. Cotter, C. E. Fenselau and O. M. Uy, *Anal. Chem.*, 1988, **60**, 2787.
32. W. A. Bryden, R. C. Benson, S. A. Ecelberger, T. E. Phillips, R. J. Cotter and C. E. Fenselau, *Johns Hopkins APL Tech. Dig.*, 1995, **16**, 296.
33. P. A. Demirev, Y.-P. Ho, V. Ryzhov and C. Fenselau, *Anal. Chem.*, 1999, **71**, 2732.
34. Z.-P. Yao, P. A. Demirev and C. Fenselau, *Anal. Chem.*, 2002, **74**, 2529.
35. R. D. English, B. Warscheid, C. Fenselau and R. J. Cotter, *Anal. Chem.*, 2003, **75**, 6886.
36. F. J. Pineda, J. S. Lin, C. Fenselau and P. A. Demirev, *Anal. Chem.*, 2000, **72**, 3739.
37. P. A. Demirev, J. S. Lin, F. J. Pineda and C. Fenselau, *Anal. Chem.*, 2001, **73**, 4566.
38. T. Ganz, *Best Pract. Res. Clin. Haematol.*, 2005, **18**, 171.
39. D. N Irani, C. Anderson, R. Gundry, R. Cotter, S. Moore, D. A. Kerr, J. C. McArthur, N. Sacktor, C. A. Pardo, M. Jones, P. A. Calabresi and A. Nath, *Ann. Neurol.*, 2006, **59**, 237.

Subject Index